...tional
Electrical Code®
Chapter-by-Chapter

About the Author

David Herres, owner and operator of a residential and commercial construction company, obtained a Journeyman Electrician's License in 1975. He has certificates in welding and wetland delineation. Beginning in 2001, Mr. Herres has focused primarily on electrical work, upgrading his license to Master status. The author of a 2006 article for *EC&M* on complimentary wiring methods, he has written nearly 40 articles on electrical and telecom wiring.

2011 National Electrical Code® Chapter-by-Chapter

David Herres

New York   Chicago   San Francisco
Lisbon   London   Madrid   Mexico City
Milan   New Delhi   San Juan
Seoul   Singapore   Sydney   Toronto

McGraw-Hill books are available at special quantity discounts to use as premiums and sales promotions, or for use in corporate training programs. To contact a representative, please e-mail us at bulksales@mcgraw-hill.com.

2011 National Electrical Code® Chapter-by-Chapter

1 2 3 4 5 6 7 8 9 0   DOC/DOC   1 7 6 5 4 3 2 1

ISBN 978-0-07-177409-3   **3 4633 00259 2853**
MHID 0-07-177409-2

This book is printed on acid-free paper.

Sponsoring Editor
Joy Evangeline Bramble

Editing Supervisor
Stephen M. Smith

Production Supervisor
Pamela A. Pelton

Acquisitions Coordinator
Alexis Richard

Project Manager
Tania Andrabi, Cenveo Publisher Services

Copy Editor
James K. Madru

Proofreader
Constance Blazewicz

Indexer
Robert Swanson

Art Director, Cover
Jeff Weeks

Composition
Cenveo Publisher Services

To my close friend Judith Howcroft, whose photographs adorn
these pages: Thanks for your encouragement, patience,
and inspiration throughout this continuing endeavor.

Contents

Introduction

In an increasingly wired world, demand for new electrical construction and for maintenance of existing structures is increasing at an accelerating rate. Simultaneously, individuals entering these fields are declining in number. Perhaps this is so because young, technologically gifted people want to design Web sites for rock groups rather than pull wire through metal raceways and terminate them in steel enclosures. Whatever the reasons, the bottom line is that the electrical workforce is underpopulated even as demand is rising. Many competent electricians are leaving the profession to putter about their retirement homes and fish for trout, a trend that adds to the labor shortage.

This social dynamic has been opposed of late by a strictly economic phenomenon. Severe dislocations in global financial markets have adversely affected the pace of new construction, and electrical workers have not been exempt. This decline has been more intense in new construction than in remodeling and repair, but it is felt throughout the industry. Overall, we see an increase in demand with some periods of instability and decreased activity. Nevertheless, the electrical segment has fared somewhat better than other sectors.

Most electrical contractors say that they are not about to fold up, but at the same time, they are not doing the volume they would like. For electricians, the answer to all this is to expand capability and in that way to leverage more high-quality work. Also, the work is more rewarding for those who are able to do installations more efficiently. The primary goal remains to create a safe, durable, and attractive product while making a fair profit.

In recent years, electrical work has expanded into areas that were unknown a few decades ago. Many types of electrical equipment—elevators, communications systems, and industrial machinery, to mention a few examples—have acquired vastly greater functionality by means of new electronic substrates. Some workers are intimidated in this environment, whereas others embrace and seem to thrive on it. Many of us are somewhere in between and aspire to greater electronic savvy.

This book is offered up to electricians who desire to advance their knowledge and expertise in the field. We will look at several areas of importance. You may be an apprentice learning the fundamentals, a journeyman with several years' experience looking to upgrade, or a master electrician always prepared to learn a new technique or gain insight into some new area.

The plan is to proceed along several fronts. The most fundamental item in every electrician's knowledge base is the National Electrical Code® (NEC®). Licensing exams are based on this book, and it is vital to the profession. State and municipal exams are fairly difficult, and as many as 50 percent of test takers do not pass the first time. Many individuals are never able to get a license. Chapter 2 addresses some of these issues and will show you how to pass the exam the first time you take it. First, we'll take a look at the history of the Code, how it developed, and its legal status today.

Besides the National Electrical Code, there are other written mandates—the *National Electrical Safety Code*, the *Life Safety Code*, the *National Fire Alarm Code*, the *Underwriters Laboratory (UL) White Book*, and others. While these mandates may not appear in electricians' licensing exams, they are of great importance in the practice of the trade. Missing out on some seemingly obscure requirement could result in time-consuming and costly rework. We shall review these and other mandates, but of course, the National Electrical Code will remain the prime focus.

The National Electrical Code (Figure I-1) was conceived in 1896 and published the following year. A couple decades earlier, Thomas

FIGURE I-1   NFPA's *NEC® Handbook* contains the complete code text plus commentary and illustrations.

Edison and his many technicians had developed the incandescent light bulb, which greatly outperformed existing gas and oil lights. Edison designed and built an electrical generation and distribution system in lower Manhattan, and within a very few years, many homes and businesses were subscribing. Edison had grasped the notions of ampacity, fusing, and wire insulation and had taken important first steps to mitigate hazards arising from the use of electricity.

Nevertheless, by the 1880s, electrical fire and injury by shock were becoming more frequent as usage increased. Insurers were losing a lot of money, and human suffering was immense.

In response, numerous codes emerged in the United States and England and on the Continent. Of course, these standards, while well meaning, were altogether out of synch. Even as some problems were confronted, others remained unresolved. The idea of grounding, for example, was highly controversial. It was noted that proper grounding increases the efficiency of overcurrent protective devices, thereby reducing the likelihood of an electrical fire. However, as used at that time, it could set the stage for electric shock, sometimes lethal. And yet, there were (and still are) many times more fatalities from electrical fire than from electrocution, especially if utility accidents are not counted.

In the first decades of widespread electrical usage regulated by numerous codes, grounding in some locations was prohibited, whereas other venues permitted or even required it. Besides grounding, other aspects of electrical construction had divergent and even contradictory mandates in various locations.

In 1896, individuals from a number of concerned organizations assembled in New York City to address these problems, and out of these meetings a draft emerged. In 1897, after extensive review, the first edition of the National Electrical Code was published. In 1911, the Code was transferred to National Fire Protection Association® (NFPA®) jurisdiction, where it remains today.

In the past 11 decades, the Code has been revised at varying intervals, stabilizing since 1975 at every 3 years. Each new edition of the National Electrical Code contains many changes. These range from a single word added, deleted, or changed for clarity to the addition or removal of whole chapters. Some of these changes mean that new methods and materials are required and that a specified installation must be done in a totally different way. Usually the trend is toward more exacting requirements in order to achieve greater safety. In some cases, though, mandates are relaxed or eliminated altogether if they are found to be excessive or unnecessary from the point of view of safety.

There is a definite protocol for creating revisions. At the back of NEC 2011 are three copies of a form that may be used to submit a proposed change. Any interested individual may participate in this process. The instructions on the form must be followed exactly. Proposals are considered by Code-making panels, put out for public comment, and ultimately voted on for inclusion in the next Code by the NFPA at an annual meeting.

Recent changes have involved ground-fault circuit interrupters (GFCIs) and arc-fault circuit interrupters (AFCIs), both of which, when originally introduced, were required only for very limited locations. Successive Code cycles have expanded usage of these devices, and it is likely that the trend will continue for these and other innovations. Such mandates significantly increase the cost to the owner of new construction, but of course, that is not important if lives are saved.

On its own, as enacted by the NFPA, the National Electrical Code has no legal standing. It is promulgated so that states, municipalities, countries, insurance companies, housing authorities, military installations, and other jurisdictions may enact it into law. They may adopt all or any part of it and amend it, strengthening or relaxing individual provisions as they see fit. As of this writing (in 2011), some jurisdictions have, for whatever reasons, not yet adopted NEC 2011 and so are still recognizing NEC 2008. When taking a licensing exam or doing a big job, it is important to be clear on this point.

Besides the National Electrical Code, the NFPA publishes many other codes and standards. Other organizations also become involved. Depending on how they are enacted in various states and venues, these codes interact in a simple and efficient way. For example, on the subject of fire alarms (as opposed to individual smoke alarms of the residential type, even when they are powered by alternating current and used for group operation):

- NFPA 101, Life Safety Code, denotes which occupancies are required to have fire alarm systems.

- NFPA 72, National Fire Alarm Code, lays out overall system design parameters, such as location and spacing of heads and pull stations, testing and maintenance procedures, minimum performance requirements, and operational protocols.

- NFPA 70, National Electrical Code, Article 760, covers the equipment and wiring of the fire alarm system, both power to the control console and zone wiring to initiating devices and to annunciators, as well as any phone lines for automatic calling. Also included are other fire alarm functions, such as guard's tour, sprinkler waterflow, sprinkler supervisory equipment, elevator capture and shutdown, door release, smoke doors and damper control, fire doors, and fan shutdown only where these functions are controlled by the fire alarm system.

- Article 725, Class 1, Class 2, and Class 3 Remote Control, Signaling, and Power-Limited Circuits, covers wiring emanating from the control panel. Where these circuits are power-limited, alternative requirements take effect for minimum wire sizes, derating factors, overcurrent protection, insulation requirements, and wiring methods and materials.

- Underwriters Laboratories (UL) or other inspecting agencies list all components such as control panel, smoke detecting heads, pull stations, batteries, and other equipment.

- NFPA 70E, Standard for Electrical Safety in the Workplace, covers on-the-job procedures and safeguards.

In addition to the main text of the Code and interspersed throughout are numerous Informational Notes. These are not characterized by the words *shall* or *shall not* like the Code's mandatory rules. Instead, they are explanatory or advisory and, as such, are not enforceable.

Additionally, the Code delineates its own jurisdiction by specifically excluding certain areas that are not covered: ships other than floating buildings, railway rolling stock, aircraft, and automotive vehicles other than mobile homes and recreational vehicles. Also not covered are installations underground in mines. Notice that this exclusion would not involve nonmine underground wiring, such as lighting in a traffic tunnel.

Other areas not covered are communications and power utilities equipment directly involved in generation and transmission. But wiring in utility office space, employees' lunchrooms, and similar areas are definitely under NEC jurisdiction.

A recurring Code phrase and key concept is *authority having jurisdiction* (AHJ). This person is usually an electrical inspector, as defined by state, municipality, or other entity. This individual must interpret Code rules and ascertain whether a given installation or plan submitted for review is in compliance. Additionally, this person must decide if installed equipment is safe to put into service. The Code does not look inside factory-made items such as a television or space heater. These items must be accepted or rejected as a unit by the authority having jurisdiction. Since it is not realistic to expect that the inspector would have the facilities or knowledge base to examine and rule on every piece of equipment that is part of an electrical installation, the Code provides that "organizations properly equipped for experimental testing" may be relied on to perform this function.

One of these organizations is UL. This body has extensive testing facilities and procedures and goes out into the field for purposes of verification. Manufacturers who introduce a new product have to get it listed by UL if they expect to market it within jurisdictions that have adopted the Code.

There have been cases of counterfeit listing, so it is important that electricians buy their materials from reputable dealers. Generally, if a product is listed, it may be presumed to meet minimum safety standards. That being said, it is still true that some products are higher quality than others. Product selection should not be made on the basis of price alone.

The National Electrical Code stresses that its role is further limited in that it addresses safety issues only. It does not contain design specifications, nor is it a manual for untrained persons. In Article 90, Introduction, the Code's purpose is stated: "The practical safeguarding of persons and property from hazards arising from the use of electricity." Principal hazards are fire and electric shock. Other hazards, mitigated by Code provisions, also could be present. For example, if an electrical installation is not supported adequately, it could fall, causing human injury and property damage. Besides design adequacy, installation requirements and sound materials, work procedures, and safety protocols are important areas of concern. In this connection, another mandate has become increasingly important: It is NFPA 70E, Standard for Electrical Safety in the Workplace. Among other material, it contains very important procedures for electrical workers' protection from injury due to arc flash.

We have seen how the Code limits itself. It is a crucial part of an electrician's knowledge base, but not the whole story. To become accomplished and truly proficient, it is necessary to develop procedures that will work when dealing with equipment or systems that are failing to function properly. When there is a malfunction in electrical distribution or end-use equipment, it is the electrician who is called. Sometimes a seemingly insignificant event such as a very slight flickering of lights is an indication that a more serious and even dangerous situation lies on the horizon. Homeowners and industrial workers trust the electrician to make an evaluation and perform necessary repairs. Knowledge and expertise necessary to do this are partly intuitive and lie outside the Code.

When a piece of electrical equipment stops working properly, an electrician is called on to make the repair. It is easy to fix a lamp, but what about an elevator, a refrigeration unit, or a submersible pump? (In these instances, there are additional licensing issues that must be considered, and we will discuss them later.) Obviously, a certain amount of knowledge concerning the inner workings is required. It is desirable to obtain schematics, repair manuals, and other documentation. In some cases this is not possible, but every effort should be made to access print or Internet resources prior to entering unfamiliar territory. Suppose that a complex piece of equipment is giving an error code on its alphanumeric display. You can type that into an Internet search engine and often instantly access manufacturer data and online professional forums that will provide the solution. Additionally, YouTube.com is a great resource. Type in the make and model of the equipment, and you will be able to view video tutorials showing how to disassemble and repair difficult units.

In a later volume we will discuss troubleshooting and repair methods and revisit them in the sections on individual types of equipment. It is a premise of this book that an electrician should endeavor to become proficient in working on anything electrical, although it may be necessary to draw a line somewhere. Most of us will never touch a particle accelerator or disassemble a digital camera, but then you never know what the future will bring.

Electricians' Licensing

Consider the enormous moral and legal implications involved every time an electrician does new or retrofit work. An improperly torqued connection or loose splice or an invalid calculation can result in property damage, personal injury, or even fatality—immediate or years down the road. Anyone doing residential electrical work should hold in mind the image of a child sleeping in a second-story bedroom. As for industrial work, the enormous available fault current can blast through a metal enclosure causing massive worker injury and death.

No individual can know everything—the purpose of the National Electrical Code is to make available a vast amount of information that, when applied properly, helps to ensure that an electrical installation will be free of hazards. However, a whole additional level of safety is required, and it is provided by the licensing of electrical workers. This ensures that these professionals are able to access and apply NEC mandates as needed on the job site. Licensing also certifies that the workers possess other, non-NEC knowledge and skills.

Many people assume that there is nationwide licensing for electricians, but this is not the case. Each state, municipality, or other venue issues its own set of licenses. Requirements and enforcement vary widely. Usually, in the United States, licensing is administered by licensing boards that are established and regulated by state legislation. A few states, notably Illinois and New York, leave the task to counties and/or municipalities. There are many variations, but a few generalizations are possible. In most cases, there is a multitier structure in place with a hierarchy of licenses available, each with its own permitted activities and requirements. To start, let's take a look at New Hampshire's well-articulated and relatively unambiguous policies. (These descriptions are valid for mid-2011, but changes occur from year to year.)

In New Hampshire, licensing mandates are established by state law enacted by the legislature and administered by the New Hampshire Department of Safety. Any high school graduate may apply for and will be granted without examination an apprentice card. The annual fee is $30, and the individual is entitled to perform electrical installations under the direct supervision of a master or journeyman electrician, who must be on the job site whenever the apprentice is working. At all times, a one-to-one ratio must be maintained, and this ratio is one of the things that an electrical inspector looks at. The apprentice, to keep his or her card current, must complete a minimum of 144 hours of electrical schooling in an approved training course during each 12-month period that the apprentice identification card is valid.

Within these limitations, the apprentice may do any type of electrical work, including calculations, heavy switchgear, and medium/high voltage under proper supervision. Of course, it is in the employer's interest to increase apprentice capabilities so that the apprentice can take on more difficult jobs.

The next step is to acquire a journeyman electrician's license, and to meet these requirements, considerable experience and knowledge are required. The basic requirement is 8,000 hours of on-the-job experience. This must be documented by means of an affidavit signed by a supervising electrician. (One year working a 40-hour week with two weeks' vacation equates to 2,000 hours, so four years is required to meet this requirement.) Besides field experience, effective June 1, 2013, 600 hours of electrical schooling including 24 hours on electrical safety are required. Any felonies within the last 10 years or adverse actions in regard to an electrician's license in another jurisdiction must be divulged and will constitute red flags.

The application fee is $50. These requirements must be met before the candidate may take the examination. The exam is in three parts, and to pass, you have to get 70 percent on each of the three parts, not 70 percent overall average. Thus you could score high on two parts, get 69 percent on one part, and fail the examination. Some of the questions are rather difficult, and a good number of applicants fail the first time. Later in this Introduction we'll consider some highly successful approaches that, if followed, will lead to passing the exam.

The first section of the journeyman's license exam consists of 50 questions based on the National Electrical Code. The second section consists of 50 questions dealing with practical electrical installations. The third part is made up of 10 questions on New Hampshire's enabling legislation and administrative rules. After passing the exam, a $150 fee is required for three years.

Requirements for a New Hampshire master electrician's license are more stringent. These are the main differences: Each applicant first must have a journeyman's license, with its 8,000-hour prerequisite. Then 2,000 more hours (one year) of field experience working as a journeyman doing electrical installations is required.

The exam is similar to the journeyman's exam with the following difference: The master exam has 25 instead of 10 questions based on the state's enabling law and administrative rules.

Besides these two licenses and the apprentice permit, New Hampshire requires a special high/medium-voltage license for anyone working with 2,001 volts or higher. There is no additional exam, but applicants must complete a state, federal, or employer certification course approved by the Electricians' Licensing Board. The fee is $270 for three years.

To complete this survey of New Hampshire electricians' licensing, a few more comments are in order:

- Utility work, upstream of the service point of connection, does not require an electrician's license. (It is also not subject to NEC coverage.)

- Individuals may do electrical installations on their own property without a license. The intent is that it is the primary residence. For example, an individual may not work on a second residence such as a summer home. A developer cannot wire an entire subdivision regardless of ownership.

- Carpenters and similar tradesmen can do incidental electrical work that they run into while plying their trade, provided that it does not involve calculations. An unlicensed individual may remove and replace light fixtures, switches, and receptacles while paneling a room but may not extend a branch circuit because that involves calculations. The defining concept is *incidental*. In other words, unlicensed trunk slammers who advertize "light electrical work" are in violation.

- The enabling legislation is for power and light wiring, and therefore, most low-voltage work (a somewhat nebulous concept) is unregulated. Residential satellite dish installers need not be licensed in New Hampshire because the modem is simply plugged into a receptacle, and current policy regards grounding back to the service (NEC mandated) as not a licensing issue.

The matter of supervision also should be reemphasized. An apprentice may do any type of electrical work but must have continuous on-site supervision by a master or journeyman. A one-to-one ratio must be maintained at all times. If there are five apprentices on the job, five master or journeyman electricians must be present. All apprentices must have legitimate state-issued cards, and visiting inspectors check for compliance.

Journeyman electricians may not engage in electrical contracting on their own. Every job must be covered by a master electrician. No ratio is required, and the supervising master electrician need not be on site at all times. Accordingly, a given job could have 15 apprentices, 15 journeymen (on site), and one master back in the office but checking the site at meaningful intervals.

Success in taking a licensing exam depends on adequate preparation. State licensing boards generally provide guidelines for exam candidates. Most exams are open-book exams. Some states allow you to take in the National Electrical Code, whereas others list additional books. Code tabs are a great help in navigating the Code quickly, which is what you have to do in an exam setting. States also permit a hand calculator, although some prohibit scientific calculators, which wouldn't be helpful anyway. A simple unit with a good, clear display is best.

Most exam questions assume copper conductors. If no mention is made to the contrary, it is safe to assume sizing is based on copper. NEC 2011 displays inch and metric values throughout, but exams usually ask questions and expect answers in inch numeration.

Exams generally are multiple choice. The best strategy is to go through the exam and answer all the questions you can without spending too much time on any one question. Then go back and tackle the more difficult ones. If questions remain that you cannot resolve, make the best educated guess possible. If a multiple-choice question has four possible answers, random guesses would give 25 percent correct answers. But you may be able to eliminate two answers, which would bring you up to 50 percent. Of 100 questions, if you know with certainty the answers to 70 percent of the questions and you get 50 percent correct on the rest, you can expect an overall score of 85 percent.

Outside the NEC Chapter Format

The fundamental premise of this book is that to perform well on an electrician's licensing exam, one must know the NEC structure, not necessarily have its contents committed to memory. Accordingly, we shall undertake a chapter-by-chapter analysis with particular attention to the order in which various elements appear. First, however, outside the NEC main body (Chapters 1 through 9), we find the following:

- Index
- Annexes
- Article 90, Introduction

(Following the Contents but before the Introduction are 11 pages listing Code committee members. These pages can be folded over to simplify navigation.)

In an open-book exam setting, the NEC Contents is helpful in finding answers quickly. For example, suppose that you have a question about a permitted use of electrical metallic tubing (EMT). You know this topic appears in Chapter 3, Wiring Methods and Materials, but rather than finding Chapter 3 and scanning through it to find the article on EMT, it is much quicker to consult the Contents, find the relevant article number, and go from there. Notice that the article numbers are in bold type at the top of each page. The numbers at the bottom of the page, in this case 70-207, indicate the NFPA document number and page, which is not what you want or need when taking an exam or doing research for job design or installation.

Thus, for finding the location within the Code when an approximate article title (but not number) is known, consult the Contents. This is the quickest way to go, and because time is of the essence in a licensing exam, proper use of the Contents definitely will result in a higher score.

In contrast, there may be cases where your knowledge is a little sketchier. Consider a question regarding locations in which ground-fault circuit interrupter (GFCI) protection is required. Such protection may be provided by placing a specialized breaker in the entrance panel to cover the branch circuit, but the more common method is to install a GFCI receptacle, which will extend this protection to other receptacles daisy-chained downstream. This question is very common in electricians' licensing exams. You will search the Contents in vain for the location of this information.

The Index, which appears at the end of the Code after the annexes, will take you where you need to go. Under "Ground-Fault Circuit Interrupters," the subtopic "Receptacles" appears, followed by a list of 20 locations with the relevant article numbers.

This example highlights a common problem in NEC-based exams. Information needed to answer a specific question may be in any one of several locations. Would GFCI mandates be found in Article 240, Overcurrent Protection, in Article 210, Part III, Branch Circuit Required Outlets, in Article 406, Receptacles, Cord Connectors, and Attachment Plugs (Caps), or perhaps in one of the many articles dealing with specific occupancies?

Other exam question topics fall into this category. Tap rules (which we shall be looking at in detail) are spread throughout the Code under various headings—Splices and Taps, Branch Circuits, Motors, Feeders, and others.

You can scan through Code pages and eventually find the answer, but valuable time will be expended. Instead, consult the keyword in the Index, and look for the word *required*. Generally, in accessing information during an open-book exam, the Index is used more often than the Contents, but reference to either as needed is far more efficient to scanning text, which should not be done.

There are various Code companion books, such as Ferm's *Fast Finder*, that many states allow to be brought into open-book exams. Generally, the best procedure is to consult the NEC Index (or Contents) first. Then, if that doesn't work, open up the *Fast Finder*, which may be a little slower but should find the NEC location.

Also outside the main NEC chapters are the annexes. These are lettered A through I and are located after Chapter 9, Tables, and before the Index. The annexes are not part of the NEC requirements and are included for informational purposes only. In this respect, they are like the fine print notes. Some of the annexes are rather arcane, whereas others are used all the time. Here is a rundown:

- Annex A is a list of product safety standards. It enumerates various electrical products such as capacitors and motors and provides the product standard number. These are mostly UL documents, and they may be consulted to get additional (beyond NEC) information that may be useful in designing a job, although this annex usually will not be consulted during an exam. Besides UL, Institute of Electrical and Electronics Engineers (IEEE) and International Society of Automation (ISA) standards are included.

- Annex B, Application Information for Ampacity Calculation, provides application information for ampacities calculated under engineering supervision. Tables similar to the ampacity tables in Chapter 3 are included, but for most exam questions, Chapter 3 tables will suffice. There is also specialized information on underground duct design and adjustment factors for more than three current-carrying conductors in a raceway or cable with load diversity. Since the contents of Annex B is applicable to Code modifications involving engineering supervision, it will not figure prominently in state electrician licensing exams.

- Annex C, Conduit and Tubing Fill Tables for Conductors and Fixture Wires of the Same Size, is used widely by electricians in design and installation phases of nonresidential work. NEC Section 300.17, Number and Size of Conductors in Raceway, provides that these conductors shall not be more than will permit heat dissipation and installation or withdrawal without damage to conductors or insulation. In order to comply with this provision, it is necessary to calculate conduit fill. Either you know the number and size of conductors in a run and wish to calculate conduit raceway size, or you have a preexisting raceway and wish to find the maximum number and size of conductors permitted. First consult NEC Table 4 for various types of raceways and find permitted fills depending on number of wires. Then find the cross-sectional area (based on conductor diameter found in Table 5), add this value for each conductor, and determine raceway size for the given conductors or number of conductors for the given raceway size. You will notice that this calculation involves more than one Code reference plus a little number crunching. The good news is that in actual practice, conductors in a single raceway are frequently all the same size and in this case may be directly accessed from the tables in Annex C, so there are no calculations and just a single lookup. The bad news is that licensing exams contain questions based on single-size or multiple-size conductors, so the more difficult calculation often must be performed. Nevertheless, Annex C is used often as a time-saver, and there will be questions based on it.

- Annex D, Examples, shows various Code calculations. This annex is helpful in learning how to do these calculations and also serves as a step-by-step guide. It may be accessed during an exam to facilitate finding correct answers.

- Annex E, Types of Construction, summarizes the five types of building construction based on fire rating. Since the main Code body references these types, it is necessary to know how to identify them. This information can be accessed easily if required by an exam question. The key is knowledge of the topics covered by each of the annexes so that they can be acquired instantly.

- Annex F, Availability and Reliability for Critical Operations Power Systems; and Development and Implementation of Functional Performance Tests (FPTs) for Critical Operations Power Systems, probably will not be used during a licensing exam, but it will be useful during a critical operations power systems design or installation.

- Annex G, Supervisory Control and Data Acquisition (SCADA), is applicable to mission-critical loads, including the fire alarm system and similar installation segments. Here again, this material may not appear on an exam, but if it does, you will know where to find it.

- Annex H, Administration and Enforcement, addresses topics that are not mandated in the Code but are provided in case states, municipalities, or other entities wish to incorporate them into their legislation.

- Annex I, Recommended Tightening Torque Tables from UL Standard 486A-B.

Another item that falls outside the main body of NEC chapters is Article 90, Introduction. Notice the article numbering. Each chapter is made up of articles with three-digit numbers with the first digit the same as the chapter number. In other words, articles in Chapter 5, Special Occupancies, are numbered in the 500s, starting with Article 500, Hazardous (Classified) Locations, Classes I, II, and III, Divisions 1 and 2, and ending with Article 590, Temporary Installations. Accordingly, Article 90, Introduction, lies outside the main body of NEC Chapters 1 through 9.

Licensing exams frequently focus on Article 90 because it creates a frame of reference for the Code and delineates its contents and legal standing. It begins with six paragraphs that state the purpose. The first sentence is very basic: "The purpose of this Code is the practical safeguarding of persons and property from hazards arising from the use of electricity." Code provisions are concerned with safety, not efficiency. It is not a design specification or instruction manual. It does not deal with electronic theory or troubleshooting techniques except where reference to them has an impact on safety, i.e., eliminating hazards. We all know that the main hazards arising from the use of electricity are fire and electric shock, although there may be other hazards, such as an improperly supported conduit that might fall and injure a person or damage property. This type of hazard also could relate to the dangers of fire and shock.

A final note in Section 90.1, Purpose, states that the NEC addresses safety principles contained in Section 131 of International Electrotechnical Commission (IEC) Standard 60364-1. These provisions include protection against electric shock, protection against thermal effects, protection against overcurrent, protection against fault currents, and protection against overvoltage. We shall compare and contrast NEC and IEC later on.

Section 90.2, Scope, is often the source for one or two exam questions. It enumerates items covered and items not covered by the Code. It will be seen that a certain logic pervades the discussion. The general public, to the extent that they are aware of the fact that there is an electrical code, assumes that it covers equipment and installations that operate over a certain voltage. This is emphatically not the case. Very low-voltage signaling cable including fiberoptic is covered. Even if the amount of electricity involved is incapable of igniting combustible material, including volatile flammable liquids, and is incapable of causing electric shock to humans, other hazards may be present. Conductor insulation in a burning building, for example, may emit a large quantity of toxic smoke that could cause fatalities. Also, it may contribute to the spread of fire that was not electrical to begin with. These issues are dealt with extensively in Chapter 8, Communications Systems, and in the other low-voltage chapters and articles.

NEC coverage, then, is defined not by voltage but by other criteria. The Code covers conductors, equipment, and raceways whether for power transmission or signaling and communication, as well as optical fiber cables and raceways. These installations are covered whether within a building or outdoors, including electric utility installations not involved directly in energy generation and distribution. Utility office space, repair shops, warehouses, employee lunchrooms, and the like are covered by the Code. Equipment on the supply side of the service point is outside the scope of the Code (Figure I-2). It is covered in American National Standards Institute (ANSI) C2, National Electrical Safety Code, published by the IEEE. Additional material on supply-side equipment appears in utility rules and protocols, not to mention publications of Occupational Safety and Health Administration (OSHA) and other regulatory bodies.

The section listing what is covered, in typical NEC logic, is followed by a section titled Not Covered in which these items are listed:

- Installations in ships other than floating buildings, railway rolling stock, aircraft, or automotive vehicles other than mobile homes and recreational vehicles.

Figure I-2 Part of a substation, this installation is usually utility-owned and not under NEC jurisdiction. This is not always the case. Large industrial facilities may purchase high-voltage power and own the transformers and distribution equipment, in which case the installation must comply with the National Electrical Code.

- Installations underground in mines and self-propelled mobile surface mining machinery and its attendant electrical trailing cable. (Note that it says "underground in mines." Therefore, underground nonmining installations, such as lighting in an underground traffic tunnel, would be covered.)

- Installations of railways for generation, transformation, transmission, or distribution of power used exclusively for operation of rolling stock or installations used exclusively for signaling and communications purposes.

- Installation of communications equipment under the exclusive control of communications utilities located outdoors or in building spaces used exclusively for such installations.

- Installations under the exclusive control of an electric utility, including service drops or service laterals and metering, wiring on private property but within easements, or wiring on property owned or leased by the utility for the purpose of communications, metering, generation, control, transformation, transmission, or distribution of electric energy (see Figure I-2).

A final piece within Section 90.2 is titled Special Permission. It describes a special exclusion from Code jurisdiction for installation of conductors and equipment not under exclusive control of the utility where such items are used to connect the supply system to the service-entrance conductors, provided that such installations are outside a building or terminated immediately inside a building wall. It is stated that this special exclusion may be granted by the *authority having jurisdiction* (AHJ). This is a frequent Code term that usually means an electrical inspector employed by the jurisdiction that has enacted the Code.

Section 90.3, Code Arrangement, is accompanied by Figure 90.3, Code Arrangement. Between the diagram and text, a clear explication of the overall Code structure is presented. It should be read, comprehended, and memorized. If you are thoroughly enlightened in this matter, a licensing exam will be much easier. It provides that Chapter 1, General, Chapter 2, Wiring and Protection, Chapter 3, Wiring Methods and Materials, and Chapter 4, Equipment for General Use apply generally to all electrical installations except as noted in the case of Chapter 8, Communications Systems.

Chapter 5, Special Occupancies, Chapter 6, Special Equipment, and Chapter 7, Special Conditions, supplement or modify Chapters 1 through 4.

Chapter 8, Communications Systems, is unique in that it is not subject to the requirements of the preceding seven chapters except where those requirements are specifically referenced in Chapter 8.

Chapter 9, Tables, is applicable as referenced within the other chapters.

Annex A through Annex H are informational only, not mandatory.

This is the structure of the National Electrical Code. It should be kept in mind throughout any exam, design, or installation work—within this framework, the entire Code makes sense.

The next provision, Section 90.4, Enforcement, contains three key provisions that serve to create a context for the Code. The first of these is discussed in Chapter 1, but we shall take a quick look at it here because it is rather important: The fact is that the Code, as enacted by NFPA vote, has no legal standing but is offered up to states, municipalities, insurance companies, and other entities that wish to adopt it into law or policy. It is further stated that the AHJ is charged with making interpretation of the rules, deciding on approval of equipment and materials, and granting the special permission contemplated in a number of the rules. The AHJ also may grant special permission to waive specific requirements if it is judged that satisfactory levels of safety will be maintained. Moreover, the AHJ may permit the use of existing electrical materials when new ones, NEC mandated, are still in the pipeline but not yet available.

It will be noted that in all cases, the AHJ, when making on-site inspections or reviewing plans prior to construction, has ultimate power to make interpretations and exceptions and to accept or reject materials and equipment. Thus the role played by UL and other testing labs is advisory in the sense that the AHJ has the option of using their listings when approving these items. It is at the discretion of the AHJ whether or not to make use of a listing, but in actual practice, it is almost always done this way. It is presumed that the AHJ does not have the facilities nor the technical expertise to pass judgment on every product that is encountered, and so the role of UL and similar organizations is not binding. That being said, it is noteworthy that the Code provides that certain products be listed by a testing lab. An example is the batteries that are used for backup power in a fire alarm system.

Section 90.8, Wiring Planning, suggests that a good practice is to provide ample space in raceways, spare raceways, and additional spaces to allow for future increases in power and communications usage. Extra distribution centers located in readily accessible locations, while not actually Code mandated, will ensure convenience and safety of operation. It is further noted that restricting the number of wires and circuits in a single enclosure minimizes the effects from a short circuit of ground fault in one circuit.

The final bit of NEC introductory material is titled Units of Measurement. In accordance with the effort to create worldwide conformity regarding units of measurement, the metric system, known as *International System of Units* (SI), is gradually becoming universal, particularly within the United States. The National Electrical Code participates in this process, but it is difficult to achieve this goal all at once. Many electricians, like other tradesmen, resist this sort of change on the theory that the inch system has worked for them and why make changes? NEC, in a sense, has had to tread a thin line. Moreover, there is the issue of soft versus hard conversion. At present, the policy is to have a dual system of units. In Code text, SI units currently appear first, and inch-pound units follow in parenthesis, except as provided in the article we are now considering.

So what are these two types of conversions? A fine print note explains that *hard conversion* is a change of dimensions into new sizes that may not be interchangeable with the original values. Hard conversion allows some inexactitude in order to avoid cumbersome decimals. *Soft conversion* is a direct mathematical translation where the actual dimension is not changed.

It is noted that trade practices are to be followed. If the actual measured size of a product is not the same as the nominal size, the trade size is to be used in all cases. When NEC material is taken from another standard, the context of the original material is to be preserved. Where industry practice is to use inch-pound units, inclusion of SI units is not required. Where any inexactitude would compromise safety, soft conversion is to be used. Since conversion to the metric system may be approximate, use of either value constitutes Code compliance. Most electricians' licensing exams use inch nomenclature and don't bother with metric. This is what we'll do in this book. Similarly, conductor sizing often assumes values specified in the Code for copper, not aluminum, and again, we will follow that convention.

This completes our review of material in Article 90, Introduction, which lies outside the main body of NEC chapters. As noted, this material is very basic and constitutes a frame of reference for the Code. Becoming familiar with this article, especially the text and figure on code arrangement (Section 90.3) will take you a long way toward the goal of performing well on an electricians' licensing exam as well as in plying our profession.

CHAPTER 1

NEC® Chapter 1, General

Article 100

The first chapter of the National Electrical Code® (NEC®) continues the task started in Article 90 of laying the groundwork for the remainder of the Code. It begins with Article 100, Definitions, which, as usual, opens with a statement of scope. At this point, it is anticipated that the reader is becoming familiar with the basic NEC article template.

Discussing scope, the article states that only Code-specific definitions are included and that commonly understood general and technical terms from related codes and standards are excluded; only terms found in two or more articles are defined herein. Terms occurring in a single article are defined in that article. Part II of the article defines terms applicable to installations and equipment operating at over 600 volts nominal. The 600-volt division is found repeatedly throughout the Code. Since it says "over 600 volts," the commonly used 600-volt system does not fall within the higher-voltage category. Also, in this context, these voltages are nominal. Actual voltages, as measured in the field, may be higher or lower depending on utility practice; line length; voltage drop within the service, feeder, and branch circuits; and amount of current being drawn at the time the measurement is made. Load current partially determines measured voltage. Nominal voltage is a constant, and this is what determines Code usage.

Most of the definitions are self-explanatory. Exam questions based on these definitions are fairly easy to answer because the terms are in alphabetical order and occur in either Part I or Part II or within the article in question, a location best ascertained by consulting the Index. In the discussion that follows, we will look at a few of the definitions that have a great impact on Code usage and understanding.

One of the really crucial definitions also comes first because it happens to be first alphabetically. We have to understand exactly the meanings and distinction between *accessible* and *readily accessible*

1

and between *inaccessible* and *not readily accessible*. This is so because a great many Code provisions depend on these distinctions. *Accessible*, in regard to equipment, means "admitting close approach." The concept is further refined by adding "not guarded by locked doors, elevation, or other effective means." *Accessible*, in regard to wiring methods, means "capable of being removed or exposed without damaging the building structure or finish or not permanently closed in by the structure or finish of a building."

Readily accessible means "capable of being reached quickly for operation, renewal, or inspections without requiring individuals to climb over or remove obstacles or resort to portable ladders and so forth." The classic example of equipment that is accessible but not readily accessible is a junction box above a suspended ceiling. (Junction boxes have to be accessible but not readily accessible.) An entrance panel, by contrast, cannot be above a suspended ceiling because it needs to be readily accessible. Interestingly, a dry transformer of 600 volts nominal or less, not exceeding 50 kVA, is permitted in hollow spaces of buildings not permanently closed in by structure, provided that it meets the ventilation requirements of Section 450.9 and separation from combustible materials requirements of Section 450.21(A). These units are not required to be readily accessible and could go above a suspended ceiling.

Concealed means "rendered inaccessible by the structure or finish of the building." Wires in concealed raceways are considered concealed, even though they may become accessible by withdrawing them.

Approved means "acceptable to the authority having jurisdiction." This may or may not be accompanied by being listed by a testing organization.

Authority having jurisdiction (AHJ) means "an organization, office, or individual responsible for enforcing Code requirements." Generally, this is the electrical inspector, but it may vary because this matter is not mandated within the Code but rather depends on state, municipal, or other legislation. Insurance companies also inspect for Code compliance and so may have personnel who constitute the AHJ, and individual property owners, managers, or government workers may perform this function. In any event, the AHJ has final say on a day-to-day basis, although in many instances there are appeal remedies in place.

Bonded means "connected to establish electrical continuity and conductivity." This concept is very important within the context of grounding. Except in limited cases, the goal is to achieve a low impedance.

Bonding jumper means "a reliable conductor to ensure the required electrical conductivity between metal parts required to be electrically connected."

Equipment bonding jumper means "the connection between the grounded circuit conductor and the equipment grounding conductor at the service."

Main bonding jumper means "the connection between the grounded circuit conductor and the equipment grounding conductor at the service." This introduces the distinction between *grounded* and *grounding*. Both are held at the same voltage potential with respect to ground, but they perform different functions. They are connected only in one location, within the service enclosure by the main bonding jumper. When they leave the service enclosure, they follow parallel paths to the power outlet, but they are emphatically not connected at any point, either intentionally or by defective equipment. We shall discuss this in greater detail when we get to Article 250, Grounding.

Exposed (as applied to live parts) means "capable of being inadvertently touched or approached nearer than a safe distance by a person." It is applied to parts that are not suitably guarded, isolated, or insulated.

Exposed (as applied to wiring methods) means "on or attached to the surface or behind panels designed to allow access."

Identified (as applied to equipment) means "recognizable as suitable for the specific purpose, function, use, environment, application, and so forth, where described in a particular Code requirement." This is generally done by UL or another testing laboratory.

Labeled means "equipment to which has been attached a label, symbol, or other identifying mark of an organization that is acceptable to the authority having jurisdiction and concerned with product evaluation, that maintains periodic inspection of production of labeled equipment or materials, and by whose labeling the manufacturer indicates compliance with appropriate standards or performance in a specified manner."

Listed means "equipment, materials or services included in a list published by an organization that is acceptable to the AHJ and concerned with evaluation of products or services, that maintains periodic inspection of production of listed equipment or materials or periodic evaluation of services, and whose listing states that either the equipment, material, or service meets appropriate designated standards or has been tested and found suitable for a specified purpose." Exam questions sometimes focus on the distinctions between identified, labeled, and listed, so these terms must be understood, and it must be known how the definitions can be accessed.

We have taken a look at some of the definitions that are germane to many Code provisions. These need to be held in mind as various locations within the Code are accessed so that misinterpretations do not occur.

Article 110

Article 110, Requirements for Electrical Installations, is very basic for all electrical work, and authors of licensing exams see it as an abundant source for questions. Following a statement of scope, the article begins with a number of general guidelines, some intended for the AHJ, some applicable to the design phase, others directed at the installer. First, a noninclusive list of eight criteria is provided for the inspector when deciding whether to approve equipment. These include

- NEC conformity, as evidenced by product marking, listing, or labeling
- Mechanical strength and durability
- Wire bending and connection space
- Electrical insulation
- Heating effect both under normal use and abnormal conditions
- Arcing effects
- Classification by type, size, voltage, current capacity, and specific use
- Other safety factors

While this material is for the AHJ and probably won't be the subject of an exam question, it nevertheless serves to put electrical equipment quality in perspective and certainly adds to an electrician's developing intuitive powers. There follow a number of general comments.

- Voltages referred to throughout the entire Code are those at which the circuit under consideration operates. Equipment must be rated at not less than the circuit nominal voltage. Generally, if equipment is operated undervoltage, it will not be damaged but will operate less efficiently. A light bulb will be dimmer than it should be but actually may have a longer life. An exception is the electric motor, which is damaged by running at less than rated voltage. It will run slower than rated rpms, and a portion of the electrical energy supplied will be converted to heat. Direct-current (dc) motors and universal motors can be run at less than rated voltage, and this property, with the proper controls, may be used intentionally to regulate speed.

- Conductor sizes throughout the Code assume that the material will be copper. Where other materials are used, notably aluminum and copper-clad aluminum, size adjustment must be made. The ampacity tables in Article 310, Conductors for General Wiring, as we shall see, give sizes for copper on the left. The right-hand side of the tables should be disregarded unless an exam question specifically refers to aluminum or copper-clad aluminum. In most respects, copper-clad aluminum conductor properties are the same as aluminum conductor properties. Conductor sizes are expressed American Wire Gauge (AWG) or in units of circular mils.

- Wiring integrity must be preserved, meaning that installations must be free from short circuits, ground faults, and connections to ground other than as required or permitted by the NEC.

- Only suitable wiring methods are included in the Code. They are permitted indoors and outdoors except as stated in the Code.

- Current-interrupting ratings must be observed. Available fault current depends on utility parameters and properties of the service and premises wiring impedance. Ironically, the better the wiring, the greater is the danger of exceeding current-interrupting ratings with possible damage to equipment and personal injury owing to arc flash. This issue becomes more important in an industrial setting, as we shall see later.

- Circuit impedance must be considered in design stages, and exam questions address this issue. Selection and coordination of overcurrent devices are necessary to protect equipment and confine any outage to a limited area. Faults may occur either between two conductors or between a conductor and an intentionally or unintentionally grounded object. Proper use of listed products is an important component of Code compliance and an important step in preventing faults.

- Conductors or equipment located in damp or wet environments must be identified for such use. Licensing exams usually focus on such properties. Similarly, gases, fumes, vapors, liquids, or any setting that has a deteriorating effect must be considered.

- Many past revisions and the current version of the Code state that electrical equipment shall be installed in a neat and workmanlike manner. The Code references American National Standards Institute (ANSI)/National Electrical Contractors Association (NECA) 1-2010, Standard Practice for Good Workmanship in Electrical Construction. This is in an Informational Note, so the document is not mandatory, but it is recommended that interested individuals consult it for more information on good work practices.

- Unused openings, unless intended, are to be closed. Where metallic plugs are used with nonmetallic enclosures, they are to be recessed ¼ inch from the outer surface. This is so because they are not grounded, and being recessed provides some protection from human contact in case they become energized owing to a ground fault within.

- The integrity of electrical equipment and connections is to be preserved. Busbars, wiring terminals, insulators, and other surfaces within electrical equipment are not to be damaged or contaminated by paint, plaster, cleaners, abrasives, corrosive residues, and similar foreign materials. Damaged, broken, bent, cut, or deteriorated parts must not be in use. Deterioration may be caused by corrosion, chemical action, or overheating.

- Wooden plugs driven into holes in masonry, concrete, plaster, and the like may not be used to mount equipment, which is to be secured firmly to the surface on which it is mounted. Entrance panels, junction boxes, and other enclosures have predrilled holes. Drywall screws may be used to mount such enclosures to wooden walls, but at least one electrical inspector has stated that only galvanized drywall screws should be used, even in an indoor setting.

- Airflow for cooling purposes must not be impeded by walls or adjacent equipment. Floor-mounted equipment must have clearance between the top surface and adjacent surfaces to dissipate rising warm air.

- Dissimilar metals must not be joined for the purpose of conducting electricity because over a period of time, corrosion will occur. This is applicable to aluminum and copper. Acid flux, used by plumbers when soldering pipes, will leave a residue that promotes corrosion. For electrical connections, use resin-based flux that is marked "Suitable for electrical connections." Connectors and terminals for conductors more finely stranded than Class B and Class C stranding, as shown in Chapter 9, Table 10, must be identified for the specific conductor class or classes. If a tightening torque is marked on equipment, it must be observed.

- Conductors must be connected to device or equipment terminals for the purpose of transferring energy to them, and the Code specifies that these connections must perform in a satisfactory manner. A thoroughly good connection must be made without damaging the conductors. This connection may be made by a pressure connector (including set-screw types), solder lugs, or splices to flexible leads.

- Elsewhere in the Code, solder connections are prohibited in certain applications. Wire-binding screws, nuts that have upturned lugs, or equivalent are permitted only for 10 AWG or smaller conductors.

- Terminals for more than one conductor and those used to connect aluminum must be identified for the purpose.

- A frequent cause of electrical fires is failure of an electrical connection. To prevent this unfortunate outcome, the Code specifies that splicing devices be identified for the purpose. Alternatively, brazing, welding, or soldering is an option, except where prohibited. (For example, Section 250.148, Continuity and Attachment of Equipment Grounding Conductors to Boxes, specifies that connections depending solely on solder may not be used.) It is further stated that all splices and joints and the free ends of conductors must be covered with an insulation equivalent to that of the conductor or with an insulating device identified for the purpose. For example, when using a wire nut (solderless connector) to join two or more wires, only enough insulation should be stripped from the end to facilitate a good connection. After the wire nut is twisted on, no copper should be visible unless the conductor is bare. Furthermore, wire connectors or splicing means for direct burial are to be listed for the application.

- Here is an NEC requirement that is of great importance and often the subject of exam questions: A conductor's temperature rating may not exceed the lowest temperature rating of any connected termination, connector, or device. Notwithstanding, conductors with temperature ratings higher than specified for terminations are permitted for ampacity adjustment and correction.

- A related requirement, also likely to appear in electricians' licensing exams: Equipment terminations for circuits rated 100 amperes or less or marked for 14 AWG through 1 AWG are to be used for conductors rated 60°C. Equipment terminations for circuits rated over 100 amperes or marked used for conductors larger than 1 AWG are to be used for conductors rated 75°C. (In both cases, higher temperature ratings are allowed if the equipment is listed and identified for such use.

- In the case of a three-phase, delta-connected system that has the midpoint of one phase grounded, it will be noticed that the opposite phase has a higher voltage with respect to ground than the other two phases. This "high leg" must be marked orange or by some other similarly effective means. This marking is to be placed at all connections throughout the system where the grounded conductor is also present.

- Switchboards, panelboards, motor control centers, meter socket enclosures, and the like in other than dwelling units must be field marked to warn against arc-flash hazard. Where high fault current is available, as in many industrial locations, severe injury may occur to workers even when current does not pass through the body, i.e., there is no electrical shock. Flash warning signs are available to comply with this requirement.

- Where arcs, sparks, flames, or molten metal may be generated within electrical equipment in ordinary operation, such equipment must be enclosed or separated and isolated from combustible material. This requirement does not refer to fault current, but instead, it addresses motors, generators, and similar equipment that may have commutators, collector rings, or a centrifugal starting switch.

- It is a Code violation to connect light or power circuits to any system that contains trolley wires with a ground return. An exception permits such usage in the case of car houses, power houses, or passenger and freight stations operated in conjunction with electric railways.

- All electrical equipment must display the manufacturer's name, trademark, or other descriptive identifying marking. Other markings, such as voltage, current, and wattage, are required elsewhere in the Code.

- Disconnecting means must be so marked unless the purpose is self-evident.

- Engineered series combination systems must be field marked to indicate that the equipment is so rated. This is to prevent improper component replacement at some time in the future that could compromise the engineering intent.

- Similarly, manufactured series combination systems also must be so marked.

- Unused current transformers associated with potentially energized circuits are to be short-circuited.

- In nondwelling units, service equipment is to be marked in the field to indicate maximum available fault current. Available fault current often runs very high in an industrial setting.

The foregoing bulleted list contains a number of requirements the Code makers have grouped together and titled Requirements for Electrical Installations. They are important points to keep in mind in the installation phase of a project, and all of them appear on electricians' licensing exams. While the points made are simple (with the possible exception of the sections on temperature limitations and equipment provisions, which have to be looked over carefully before they make sense), they may be a little bit elusive because it is not immediately evident where this content may be found. This is particularly true of Chapters 1 and 2. It is suggested that the NEC student make outlines of these two chapters and read them over until they become very familiar.

Part II of Chapter 1, General, is titled 600 Volts, Nominal, or Less. Here again, this lower-voltage category includes the frequently used

600-volt nominal system. The most important item in Part II is Section 110.26, Spaces about Electrical Equipment. It is important from the standpoint of licensing exams, which always contain working-space and height questions, and also because, on a design/installation level, if a mistake is made and the inspector picks up on it, very costly rework may be triggered. And if the inspector does not catch it and injury or fatality results, the implications are unthinkable. There is a simple remedy—learn Section 110.26, and do not neglect to apply it. On an exam level, the provisions are easy to grasp, and there should never be a problem getting the correct answer, but it may be difficult to access this material in the first place if you do not know where to look. One might search around in Chapter 2, Wiring and Protection, Chapter 3, Wiring Methods and Materials, or Chapter 4, Equipment for General Use or even an annex or table.

Now that we have firmly in mind that working space, dedicated equipment space, and related rules are in Part II (for up to and including 600 volts) of Article 110, Requirements for Electrical Installations, let us take a look at the specific mandates. Section 110.26, Spaces about Electrical Equipment, mandates two separate and discrete spaces—working space and dedicated equipment space. The first of these requires a protective zone around electrical equipment so that in the event of arc flash or other emergency, a worker will not be trapped and unable to escape. Dedicated equipment space, in contrast, serves to isolate equipment and protect it from damage. These are two sets of spaces with separate requirements, both of which must be observed.

Bearing in mind that we are currently examining Part II, which deals with installations 600 volts nominal and less, working spaces are given in Table 110.26(A)(1) (Figure 1-1). This is the place to start. Notice that there are two voltage categories, listed in the first column. These are nominal voltages to ground, not leg-to-leg, so a 240-volt, single-phase system would be in the 150-volt category. Minimum clear distances are given for each of three conditions, and these conditions are spelled out in the note appended under the table. This is all quite clear. It just remains to be ascertained how these numbers are applied, particularly in regard to the geometric layout of these spaces, and also what must be excluded from these spaces. Moreover, there are some exceptions. Throughout we must keep in mind that these are working spaces. Dedicated equipment spaces are different from and in addition to working spaces.

The first thing to consider is which types of equipment require working spaces. In this category is equipment that is likely to require examination, adjustment, servicing, or maintenance. Examples are entrance panels, load centers, motor control centers, and heating and air-conditioning controls. There may be gray areas. If in doubt, provide working space. If the equipment does not fall within this category, the minimum clearances in Table 110.26(A)(1) do not apply.

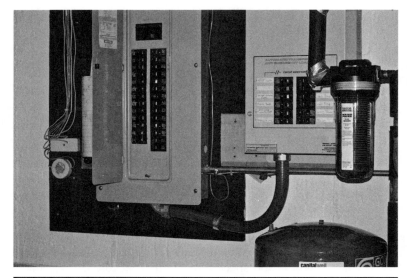

FIGURE 1-1 Water piping infringes the working space in front of this transfer switch in violation of Section 110.26, Spaces about Electrical Equipment. Also, there should be a bonding jumper around the plastic water filter housing to provide ground continuity.

However, an unspecified amount of access and working space is required (in the first paragraph of 110.26) for all electrical equipment.

We normally think of working space as extending in front of equipment, but the same requirements apply to the sides and back if these surfaces need access for examination, adjustment, servicing, or maintenance of electrical equipment. If these surfaces require access to work on nonelectrical equipment, the minimum clear distance for working space is reduced, in Section 110.26(A)(1)(a), Dead-Front Assemblies, to 30 inches. A good way to remember this is to recall that the original working space minimum clear distances in Table 110.26(A)(1) are expressed in feet and inches, whereas the relaxed requirement for nonelectrical work is just in inches, 30 to be precise. We should discuss the geometry of working space: The width of the working space is equal to the width of the equipment, but never less than 30 inches. This, of course, is so that it can be approached by a person. Moreover, the work space must permit equipment doors or hinged panels to open at least 90 degrees, so in some cases the work space must be increased. The 30 inches, where applicable, can extend from either the left edge or the right edge of the equipment; it does not have to be centered. Moreover, working spaces of adjacent pieces of equipment may overlap. This is actually a good thing because it usually means greater than minimum space is provided. In the case of overlapping working spaces, the adjacent equipment cannot be deeper or the working space minimum clearance would be compromised.

As for height, the working space is to be 6½ feet or the height of the equipment, whichever is greater, measured from grade, floor, or platform. Other related electrical equipment above and below the unit in question is permitted as long as it does not extend more than 6 inches from the front surface. There are two exceptions to these rules.

- In existing dwelling units, working spaces for service equipment or panelboards up to 200 amperes are permitted to have less height (not less clear distance).

- Meters in meter sockets may extend beyond the other equipment. (The meter can extend beyond the 6 inches permitted in Section 110.26(A)(3), but the meter socket cannot do so.

We have seen what is needed for working space. There are also related requirements, some simple to understand and implement, some more complex. First, and this should be obvious, the working space is not to be used for storage. This is likely to be beyond the electrician's control because it may happen after the installation has been completed. In an industrial setting, it may be appropriate to outline the working space by painting a yellow line on the floor and appending a warning sign. In the same section, 110.26(B), Clear Spaces, it is further mandated that when normally enclosed live parts are exposed for inspection or servicing, the working space, if in a passageway or general open space, is to be suitably guarded. Another section, 110.26(D), addresses the issue of illumination. It states that illumination is to be provided for the working spaces about service equipment, switchboards, panelboards, and motor control centers that are installed indoors and that this illumination is not to be controlled solely by automatic means. Automatic means, such as a motion detector, are permitted, but a manual bypass is needed as well. These are the simple requirements. The others, Section 110.26(C), Entrance to and Egress from Working Space, which includes (3) Personnel Doors, are more complex, but they need to be dealt with carefully because they appear frequently on exams.

This section requires at least one entrance of sufficient area to provide access to and egress from working space about electrical equipment. Even though the word *access* is used, the main intent is worker safety, i.e., egress. In the event of an arc-flash event and possible subsequent fire, the object is to provide a way out for threatened and/or injured workers. For large equipment rated 1,200 amperes or more and over 6 feet wide that contains overcurrent, switching, or control devices, there must two entrances to and egresses from the working space, and they must be at least 24 inches wide and 6½ feet high at each end—in other words, two doors. This requirement may be reduced to one entrance provided either

- There is continuous and unobstructed means of egress.
- The minimum working clearance is doubled.

If there are personnel doors less than 25 feet from the nearest edge of the working space, they must be equipped with panic hardware so that they open under simple pressure. The idea is that an arc-flash event or electrical fire could severely burn a worker's hands, preventing operation of a doorknob.

These are the basic working space requirements. You can expect exam questions on this and on the following material in 110.26(E), Dedicated Equipment Space, which is less complex. You should keep in mind that both sets of requirements are in Article 110, Requirements for Electrical Installations, and that they are two entirely discrete types of spaces. They serve separate purposes and have different locations.

Working space is concerned primarily with worker safety, whereas dedicated equipment space has to do with equipment integrity. The latter applies to switchboards, panelboards, and motor control centers. Rather than clearance in front of equipment, it begins on grade, platform, or floor; encompasses the height and width (footprint) of the equipment; and extends upward to a height of 6 feet above the equipment or to the structural ceiling, whichever is lower. A suspended ceiling is permitted within this zone. This space is for related electrical equipment, but no piping, ducts, leak-protection apparatus, or other items foreign to the electrical installation may be present. Notice that the dedicated equipment space may have related electrical equipment, but the working space may not. This is so because electrical equipment, related or not, would interfere with the clearances necessary for worker safety.

The area above the dedicated equipment space is allowed to contain foreign systems, provided that protection against water damage is installed. Sprinkler protection is permitted for the dedicated equipment space, but the piping that affords this protection must be outside that zone.

We have examined NEC requirements for working space and dedicated equipment space. Finding the answers for the inevitable exam questions on these topics is fairly straightforward as long as you can instantly locate the requirements in Chapter 1.

Part II of Article 110 concludes with a few simple requirements for electrical installations. Live parts operating at 50 volts or more must be guarded against accidental contact by suitable means, which are enumerated. Section 110.28 lists enclosure types in Table 110.28, Enclosure Selection. Exam questions may concern environmental conditions. Protection is afforded by various enclosure types. Environmental conditions include rain, snow, sleet, corrosive agents, and similar. The table has two parts, titled For Outdoor Use and For Indoor Use. Remember the Code location!

Part III, Over 600 Volts, Nominal, is like Part II, 600 Volts, Nominal, or Less, but the requirements are more stringent for the higher voltage. Construction specifications for transformer vaults are supplied, augmenting those found in Article 450, Transformers and Transformer Vaults (including Secondary Ties). Table 110.31 provides minimum distances from fence to live parts for various nominal voltages. Table 110.34(A), Minimum Depth of Clear Working Space of Electrical Equipment, parallels the table in Part II, but there are five voltage levels, and they are higher. The three conditions are the same.

Table 110.34(E), Elevation of Unguarded Live Parts above Working Space, addresses three voltage levels, which are between phases, not to ground. Section 110.40 states that conductors may be terminated based on the 90°C temperature rating and ampacity unless otherwise identified.

Higher voltage level requirements are rarely the subject of electricians' licensing exams, but if they do appear, you will know where to find the answers. Similarly, Part IV, Tunnel Installations over 600 Volts, Nominal, is outside the scope of most exams. Part V, Manholes and Other Electrical Enclosures Intended for Personnel Entry, All Voltages, contains important provisions regarding size, strength, cabling work space, equipment work space, conductor installation, access, ventilation, and guarding. Fixed ladders must be corrosion-resistant. Exam performance will be enhanced by knowing where this information is located, but it is not realistic to think that all requirements need to be memorized.

This is the content of Article 110, Requirements for Electrical Installations. It is a repository for various mandates that may be difficult to find. After all, the entire Code could be titled Requirements for Electrical Installations. As we have seen, the Index is an excellent gateway, and the Contents is useful in quickly finding the answers to exam questions. In any event, you must have an organized plan for getting to the right Code location without Code hopping or getting bogged down.

NEC® Chapter 2, Wiring and Protection

C hapter 2 is also basic to the Code. The title is very general, so it may be difficult to access its contents under the time constraints of an electricians' licensing exam. Careful preparation and advance consideration are called for. The chapter opens with Article 200, Use and Identification of Grounded Conductors. Exams generally have questions on this article. Since the material is simple and its organization straightforward, you need only recall where to access it. Remember: It is the first article of Chapter 2. Section 200.1, Scope, lists three topics:

- Identification of terminals

- Grounded conductors in premises wiring systems

- Identification of grounded conductors

Identification and correct termination are necessary in any wiring job to ensure that it is free of hazards, the stated goal of the Code. For this reason, examiners are going to focus on the article. It is important in this context to keep in mind the difference between grounded and grounding conductor. Both are connected to the main bonding jumper in the entrance panel (or for a separately derived system, system bonding jumper) and ultimately to the ground electrode, so ideally they are at the same voltage potential. However, the grounded conductor is a current-carrying conductor and, as such, may acquire a different potential, depending on circuit loading and impedance back to the source. For this reason, they must be differentiated. Remember that grounded conductors are treated in Article 200, whereas identification of equipment grounding conductors appears in an entirely different place, Section 250.119. The Index is a reliable guide in this matter.

Before initiating a detailed discussion of color coding, four brief but essential principles are set forth. The first is that aside from color, the grounded conductor, where insulated, is to have insulation suitable for the same circuit if less than 1,000 volts. For over 1,000 volts, if not impedance grounded (which we will discuss later) and if a solidly grounded neutral system, 600-volt insulation will suffice.

The second mandate is that continuity must not depend on connection to a metallic enclosure, raceway, or cable armor. As we shall see when we get to Article 250, Grounding and Bonding, the grounding conductor, with limitations, may use these elements for its purpose, but in many instances an extra wire is required. Details are in Chapter 3.

The third point is that premises wiring is to be connected to a supply system that includes a corresponding grounded conductor. The connection has to be solid as opposed to connection through electromagnetic induction, i.e., induced through magnetic action in the core of a transformer. An exception excludes from this requirement listed utility-interactive inverters found in photovoltaic and fuel-cell power systems.

The fourth stated principle before addressing grounded conductor identification is a brief statement that a neutral conductor may not be used for more than one branch circuit, for more than one multiwire branch circuit, or for more than one set of ungrounded feeder conductors unless permitted elsewhere in the Code. This might appear at first to preclude a multiwire branch circuit, but it does not in part because a multiwire branch circuit is a single circuit, not more than one branch circuit. Later on, as we shall see, the Code takes up the subject of multiwire branch circuits, which have acquired increasingly stringent requirements in recent Code cycles.

Section 200.6, Means of Identifying Grounded Conductors, gets to the heart of the matter in this first segment of Chapter 2, and there is considerable detail. For purposes of a licensing exam or job-specific installation questions, the material does not have to be memorized as long as you know exactly where to access it, specifically not in Article 250, Grounding and Bonding, but in Article 200, Use and Identification of Grounded Conductors. The first thing we have to realize is that these means, i.e., the color-coding rules, differ depending on the size of the conductor. Size 6 American Wire Gauge (AWG) wire and smaller come under a different set of requirements than larger sizes. Throughout the Code, various cutoff figures are used in connection with stated mandates. Sometimes the wording is *less than or equivalent* as opposed to *and smaller*. Notice that the crucial difference is whether the stated cutoff is in the higher category or the lower category. This difference in wording makes a difference in how exam questions are answered, and the examiners often compose questions that test your ability to make the necessary distinction. Carefully read all questions with this in mind because it may make a difference in percentage points on your final score.

In the section under consideration, size 6 AWG and smaller are to be identified as follows:

- A continuous white outer finish
- A continuous gray outer finish
- Three continuous white stripes along the conductor's entire length on other than green insulation
- An outer covering finished to show a white or gray color with colored tracer threads in the braid identifying the source of manufacture
- Field-marked grounded conductor terminations for mineral-insulated, metal-sheathed cable
- Similar field markings for photovoltaic-system single grounded conductors
- Any of the preceding for aerial cable or a ridge on the exterior of the cable (In this and other NEC contexts, a *ridge* means grounded conductor.)

Fixture wire grounded conductor identification is covered in 402.8, Grounded Conductor Identification. An exam question covering this matter could involve some Code hopping.

At one time, gray was used for ungrounded conductors, and Informational Notes stress that care must be taken in working on old wiring lest one get zapped or make a wrong connection. In the past, it was called *natural gray*, but the adjective was dropped as meaningless. (All gray is natural.)

For conductor size 4 AWG and larger, white, gray, or three white stripes the entire length of other than green are permitted, but (here's the difference) in the field, white or gray marking at terminations is an alternative. The marking must entirely encircle the conductor, applied to the insulation. Often electricians make three circles with paint or tape, but the Code requires only one, as long as the circle is complete.

Why are the requirements less stringent in regard to reidentifying larger conductors? It is because in conductor size 4 AWG and larger, it is not realistic to suppose that electricians would stock extra reels of this very expensive wire in different colors in order to comply with color-coding rules. Furthermore, in the larger sizes, reidentification banding is more prominent and durable. Also, there is less chance of error than would be the case with a huge number of branch circuits.

Section 200.6(D), Grounded Conductors of Different Systems, states the principle that grounded conductors of different systems (i.e., voltage, phase, frequency, alternating current [ac], or direct current [dc]) must be distinguished if they occupy the same raceway, cable, box, auxiliary gutter, or other type of enclosure. Methods for complying with this provision are set forth. All grounded conductors of one system are to be identified in the usual way. Other system grounded conductors are identified differently but also comply with 200.6(A) or (B). The means of identification must be posted where the conductors of different systems originate. It is industry practice, not specifically mandated by the NEC, that the higher-voltage system, such as a 480Y/277-volt system, use gray for the grounded conductor, and the lower-voltage system, such as a 208Y/120-volt system, employ white for the grounded conductor.

Section 200.6(E), Grounded Conductors of Multiconductor Cables, contains the same basic mandates but allows exceptions for multiconductor flat-cable size 4 AWG or larger (external ridge permitted), for installations where conditions of maintenance and supervision ensure that only qualified persons service the installation (field-marked terminations permitted), and for varnished-cloth-insulated cable (field-marked terminations permitted).

Section 200.7, Use of Insulation of a White or Gray Color or with Three Continuous White Stripes, opens a different perspective on the whole matter of color coding. Whereas previous sections have approached the subject from the point of view of what means are permitted to identify grounded conductors, this section discusses what uses are permitted for white and other grounded conductor identifiers. Here it is possible to be misled if you don't scrutinize the wording. Section 200.7(A), General, states that the following are to be used to identify only the grounded conductor unless otherwise permitted in (B) or (C):

- White or gray covering
- Three continuous white stripes on other than green insulation
- A marking of white or gray at terminations

Notice that this list of items is restrictive. It is stated that the preceding items may be used to identify only the grounded conductor and for no other purpose. From this it is not possible to conclude that in all cases all these means, notably the last, are permitted.

Section 200.7(B), Circuits of Less Than 50 Volts, says that white (and related) may be used for ungrounded conductors but subject to restrictions in Section 250.20(A). So we must flip ahead to the cited passage. Section 250.20 is titled Alternating-Current Systems to be Grounded, and (A) Alternating-Current Systems of Less Than 50 Volts lists three under-50-volt systems that must be grounded, and from these we may determine which ones are not required to be grounded and so may use white as an identifying means for ungrounded conductors. Circuits requiring a grounded conductor include those

- Supplied by transformers if the transformer supply system exceeds 150 volts to ground

- Supplied by transformers if the transformer supply system is ungrounded

- Installed outside as overhead conductors

Accordingly, indoor or outdoor nonoverhead conductors, sec-ondaries of transformers whose primary does not exceed 150 volts to ground, and secondaries of transformers whose primaries include a grounded conductor may use white (or related) as ungrounded con-ductors. An example would be the commonly used 24-volt thermo-stat wire, which does not need a grounded conductor. Section 200.7(C), Circuits of 50 Volts or More, contains an important NEC 2011 change that will be appearing on many electricians' licensing exams. White and related conductors are permitted to be used as ungrounded con-ductors if reidentified, but only with limitations. It can happen when these conductors are part of a cable assembly such as type NM (trade named Romex). It is not permitted and not useful to reidentify a white conductor for use as an ungrounded conductor in a raceway—you could just pull the correct wire to begin with. Commonly available cable assemblies contain white wires, and the Code recognizes that you have to deal with that as a given. For example, a 240-volt, single-phase ac motor needs two hot wires but no neutral (and, of course, the bare or green equipment ground). Standard NM cable or another cable assembly may be used, reidentifying the white.

Another case where this is called for is the switch loop. Prior to NEC 1999, it was not necessary to reidentify the white wire because it was felt to be self-evident that it would be an ungrounded conductor owing to the fact that it was connected to a switch. For greater clarity, the reidentification requirement was added. Here's the NEC 2011 change: The reidentified wire has to be for the switch supply. The output must be black (or other color permitted for ungrounded con-ductors, i.e., not white or related and certainly not green or bare). Similar use of white for ungrounded conductors is permitted for fix-ture cords, with restrictions.

Sections 200.9 and 200.10 lay out some fairly obvious require-ments regarding terminal identification. For the most part, these apply more to manufacturers than to installers. Notice that terminal identification is not required in panelboards, where it is self-evident. Article 200 closes with 200.11, Polarity of Connections, which states, by way of summation, that no grounded conductor may be connected to any terminal or lead so as to reverse the designated polarity.

Article 210

Article 210, Branch Circuits, is lengthy and contains numerous very basic provisions that appear on all electricians' licensing exams. Preparation for an exam should include thorough familiarity with the contents of this article so that answers can be accessed quickly.

Since the term *branch circuit* appears in many locations throughout the Code, its definition appears in Article 100, Definitions. It consists of the circuit conductors between the final overcurrent device protecting the circuit and one or more outlets.

Part I, General Provisions, contains basic mandates that are easy to understand but may be difficult to find quickly during an open-book exam. For example, if you were asked a question on ground-fault circuit-interrupter requirements, would you know to look under Branch Circuits? Article 210 also covers multiwire branch circuits, more on means of conductor identification, branch circuits required, arc-fault circuit-interrupter protection, overcurrent protection, permissible loads, and required outlets. Interspersed there is coverage of various specific types of equipment, such as heating, air-conditioning, and refrigeration outlets. If you went to Chapter 4, Equipment for General Use, you would be Code hopping, and valuable exam minutes would be lost.

Here we shall discuss specific components of Article 210, with a view not to memorize the specific mandates but instead to create a mental inventory of the contents so that answers can be found as needed.

As usual, Article 210 opens with a statement of scope. It is stated that the article does not cover branch circuits that supply only motor loads. Those provisions are found in Article 430, Motors, Motor Circuits, and Controllers. Provisions of both articles apply to branch circuits with combination loads. Section 210.3, Rating, makes the point that branch-circuit ratings depend on the overcurrent rating or setting. The conductor size does not determine the rating. It may be larger, for example, to control voltage drop or because the installer wanted to overbuild for whatever reason. This section also lists standard ratings: 15, 20, 30, 40, and 50 amperes.

Section 210.4, Multiwire Branch Circuits, has been a contentious subject for many years. Although multiwire branch circuits are permitted by the Code, many electricians don't like them, and their reasons are valid. Because the term appears in many locations throughout the Code, *multiwire branch circuit* is defined in Article 100. It is a branch circuit consisting of two or more ungrounded conductors that have a voltage between them and a grounded conductor that has equal voltage between it and each ungrounded conductor of the circuit. It is connected to the neutral or grounded conductor of the system. The usual scenario is two hot wires from the two legs of a single-phase 120/240-volt panel and a shared neutral. Thus three wires (not counting equipment-grounding conductors, which are always present) are able to do the work of four.

The basic concept is that the two hot wires must be connected to different phases. The neutral carries only the unbalanced current, the difference between two legs. The neutral carries maximum current when one leg is loaded slightly less than the circuit rating so that the breaker does not trip, and the other leg has zero current. However, if you were to increase the current flow on the other leg by loading it more heavily, the neutral actually would carry less current. In cable or raceway, there is less heat than if both loads were on separate two-wire circuits. Significant cost savings result from less wire and smaller conduit. Also, 12-3 type NM is easier to install because it is round. It fishes through hollow spaces easier, and there is less problem with twisting. These are the advantages of using multiwire branch circuits.

Is there a downside? Yes, absolutely. The neutral has between zero and rated circuit current depending on how the legs are balanced. If both hot wires were to be connected to the same phase at the point where they receive power, however, the neutral could be loaded to twice its ampacity. It could become hot enough to ignite nearby combustible material. An error in initial installation or rework at some point in the future could result in property loss or fatality. We will discuss how NEC attempts to keep this from happening, but the fact is that homeowners or untrained workers could shift an ungrounded conductor to the wrong phase.

Section 210.4 begins by stating that multiwire branch circuits, as defined previously, are permitted, stipulating that all conductors of any given multiwire branch circuit must originate from the same panelboard or similar distribution equipment. An Informational Note states that three-phase, four-wire, wye-connected systems may experience heating of the neutral if connected to nonlineal loads owing to high harmonic currents. This is a separate but related problem that may be remedied by altering the load, upsizing the neutral, or creating additional branch circuits. Use of multiwire branch circuits actually may be helpful in this regard, but this does not address the problem of miswired ungrounded conductors.

(B) Disconnecting Means mandates that all ungrounded conductors in each multiwire branch circuit must have provisions for simultaneous disconnection, and this must be at the point where the branch circuit originates. In single-phase installations, the most common means of complying with this requirement is use of a double-pole breaker. It is also permissible to employ two single-pole breakers with a handle tie. The handle tie must be identified. The trunk slammer's expedient of driving a piece of wire through the holes in two adjacent breaker handles is not permitted. The wire could bend or slide out at some time in the future, producing a situation where the load would be turned off but still partially energized, resulting in shock to a person working on the equipment.

(C) Line-to-Neutral Loads provides that multiwire branch circuits may supply only line-to-neutral loads with two exceptions:

- Where only one utilization equipment is supplied

- Where a branch-circuit overcurrent device opens all ungrounded conductors simultaneously

(D) Grouping is a key requirement because it is NEC's answer to the problem of inadvertent miswiring. It requires that all conductors (grounded and ungrounded) of each multiwire branch circuit be grouped by means of cable tie or similar within the panelboard or other point of origin. An exception allows this grouping to be omitted if a cable or raceway entry makes the grouping self-evident. It is likely that this section will be the subject of an exam question. If you know that the relevant NEC language occurs early in Chapter 2, you certainly will gain time on this question.

Section 210.5, Identification for Branch Circuits, recapitulates identification requirements in 200.6, Means of Identifying Grounded Conductors, and looks ahead to 250.119, Identification of Equipment Grounding Conductors. An important new requirement is added: If there is more than one nominal voltage system, ungrounded conductors must be identified by phase or line and system. This identification must be present at all termination, connection, and splice points. The usual means of identification is color coding, but other approved methods are permitted. The method of identification must be posted at each branch-circuit panelboard or distribution equipment.

Section 210.6, Branch-Circuit Voltage Limitations, is a little bit complex and requires scrutiny. It opens with (A) Occupancy Limitation and goes on to stipulate permitted usages for four different voltage levels. The occupancy limitation refers to dwellings, guest rooms, or guest suites of hotels, motels, and similar occupancies. It is stated that voltage between conductors may not exceed 120 volts, nominal, connected to luminaires (at one time referred to as "light fixtures") or to cord-and-plug-connected loads 1,440 volt-amperes, nominal, or less, or less than 1/4-horsepower motors. Larger cord-and-plug-connected equipment such as dryers and cooking equipment (1,440 volt-amperes and over) may be connected to over 120-volt circuits, i.e., 208 or 240 volts.

The voltage-level limitations apply generally. This means that we are no longer talking about the occupancy limitation in (A). Where the voltage does not exceed 120 volts, nominal, these types of equipment may be supplied:

- Lampholders within their voltage rating
- Auxiliary equipment (ballasts and starting equipment) of electric-discharge lamps
- Cord-and-plug or hard-wired utilization equipment

Where the voltage exceeds 120 volts, nominal, between conductors and does not exceed 277 volts to ground, these types of equipment may be supplied:

- Listed electric-discharge or listed light-emitting diode–type luminaires

- Listed incandescent luminaires supplied at 120 volts or less from the output of an integral stepdown autotransformer where the outer shell terminal is electrically connected to a grounded conductor of the branch circuit

- Luminaires that have mogul bases

- Nonscrew shell lampholders applied within voltage ratings

- Auxiliary equipment of electric-discharge lamps

- Cord-and-plug-connected or hard-wired utilization equipment

Notice that 277 volts may not supply a medium-base (common household) screw shell lampholder.

Where the voltage exceeds 277 volts, nominal, to ground but does not exceed 600 volts, nominal, between conductors, these types of equipment may be supplied:

- Auxiliary equipment of electric-discharge lamps in permanently installed luminaires not less than 22 feet high on poles or the like for outdoor illumination (18 feet high in tunnels)

- Cord-and-plug-connected or hard-wired nonluminaire equipment

- Dc–powered luminaires if there is a listed dc ballast that provides isolation between the dc power source and the lamp circuit and protection from electric shock is provided when relamping (Exceptions are provided for industrial infrared heating appliances and railways properties.)

Where the voltage is greater than 600 volts, nominal, between conductors, conditions of maintenance and supervision must ensure that only qualified persons service the installation.

Section 210.7, Multiple Branch Circuits, provides that where two or more branch circuits are connected to the same yoke, there has to be a means to disconnect all ungrounded conductors simultaneously at the entrance panel or other point of origin. The purpose of this requirement is to prevent a situation where a worker deenergizes only part of a device and assumes that it is safe to touch.

Section 210.8, Ground-Fault Circuit-Interrupter Protection for Personnel, is perhaps the most frequently accessed portion of the Code in part because it has been changed to provide expanded coverage every Code cycle since the introduction of ground-fault circuit interrupters (GFCIs) in 1971. These devices are tremendous lifesavers. When a circuit is loaded, a certain amount of current flows through the branch-circuit ungrounded conductor from the point of distribution such as an entrance panel to the load. This translates to a certain finite number of electrons passing a given point along that conductor per second. The same amount of current should flow back through the grounded conductor to the neutral bar at the service. These two quantities should be the same, except for a very small amount of leakage that may take place within a tool or appliance owing to imperfect insulation. A GFCI compares the incoming current in the black wire with the outgoing current in the white wire. If the difference becomes greater than 5 mA, the device trips like a regular circuit breaker and opens the circuit. (For the purpose of listing, this figure may range from 4 to 6 mA.)

Premises wiring may have insulation that has deteriorated owing to age, repeated heating, penetration by an errant nail or nick, or abrasion acquired during initial installation. If fault current in excess of 5 mA flows between the hot side and the equipment-grounding conductor or any normally non-current-carrying conductive body, the device will trip. In a motor, the insulation can break down owing to age or repeated overheating. Sometimes a motor will still function but will cause a GFCI to trip. This is particularly true of a sump pump, submersible motor, or hermetically sealed refrigeration pump/motor. It is sometimes said that GFCIs and refrigeration are incompatible. A freezer in a basement can have its GFCI supply trip out with disastrous consequences. As we shall see, there are certain ways to work around the GFCI rules to prevent this from happening.

There are several variations on the basic GFCI design: receptacle type, attachment plug cap type, circuit-breaker type, plug-in type, and portable. To be in compliance, any of these must be installed in a readily accessible location. The circuit-breaker-type GFCI is easy to use—just install it in place of a conventional circuit breaker in the entrance panel or load center. The only problem is that these units are very expensive. The receptacle-type GFCI is used widely. It has feed-through capability. Supply power goes to the input terminals, and duplex receptacles daisy-chained downstream are all GFCI protected. Included with the GFCI are stickers to be affixed to the protected receptacles. It is common practice to protect an outdoor receptacle by feeding it from a bathroom GFCI, and this is NEC compliant, but the only problem is that a homeowner, finding such an outdoor receptacle dead, may not know that an indoor GFCI needs to be reset. Also, the indoor GFCI may experience nuisance tripping owing to moisture on the outside.

Section 210.8 includes a list of required locations within the context of Article 210, Branch Circuits. Additional required locations, such as health care facilities, are mandated throughout the Code. These can be located by referring to the Index.

The required locations in 210.8 are broken down into (A) Dwelling Units and (B) Other Than Dwelling Units. In this article, the requirements are for 125-volt, single-phase, 15- and 20-ampere receptacles. (Elsewhere, the focus is a little different. For example, Section 422.49, High-Pressure Spray Washers, provides that this equipment, if single-phase, cord-and-plug-connected, and rated at 250 volts or less, must have a factory-installed GFCI as an integral part of the attachment plug or located in the supply cord within 12 inches of the plug. It doesn't specify amperes.)

In dwellings, the locations are bathrooms, garages and accessory buildings, outdoors (except for non–readily accessible receptacles dedicated to snow melting, deicing, or pipeline and vessel heating), crawl spaces at or below grade level, unfinished basements, kitchen countertop surfaces, sinks outside kitchens where receptacles are within 6 feet of the outside edge, and boathouses.

In nondwellings, the locations are bathrooms, kitchens, rooftops, outdoors (with exceptions), and sinks—same zone requirement as in dwellings, indoor wet locations, locker rooms with showers, and garages with electrical diagnostic equipment, electrical hand tools, or portable lighting equipment.

(C) Boat Hoists requires GFCI protection for outlets not exceeding 240 volts. This requirement is for dwellings and is separate from the requirement for boathouses associated with dwellings and applies to both cord-and-plug-connected and hard-wired boat hoists.

Section 210.9, Circuits Derived from Autotransformers, provides that these circuits must have a grounded conductor that is electrically connected to a grounded conductor of the system supplying the autotransformer. There are two exceptions. Exception number 1 exempts stepping up from 208 to 240 volts or stepping down from 240 to 208 volts. You don't need the connection to a grounded conductor when going either way between these two voltage levels only. Similarly, exception number 2 exempts going either way between 600 and 480 volts, but only in industrial occupancies where conditions of maintenance and supervision ensure that only qualified persons service the installation.

It is to be emphasized that the connections required in this article must be solid electrical connections, not connections through electromagnetic induction. A conventional transformer has the primary and secondary windings insulated from each other. An autotransformer has a single winding with primary and secondary circuits tapped at different points to achieve voltage transformation. One connection is common to both circuits, so they are in fact electrically connected, where the other primary and secondary connections are tapped at different points. Care must be used in working with an autotransformer to ensure that high voltage from the primary of a step-down transformer does not infiltrate a secondary circuit. This being said, autotransformers are smaller, less expensive, and more efficient, and so their use is appropriate at times.

Section 210.10, Ungrounded Conductors Tapped from Grounded Systems, is an example of the fact that NEC tap rules occur in various locations throughout the Code, and for this reason, under time constraints of an exam or in design work of an actual project, difficulty may be experienced in accessing the right answers. It would be a good idea to make an index of all NEC tap rules, perhaps written in some blank space in your Code book. However, this could be problematic if your state does not allow such notes to be brought into the exam. The best approach would be to check with your licensing board in advance.

This section allows ungrounded conductors, in the context of a branch circuit, to be tapped from a grounded system. It must be emphasized, however, that even a totally ungrounded circuit must have an equipment-grounding conductor of necessity connected to ground at the entrance panel, load center, or other point of origin.

Ungrounded conductors may be tapped from grounded systems subject to an important condition—that any switching devices in these ungrounded circuits have poles in each ungrounded conductor. All such poles of multipole switches must manually switch together if these switching devices count as the required switching device in the seven types of equipment that are listed in 210.10. Prominent among them is the simple 240-volt motor. Beginning electricians often assume that a grounded neutral should be run for this load and are surprised to find there is nowhere to connect it. The answer is in 210.10, a very brief article with wide application.

Section 210.11, Branch Circuits Required, is another frequently consulted area of the Code. It is simple and concise, but you have to know where to find it. In terms of exam questions, Chapter 2 is of great importance, and Article 210, Branch Circuits, always plays a role, so memorization is called for—not necessarily the exact requirements but rather the location and inner structure so that you can find answers instantly.

(A) Number of Branch Circuits states that the minimum number of branch circuits must be determined from the total calculated load (discussed later in this chapter in great detail) and the size or rating of the circuits used. Remember that the rating of a circuit is determined by the rating in amperes of the overcurrent device, which, in turn, determines the minimum-size conductor. If the conductor size is increased to minimize voltage drop in a long run or for any other reason, it doesn't matter. The circuit rating remains dependent solely on the overcurrent device. It is further stated that in all installations, the number of circuits is to be adequate for the load.

(B) Load Evenly Proportioned Among Branch Circuits presupposes material covered later in Chapter 2. If the load is calculated on a square-foot basis, the wiring from initial service to branch-circuit panelboards must be sufficient to serve that calculated load. This load is to be evenly proportioned among multioutlet branch circuits emanating from the panelboard. What's all this about calculated loads? We shall see when we get to Table 220.12, General Lighting Loads by Occupancy, several pages down the road.

(C) Dwelling Units presents three important branch-circuit requirements that are likely to appear on any electricians' licensing exam and that are also important on a daily basis for those who do residential work.

(1) Small-Appliance Branch Circuits provides that at least two 20-ampere small-appliance branch circuits are required for all receptacle outlets specified by 210.52(B), Small Appliances. This is a forward-looking requirement usually thought of as pertaining to kitchens but actually broader than that. We shall discuss the details when we get to that section.

(2) Laundry Branch Circuits provides that at least one 20-ampere branch circuit is required to supply laundry receptacle outlet(s) required by 210.52(F), Laundry Areas. This circuit must not supply other outlets. Again, this is a forward-looking requirement (to be discussed later).

(3) Bathroom Branch Circuits provides that at least one 20-ampere branch circuit is required to supply bathroom receptacle outlets. This circuit must not supply other outlets. (An exception permits it to supply other outlets within the same bathroom as long as a second bathroom is not supplied.) These branch-circuit requirements for dwelling units are a preview of material appearing later in this chapter.

Section 210.12, Arc-Fault Circuit-Interrupter Protection, mandates sweeping changes in the way residential wiring is done today. Just as GFCIs undoubtedly protect people from shock, arc-fault circuit interrupters (AFCIs) prevent fire. Both types of protection have saved countless lives. First, we'll take a look at how AFCIs work, and then we will examine installation details and where they are required according to NEC 2011.

An arc fault occurs when wiring is damaged by nail penetration or other mishap or as a result of improper installation such as insufficiently torqued terminations or nicked wire, which, owing to reduced ampacity at some location, may burn through but remain close enough to arc. Arc faults may be series or parallel. In a parallel arc fault, the current draw may or may not be sufficient to trigger overcurrent protection shutdown. In a series arc fault, the current is less, and a conventional breaker will not sense that there is a problem. In a circuit that is powered down, an ohmmeter may not detect the fault. This is due to the nature of arcing. Often a parallel arc fault will remain inactive until a certain voltage is applied. At this point, the insulation, either the remnant of damaged conductor insulation or simply an air gap, will ionize and suddenly become conductive. This is like lightning in the sky, where the voltage potential increases until an ionized path is created, at which point that path suddenly becomes conductive, accompanied by release of heat and light. The arc blast may extinguish itself, i.e., blow out the ionized material, whereupon the voltage begins to build again. An arc fault may clear itself by burning out conductive material so as to increase the gap, but before this happens, it may ignite nearby combustible material.

By its nature, an arc fault is not like a steady overload, nor is it like a 60-cycle hum. It has an intermittent, spiky, sputtery quality characterized by fast rise and decay times in the waveform. The AFCI, by virtue of internal circuitry, is able to detect these characteristic signs that there is an arc fault, and it opens the circuit. It is up to the electrician, at this point, to find and correct the fault rather than just swinging the circuit over to a non-AFCI breaker. The potential for preventing loss of life and property far outweighs the downside that includes added cost of initial installation, nuisance tripping, and possible interruption of electrical supply.

AFCIs are required in all 120-volt, single-phase 15- and 20-ampere circuits supplying outlets installed in dwelling-unit family rooms, dining rooms, living rooms, parlors, libraries, dens, bedrooms, recreation rooms, closets, hallways, or similar rooms or areas. Notice that whereas GFCIs are mandated for receptacles, AFCI protection is extended to branch circuits supplying outlets in general. On the other hand, GFCI protection is broader in that it is required for many non-dwelling locations such as commercial kitchens, garages, and outside. In order to protect the entire branch circuit, AFCI devices must be of the breaker type within the entrance panel or load center where the branch circuit begins. Two exceptions permit AFCI protection to begin at the first outlet if upstream wiring is in metal raceway or non-metallic conduit or tubing encased in at least 2 inches of concrete. A third exception exempts individual branch circuits supplying a fire alarm system provided that the supply wiring is in metal raceway or steel-sheathed cable.

Section 210.12(B), Branch-Circuit Extensions or Modifications—Dwelling Units, states that in an area where AFCI protection is required, if existing wiring is modified, replaced, or extended, it must have AFCI protection at the origin of the branch circuit or at the first receptacle outlet. If the second option is chosen, it is not necessary to provide raceway protection upstream. If it were not for this option, the electrician would have to replace an entire entrance panel if it happened to be a fuse box.

Section 210.18, Guest Rooms and Guest Suites, states that these locations, if provided with permanent cooking provisions, must have branch circuits complying with dwelling-unit rules.

Part II, Branch-Circuit Ratings, contains a variety of material pertaining to branch-circuit tap rules, overcurrent protection, receptacle requirements, and much more. It begins by stating that for branch circuits not more than 600 volts, branch-circuit conductors must have ampacity not less than the load to be served. Additionally, circuit conductors supplying continuous loads must have capacity of 125 percent of the load, and in combination, circuits must have capacity of

100 percent of the noncontinuous load plus 125 percent of the continuous load. (An exception allows the extra 25 percent to be waived if the assembly including overcurrent device is listed for 100 percent of its rating.) We recall that a continuous load is defined in Article 100 as one in which the maximum current is expected to continue for 3 hours or more. Fixed electric space heating is an example of a continuous load and, as such, requires 125 percent larger branch-circuit conductors and overcurrent device, which determines the rating of the circuit. This also influences feeder and service sizes.

There are four Informational Notes. Note 1 references Section 310.15 for ampacity ratings of conductors. Here we see the interaction between NEC Chapter 2, Wiring and Protection, and NEC Chapter 3, Wiring Methods and Materials. These two most basic Code chapters go together to form the electrician's fundamental protocol. Chapter 2 states the basic principles, and Chapter 3 tells how to implement them. Note 2 also points ahead to Chapter 4, referring to Part II of Article 430, Motors, Motor Circuits and Controllers, for minimum rating of motor branch-circuit conductors. This introduces us to the fact that motor ampacity calculations are done in an entirely different way than for other types of wiring and have to be considered separately. For now, we merely make mention of this fact, which will be treated when we get to this most important (and lengthy) article. Note 3 again looks ahead to Section 310.15(A)(3) for temperature limitation of conductors. Note 4 mentions the 3 and 5 percent figures for recommended voltage drop—at the farthest outlet of power, heating, and lighting loads and including feeders in addition to branch circuits, respectively. These Informational Notes are not mandatory Code rules. Voltage drop is not considered a safety issue, so these recommended percentages are advisory only.

(2) Branch Circuits with More Than One Receptacle states that the conductors for these circuits are to have an ampacity of not less than the branch circuit. Don't be misled into thinking that the circuit rating or conductor ampacity has something to do with the number of receptacles. Providing lots of receptacles, placed even closer than the maximum spacing required for dwellings, merely guarantees that fewer extension cords will be needed but does not necessarily affect the loading of the circuit.

(3) Household Ranges and Cooking Appliances presents two basic principles regarding sizing of circuit conductors and branch circuits for certain electric cooking appliances. Household ranges, wall-mounted ovens, counter-mounted cooking units, and other household cooking appliances require branch-circuit conductors with an ampacity not less than the rating of the branch circuit and not less than the maximum load to be served. For ranges of 8¾ kW or more rating, the minimum branch-circuit rating is 40 amperes. This 8¾-kW cutoff point becomes very important later (Table 220.55, Demand Factors and Loads for Household Electric Ranges, Wall-Mounted Ovens,

Counter-Mounted Cooking Units, and Other Household Cooking Appliances Over 1¾ kW Rating). At this point, the basic principles are being introduced. Detailed calculation instructions come later, in Article 220, Branch-Circuit, Feeder, and Service Calculations. It is there that the answers to actual Code exam questions will be found.

Exception number 1 contains important tap rules. It is permitted to go to a smaller size when tapping conductors for electric ranges, wall-mounted electric ovens, and counter-mounted electric cooking units. Conductors tapped from a 50-ampere branch circuit must have 20 amperes of minimum ampacity, and they must be sufficient for the load to be served. Included are loads supplied with the appliance when these loads are smaller than the branch-circuit rating. These taps must be no longer than necessary for servicing purposes.

Exception number 2 allows neutral reduction for these same types of cooking equipment. Where the maximum demand of a range has been calculated on the basis of Column C in the Table 220.55 (mentioned above) and is 8¾ kW or more, the neutral may be reduced to 70 percent of the branch-circuit rating, but not smaller than 10 AWG.

(4) Other Loads provides that branch-circuit conductors (other than certain enumerated exceptions) must be at least 14 AWG.

(B) Branch Circuits Over 600 Volts again references those ampacity tables in Chapter 3. Section 310.15, Ampacities for Conductors Rated 0–2,000 Volts, and Section 310.60 are the references. Notice that the cutoff is now 2,000 volts.

(1) General specifies that for over 600 volts, the ampacity of branch-circuit conductors must be at least 125 percent of the designed potential load of equipment that will be operated simultaneously. An alternative is that branch-circuit sizing may be determined by qualified persons under engineering supervision only for supervised installations.

Section 210.20, Overcurrent Protection, emphasizes the regulations discussed earlier regarding continuous and noncontinuous loads.

Section 210.21, Outlet Devices, talks about lampholders and receptacles.

(A) Lampholders states that only heavy-duty lampholders (having a rating of 660 watts or greater if of the admedium type, 750 watts or greater if of any other type) may be connected to a branch circuit rated over 20 amperes. The bottom line is that medium-base screw shell lampholders are not to be connected to branch circuits over 20 amperes. Additionally, ballast circuits of fluorescent lights are restricted to 20 amperes maximum.

(B) Receptacles states in (1) Single Receptacle on an Individual Branch Circuit that such a device must have an ampere rating not less than the branch circuit. It must be remembered that a single receptacle is a yoke that will accommodate only one plug, as opposed to a duplex receptacle. So this section does not apply if there are two or more single receptacles or one or more duplex receptacles. Exceptions are for a portable motor of 1/3 horsepower or less and for arc welders, which have special rules because of their duty cycle.

(2) Total Cord-and-Plug-Connected Load provides that a receptacle connected to a branch circuit supplying two or more receptacles or outlets must not supply a total cord-and-plug-connected load greater than

- 12 amperes for a 15-ampere receptacle connected to a 15- or 20-ampere circuit
- 16 amperes for a 20-ampere receptacle connected to a 20-ampere circuit
- 24 amperes for a 30-ampere receptacle connected to a 30-ampere circuit

The first item presupposes that a 15-ampere receptacle may be connected to a 20-ampere circuit. We normally think of an overcurrent device as protecting the weakest link in the chain, but this is not the case for a 15-ampere receptacle on a 20-ampere circuit, provided that it is not a single, nonduplex receptacle.

(4) Range Receptacle Rating permits the ampere rating of a range receptacle to be based on a single range demand load as specified in Table 220.55 (referenced earlier), which we shall discuss soon.

Table 210.24, Summary of Branch-Circuit Requirements, gives conductor minimum sizes, lampholders permitted, receptacle ratings, and permissible loads for various circuit ratings, overcurrent protection levels, and maximum loads. These range from 15 to 50 amperes.

All of this is fairly straightforward, and answers to exam questions are easy to extract, but would you know that receptacle and lampholder sizing appears in an article titled Branch Circuits?

Section 210.23, Permissible Loads, lays out certain basic rules, organized by circuit rating. First, a fundamental principle is repeated: No load may exceed its branch-circuit rating. (Remember that the branch-circuit rating is set by the overcurrent device.) The remainder of the article applies to branch circuits supplying two or more outlets or receptacles. Many nonelectricians think that an outlet is another name for a receptacle. They will go into a hardware store and ask for 10 outlets, and the sales clerk will know what they are talking about. Actually, an outlet is a point where any utilization equipment receives its supply of electrical energy, including electric motors, smoke detectors, receptacles, and so on. A receptacle sitting on a shelf in a warehouse or store is not an outlet. It becomes an outlet when it is installed.

(A) 15- and 20-Ampere Branch Circuits may supply lighting units, other utilization equipment, or a combination of both, subject to the following:

(1) Cord-and-Plug-Connected Equipment Not Fastened in Place states that such equipment may not exceed 80 percent of the branch-circuit rating.

(2) Utilization Equipment Fastened in Place states that the total rating of such equipment may not exceed 50 percent of the branch-circuit ampere rating if lighting units and/or cord-and-plug-connected utilization not fastened in place are present.

(B) 30-Ampere Branch Circuits permits such circuits to supply fixed lighting with heavy-duty lampholders in nondwellings and utilization equipment in any occupancy. Any one utilization equipment may not exceed 80 percent of the branch-circuit rating.

(C) 40- and 50-Ampere Branch Circuits permits such circuits to supply cooking appliances that are fastened in place in any occupancy. In nondwellings, these circuits may supply fixed lighting with heavy-duty lampholders and other utilization equipment. Thus 240-volt equipment, such as dryers, accordingly, in dwellings is limited to 30 amperes.

(D) Branch Circuits Larger Than 50 Amperes provides that only nonlighting loads may be supplied by these circuits.

A reasonable exam strategy would be to have (A)(1) and (A)(2) committed to memory and to know where to locate (B), (C), and (D).

Section 210.25, Branch Circuits in Buildings with More Than One Occupancy, is very brief but should be good for one exam question.

(A) Dwelling Unit Branch Circuits states the important principle that such circuits in a single dwelling unit may supply loads only within or associated with that dwelling.

(B) Common Area Branch Circuits prohibits common-area loads from being fed from an individual dwelling unit or tenant space. Common areas must have their own panel so that it is possible to disconnect common-area branch circuits without entering a tenant's living area.

Part III, Required Outlets, is one of the really key locations in the Code. This is where we find receptacle layout rules for kitchen and other rooms in dwellings as well as numerous fine points regarding baseboard heat receptacles, show-window mandates, and requirements for heating, refrigeration, and air-conditioning servicing outlets.

Section 210.50, General, contains some specifications, the third of which could appear on a licensing exam and may be hard to find if you don't know the exact figure, although among multiple choices it probably would be the logical choice.

(A) Cord Pendants states that a cord connector supplied by a permanent cord pendant is a receptacle outlet.

(B) Cord Connections states that a receptacle outlet must be installed for any flexible cords with attachment plugs. Receptacles are not required where flexible cords are permanently connected. A screw-plug adapter screwed into a lampholder to create a means for plugging in an attachment plug is prohibited.

(C) Appliance Receptacle Outlets in dwellings must be within 6 feet of the appliance.

Section 210.52, Dwelling Unit Receptacle Outlets, has many requirements that come up all the time in residential work. It is relevant to 125-volt, 15- and 20-ampere receptacle outlets.

These required receptacles are in addition to receptacles that are part of a luminaire or appliance controlled by a wall switch located inside a cabinet or located more than 5½ feet above a floor. This is to say, a receptacle may be permitted higher than 5½ feet above a floor, but it does not count toward fulfilling the minimum required number of receptacles. Throughout this discussion, remember that these requirements are applicable to dwellings only. In a commercial setting, there is not a required number of outlets, and other mandates herein are also not applicable.

It is further stated that factory-installed receptacle outlets that are part of permanently installed electric baseboard heaters may be counted toward the requirement in Section 210.52(A)(1), Spacing. (Needless to say, these receptacles are not to be bugged off the 240-volt heat circuits) (Figure 2-1).

(A) General Provisions states that the following provisions are applicable to every kitchen, family room, dining room, living room, parlor, library, den, sunroom, bedroom, recreation room, or similar room or area of dwelling units:

(1) Spacing provides that receptacles are to be placed so that no point measured horizontally along the floor line of any wall space is more than 6 feet from a receptacle outlet. This would seem to mean that you have to have one receptacle every 12 feet, but actually, it is a bit more complicated than that.

(2) Wall Space defines that term to mean

- Any space 2 feet or more in width (including space measured around corners) and unbroken along the floor line by doorways and similar openings, fireplaces, and fixed cabinets

- The space occupied by fixed panels in exterior walls, excluding sliding panels

- The space afforded by fixed room dividers, such as freestanding bar-type counters or railings

(3) Floor Receptacles states that floor receptacles are not to be counted toward meeting the requirement in (1) unless located within 18 inches of the wall. Remember that face-up receptacles are permitted in a floor but not in a countertop. Generally, these are not used unless there is a problem getting the branch circuit through a wall in retrofit work.

(4) Countertop Receptacles makes the point that receptacles installed for countertop surfaces required by 210.52(C) are not to be counted toward the requirement in Section 210.52(A). This simply means that if a receptacle is at the end of a countertop where a wall space begins, it is of no help in fulfilling the wall-space requirement,

FIGURE 2-1   The Code does not explicitly prohibit receptacle placement over a baseboard heater unit. What it says is that equipment must be installed in accordance with the manufacturer's instructions, and listing requirements for electric baseboard heaters specify that they not be installed below receptacles. The problem is that a cord may get caught inside a slot and become overheated. This baseboard heater is not electric; it is forced hot water, so it is in compliance with the letter if not the spirit of the NEC mandate.

and therefore, the first receptacle along the floor line would have to be 6 feet, not 12 feet, from the end of the counter.

(B) Small Appliances is central to Article 210. It refers back to 210.11(C)(1), which requires only two small-appliance branch circuits in dwelling units. The mandate appeared earlier, and now we come to the details:

(1) Too Much Space states that in the kitchen, pantry, breakfast room, or similar area of a dwelling, two 20-ampere small-appliance

branch circuits are to be provided for wall, floor, and countertop receptacles and receptacle outlets for refrigeration. Notwithstanding this language, the refrigeration receptacle may be supplied by an individual 15-ampere branch circuit. The refrigerator can go either way. Refrigeration equipment, moreover, is exempt from the GFCI requirement. The 20-ampere small-appliance branch-circuit receptacles where they are located along the countertop, of course, must be GFCI protected. Recall that in a nondwelling such as a restaurant, all kitchen receptacles must be GFCI protected, but in a dwelling kitchen, this requirement applies only to countertops.

(2) No Other Outlets specifies that the two small-appliance branch circuits may supply no other outlets, with two exceptions. Exception number 1 permits the 20-ampere small-appliance branch circuits to supply an electric clock. Exception number 2 allows them to supply receptacles for supplemental electric power and lighting on gas-fired ranges, ovens, or counter-mounted cooking units (Figure 2-2). These two exceptions recognize that the loads specified would draw very little current and so would not detract from the main function of the 20-ampere small-appliance branch circuits, which is to power coffee makers, toasters, skillets, mixers, and other common kitchen equipment for which 15-ampere branch circuits might not suffice.

FIGURE 2-2 Countertop spaces separated by range tops, as illustrated here, are considered separate countertop spaces when figuring receptacle spacing. The 20-ampere small-appliance branch circuits are not permitted to serve other loads. An exception allows power for the gas igniter in this gas range because it consumes a minute amount of power.

(3) Kitchen Receptacle Requirements states that the two 20-ampere small-appliance branch circuits must supply the countertop, but they also may supply other receptacles in the rooms enumerated in 210.52(B)(1). If there is a second kitchen in the dwelling, it must have separate 20-ampere small-appliance branch circuits.

(C) Countertops provides the rules for receptacle spacing, clarifies island and peninsular countertop spacing, and vertical location. (All this in an article titled Branch Circuits!)

(1) Wall Countertop Spaces indicates that such spaces are like receptacle spacing along a wall in a room (in dwellings) except that the key figure is 24 inches, not 6 feet. Receptacle outlets are to be placed so that no point along the wall line is more than 24 inches from a receptacle outlet. An exception exempts the space directly behind a range, counter-mounted cooking unit, or sink. However, if the space behind one of these objects is 12 inches or greater (18 inches or greater from the wall intersection for a corner unit), there is no exemption.

(2) Island Countertop Spaces indicates that such spaces with a long dimension of 24 inches and a short dimension of 12 inches need at least one receptacle.

(3) Peninsular Countertop Spaces indicates that such spaces of the same dimensions also need at least one receptacle. Measurements are taken from the connecting edge.

(4) Separate Spaces provides that range tops, refrigerators, or sinks serve to divide a countertop into two separate countertop spaces, each of which must be in compliance with receptacle spacing requirements.

(5) Receptacle Outlet Location provides that receptacles may be on or above, but not more than 20 inches above, the countertop to qualify. Receptacle outlets rendered not readily accessible by appliances fastened in place, appliance garages, sinks, or range tops do not fulfill the requirement. The outlet must be not more than 12 inches below the countertop to qualify (Figure 2-3).

(D) Bathrooms provides that in dwellings, there must be at least one receptacle within 3 feet of the outside edge of each basin (Figure 2-4). Two adjacent basins may share a receptacle. These have to be GFCI protected. They also (like the kitchen) must be not more than 12 inches below the countertop to qualify.

(E) Outdoor Outlets indicates that outdoor outlets are to be installed as follows:

(1) One-Family and Two-Family Dwellings provides that such dwellings at grade level must have for each unit one receptacle outlet accessible while standing at grade level not more than 6½ feet above grade in front of the building and one in back. These must be GFCI protected.

(2) Multifamily Dwellings provides that for each such unit where the dwelling unit is located at grade level and provided with individual exterior entrance/egress, at least one receptacle outlet

FIGURE 2-3   A kitchen receptacle under the countertop counts toward the required maximum spacing as long as it is not more than 12 inches below the countertop or recessed more than 6 inches from the edge of the countertop.

accessible from grade level and not more than 6½ feet above grade is to be installed.

(3) Balconies, Decks, and Porches provides that such appurtenances, if they are accessible from inside the building, must have at least one receptacle outlet not more than 6½ feet above the deck. GFCI protection is necessary.

(F) Laundry Areas provides that in dwellings, such areas must have at least one receptacle supplied by a 20-ampere branch circuit. The circuit can have no other outlets. There are two exceptions: Exception number 1 excludes a dwelling unit that is an apartment in a multifamily building where laundry facilities on the premises are available to all building occupants. Exception number 2 excludes other than one-family dwellings where laundry facilities are not to be installed or permitted.

(G) Basements, Garages, and Accessory Buildings contains two provisions for a one-family dwelling:

(1) In addition to those for specific equipment, at least one receptacle is to be installed in each basement, attached garage, and each detached garage or accessory building that has electric power. These are to be GFCI protected.

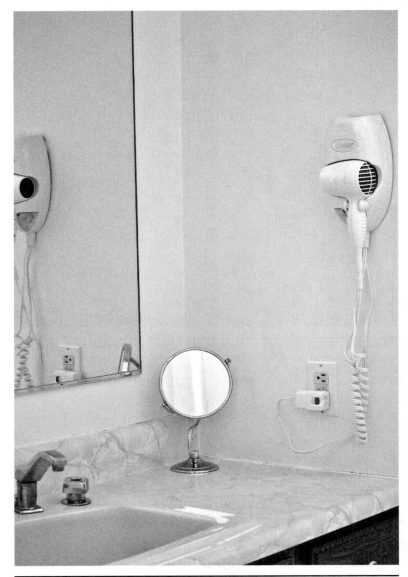

Figure 2-4 Section 210.52, Dwelling Unit Receptacle Outlets, provides in (D), Bathrooms, that in dwelling units at least one receptacle outlet is to be installed within 3 feet of the outside edge of each basin. Section 210.8, Ground-Fault Circuit-Interrupter Protection for Personnel, provides in (A) that all dwelling-unit bathroom receptacles be GFCIs.

(2) If a portion of a basement is finished to create living space, each discrete unfinished portion must have a receptacle outlet.

(H) Hallways provides that hallways 10 feet or more in length must have one or more receptacle outlets.

(I) Foyers indicates that foyers that are not part of a hallway having an area over 60 square feet are required to have receptacles in each wall space 3 feet or more in width unbroken by doorways, floor-to-ceiling windows, and similar openings.

Section 210.60, Guest Rooms, Guest Suites, Dormitories, and Similar Occupancies, provides that such areas are required to have receptacle outlets as if they were dwellings. If these occupancies have permanent provisions for cooking, applicable dwelling requirements kick in. A portable microwave oven is not a permanent provision for cooking. Frequently, these occupancies have permanent furniture layouts. A bed, for example, may be bolted to the wall. Accordingly, the Code allows receptacle locations to be altered so that they are not blocked. Nevertheless, the total number of receptacles must not be reduced from the number required for a dwelling. Receptacles behind a bed must be located to prevent damage to attachment plugs or a suitable guard must be provided.

Section 210.62, Show Windows, indicates that show windows are for displays set up inside a store or commercial establishment that can be viewed by passersby on a sidewalk in the hope of drawing them in to make a purchase. Typically, they consist of a raised platform elevating them from the floor inside and extend to or close to the ceiling. NEC requires that at least one receptacle outlet be installed within 18 inches of the top of the show window for each 12 linear feet of show window area measured horizontally at its greatest width. The purpose of this mandate is to discourage the use of extension cords and floor receptacles.

Section 210.63, Heating, Air-Conditioning, and Refrigeration Equipment Outlet, requires that a 125-volt, single-phase, 15- or 20-ampere receptacle outlet be placed within 25 feet of heating, air-conditioning, and refrigeration equipment. It must be in an accessible (not necessarily readily accessible) location for the purpose of servicing the equipment. The outlet must not be on the branch circuit that powers the equipment so that the equipment can be powered down during servicing. Depending on location, the outlet may have to be GFCI protected. This section appears frequently on electricians' licensing exams. Expect to see a question on the 25-foot figure.

Section 210.70, Lighting Outlets Required provides the following:

(A) Dwelling Units indicates that dwelling units must have lighting outlets in the following areas:

(1) Habitable Rooms indicates that at least one wall switch–controlled lighting outlet is necessary in every habitable room and bathroom (Figure 2-5). An exception allows, in other than kitchens and bathrooms, one or more wall switch–controlled receptacles instead.

FIGURE 2-5 Section 210.70 requires at least one wall switch–controlled lighting outlet in every habitable room and bathroom. The switch should be adjacent to the point of entry, and if there is more than one entry, three-way and possibly four-way switches are needed.

A second exception permits occupancy sensors instead, either in addition to wall switches or located at a customary wall switch location and equipped with manual override.

(2) Additional Locations provides the following:

- At least one wall switch–controlled lighting outlet is required in hallways, stairways, attached garages, and detached garages with electric power.

- For dwelling units, attached garages, and detached garages with electric power, at least one wall switch–controlled lighting outlet is required to provide illumination for outside entrances or exits with grade level access. This requirement does not apply to vehicle doors.

- If one or more lighting outlets are installed for interior stairways, a wall switch is required at each floor level and landing that includes an entrance. The purpose is to control lighting outlets where the stairway between floor levels has six risers or more.

An exception to these three items allows automatic central or remote control in lieu of wall switches.

(3) Storage or Equipment Spaces requires that in attics, under-floor spaces, utility rooms, and basements, at least one switch-controlled lighting outlet be provided if these spaces are used for

storage or if they contain equipment requiring servicing. The light is to be at or near the equipment requiring servicing. An example would be a basement with a furnace and/or water pump.

(B) Guest Rooms or Guest Suites moves away from the subsection on dwellings, although the requirement repeats that in 210.70(A)(1): Every habitable room and bathroom has to have a wall switch–controlled light.

(C) Other Than Dwelling Units requires a lighting outlet in attics and underfloor spaces containing equipment that may require servicing. A switch is to be provided at the point of entry.

Thus ends the lengthy Article 210, Branch Circuits. Most of us cannot retain every provision. Knowledge of the principal provisions and a clear mental map of the article are necessary tools of the trade, especially when taking a licensing exam. You need to know the contents of the article and where the point you are searching for is located in order to avoid Code hopping in the heat of an exam or on the job, while discussing design points with the owner.

Article 215

Article 215, Feeders, continues our tour upstream in the direction of the service. We should revisit Article 100, Definitions, to find out how NEC defines a feeder. A *feeder* consists of all circuit conductors between the service equipment, the source of a separately derived system or other power-supply source, and the final branch-circuit overcurrent device. So the feeder is made up of circuit conductors and does not include the overcurrent devices, disconnects, or loads. The article is much shorter than Article 210, Branch Circuits, but it does contain a few key points that should be comprehended—notably overcurrent protection and common neutrals.

Section 215.1, Scope, states that this article covers installation requirements, overcurrent protection requirements, minimum size, and ampacity of conductors for feeders supplying branch-circuit loads.

Section 215.2, Minimum Rating and Size, is subdivided into sections on feeders not more than 600 volts and feeders over 600 volts. Notice that a 600-volt, nominal, system would fall into the lower category.

(A) Feeders Not More Than 600 Volts:

(1) General states that feeder conductors are to have an ampacity not less than required to supply the load as calculated in Parts III, IV, and V of Article 220. What does this mean? It introduces the whole concept of permitted downsizing based on demand factors. When we get to Article 220, Branch-Circuit, Feeder, and Service Calculations, we will see how these calculations are performed. For now, suffice to say that this article permits us to use smaller conductors than would be necessary for the total connected load. This is so because, owing to

load diversity, there is little chance that the entire connected load would be turned on simultaneously. Even if that happened, there should be no dangerous overheating—the overcurrent device is downsized along with the circuit rating, and it would cut out, serving its intended function. When we get to our treatment of Article 220, we will see that the more units such as laundry equipment there are in a large apartment building, the less statistical chance there is that they will all operate simultaneously, so the more they can be derated. It is also provided that feeder conductors must adhere to the general principle that they are sized not less than the noncontinuous load plus 125 percent of the continuous load. This calculation is to be performed before the application of adjustment or correction factors found in Table 310.60(C)(4), Ambient Temperature Correction Factors. This can be a bit confusing. Requirements in Chapter 2 refer to material in Chapter 3. Complete understanding of an earlier chapter presupposes knowledge of a later chapter. We shall just have to suspend that discussion until we get to it. Once the entire Code has been covered, it will be possible to look ahead and back with great confidence.

An exception permits grounded conductors not connected to an overcurrent device to be sized at 100 percent (not 125 percent) of the continuous load.

(2) Grounded Conductor provides that the size of the feeder-circuit grounded conductor is to be no smaller than required by Section 250.122, Size of Equipment Grounding Conductors, and its associated table, which, despite its title, has wide use throughout the Code for sizing other conductors. Use of this table will be discussed further when we come to consider the very important Article 250, Grounding and Bonding.

(3) Ampacity Relative to Service Conductors states that feeder conductor ampacity is not to be less than that of service conductors when feeder conductors carry the total load supplied by service conductors with an ampacity of 55 amperes or less.

(4) Individual Dwelling Unit or Mobile Home Conductors provides that such feeder conductors never have to be larger than service conductors. Of course, they are allowed to be larger if, for example, the designer/installer decided to upsize them to control voltage drop.

(B) Feeders Over 600 Volts provides that ampacity of feeder conductors supplying a load consisting entirely of transformers is to be the sum of transformer nameplate ratings and that for combination loads consisting of transformers and utilization equipment, the ampacity of feeder conductors supplying these loads is to be the sum of transformer nameplate ratings plus 125 percent of designed potential load of utilization equipment that will be used simultaneously.

Section 215.3, Overcurrent Protection, states that feeders must conform to the overcurrent rules in Article 240, Overcurrent Protection, and furthermore, the continuous-load mandate is in effect.

Section 215.4, Feeders with Common Neutral Conductor, allows up to three sets of three-wire feeders or two sets of four- or five-wire feeders to have a common neutral. This permits a substantial savings in wire and conduit. (If located in a metal raceway or metal enclosure, all conductors sharing a common neutral must be in the same raceway or enclosure. This is to avoid inductive heating.)

Section 215.5, Diagrams of Feeders, states that a diagram showing feeder details must be provided prior to installation if required by the authority having jurisdiction (AHJ). Details are given in the Code text. This mandate applies to dwellings as well as nondwellings, although it is hard to imagine that an inspector would require a diagram for a simple feeder from a main disconnect to an entrance panel. In any event, compliance would be a simple task.

Section 215.6, Feeder Equipment Grounding Conductor, provides that equipment-grounding conductors must be supplied. A common scenario for a residential single-phase, 120/240-volt system is the main disconnect fed by three-wire service conductors from the meter. After the main disconnect, if located in a separate enclosure from the entrance panel, the fourth wire, an equipment-grounding conductor, must be installed and terminated at the entrance panel, which may or may not have a main breaker. Trunk slammers might install three wires and call it good.

Section 215.7, Ungrounded Conductors Tapped from Grounded Systems, allows two-wire dc circuits or ac circuits of two or more ungrounded conductors to be tapped from the ungrounded conductors of circuits having a grounded neutral. Switching devices in the tapped circuit must have a pole in each ungrounded conductor. It is necessary, to protect the maintenance worker, to cut out all legs simultaneously. Of course, in these ungrounded circuits, you still need an equipment-grounding conductor.

Section 215.9, Ground-Fault Circuit-Interrupter Protection for Personnel, states that feeders supplying 15- and 20-ampere receptacle branch circuits are allowed to be protected by a GFCI instead of complying with 210.8 and 590.6(A), which require GFCI protection for workers on construction sites. It is necessary to ensure that lighting will not be affected. It is permissible to GFCI protect a feeder, and this will meet requirements for all branch circuits. Cost savings are possible, but the downside is that in the event of a line-to-ground fault, all branch circuits emanating from that feeder will be taken out.

Section 215.10, Ground-Fault Protection of Equipment, provides that a feeder disconnect rated 1,000 amperes or more installed on solidly grounded wye electrical systems of more than 150 volts to ground but not more than 600 volts phase to phase must have ground-fault protection of equipment. This provision applies to 480Y/277-volt systems.

Section 215.11, Circuits Derived from Autotransformers, requires, in such circuits, a direct (as opposed to inductive) connection to the grounded conductor in the primary. An exception exempts 208- to 240-volt and 240- to 208-volt transformer connections.

Section 215.12, Identification of Feeders, mirrors requirements for branch circuits.

As we have seen, this article is much simpler than Article 210 because there is not the lengthy discussion of loads. The only curve ball is that tap rules are not included. Feeder tap rules are found in Section 240.21(B), Feeder Taps, Section 240.92, Location in Circuit, and Section 430.28, Feeder Taps, which is part of Article 430, Motors, Motor Circuits, and Controllers. This takes us to a lengthy and very challenging topic, Article 220.

Article 220

Article 220, Branch-Circuit, Feeder, and Service Calculations, is made up of five parts. Part I, General, contains some basic introductory material that applies to all the remaining parts. Part II, Branch-Circuit Load Calculations, as the title implies, begins downstream with the basic loads seen in dwellings and nondwellings. Part III, Feeder and Service Load Calculations, is called by electrical designers *the standard calculation*. It is used most frequently. Part IV, Optional Feeder and Service Load Calculations, is a simpler way of figuring these loads, and there can be significant savings, especially on large jobs. Some designers perform calculations using both methods and apply the smaller of the two, whereas others stick to Part III. In an exam setting, you may be asked to perform calculations based on Part III or Part IV, so it is necessary to learn both. Part V, Farm Load Calculations, is different from the other parts because of characteristic farm-load diversity parameters.

Part I, General, states the rules of the game.

Section 220.3, Application of Other Articles, refers to Table 220.3, Additional Load Calculation References, which provides a list of 28 other articles that have provisions that are in addition to or modifications of those within Article 220. Examples are electrolytic cell lines and elevator feeder demand factors. Many of these will be found in Chapter 4, Equipment for General Use.

Section 220.5, Calculations, provides the following:

(A) Voltages. Unless other voltages are specified, for purposes of this article, nominal voltage systems are

- 120
- 120/240
- 208Y/120
- 240

- 347
- 480Y/277
- 480
- 600Y/347
- 600

These nominal voltages are to be used in all calculations. Actual measured voltages may be slightly higher or lower depending on utility specifications and premises wiring impedance and loading.

(B) Fractions of an Ampere states that calculations may be rounded to the nearest whole ampere, dropping decimal fractions smaller than 0.5 ampere. Loads are calculated in volt-amperes (VA) and kilovolt-amperes (kVA) rather than watts and kilowatts, which are often given on the nameplate. These are the same for lineal loads. The first step in calculating loads, circuit ratings, conductor ampacity, and wire size is to convert these load values to amperes, which involves dividing by nominal voltage and rounding off as provided.

Part II, Branch-Circuit Load Calculations, is the basis for the rest of the article. From years of experience and the application of logic, certain typical load figures have emerged.

Section 220.12, Lighting Load for Specified Occupancies, refers to Table 220.12, General Lighting Loads by Occupancy. This is one of the most accessed tables in the entire Code. Twenty occupancies are listed with general lighting load for each. Values range from highs of 3½ VA per square foot for banks and office buildings to lows of ¼ VA per square foot for warehouses and storage spaces. Note that the last three entries are for spaces within a larger occupancy, excluding dwellings, in which the spaces are not differentiated. A very frequently accessed figure is 3 VA per square foot for dwelling units. Stores and schools are rated the same as dwelling units. Some commercial occupancies carry a higher characteristic load, whereas others are rated lower. It is specified that floor areas are calculated from the outside dimensions of the building. For dwelling units, do not include open porches, garages, or unused or unfinished spaces not adaptable for future use. Branch-circuit, feeder, and service loads are based on this table, but this is just the beginning. As we shall see, demand factors and other site-specific considerations have to be considered before a final design can be generated. An Informational Note states that these values are based on minimum-load conditions and 100 percent power factor and may not provide sufficient capacity for the installation contemplated. In view of this warning, the best approach may be to use the standard calculation in all cases. For exam purposes, as mentioned earlier, the optional feeder- and service-load calculations in Part IV must become familiar as well.

Section 220.14, Other Loads—All Occupancies, states that the minimum loads for general-use receptacles and outlets not used for general illumination are as follows:

(A) Specific Appliances or Loads provides that loads not covered in the list that follows are to be based on the ampere rating of the appliance or load served.

(B) Electric Dryers and Electric Cooking Appliances in Dwelling Units permits load calculations specified in Section 220.54 for electric dryers and Section 220.55 for electric ranges and other cooking appliances. Both these sections refer to demand-factor tables found in Part III, Feeder and Service Load Calculations, which we will be discussing soon.

(C) Motor Loads—These are in a category apart from other loads, and calculations are discussed in 430.22, Single Motor, 430.24, Several Motors or a Motor(s) and Other Load(s), and 440.6, Ampacity and Rating (for Air-Conditioning and Refrigeration Equipment).

(D) Luminaires provides that an outlet supplying luminaire(s) is to be calculated based on the maximum volt-ampere rating of the equipment and lamps for which the luminaire(s) is rated. In the case of a dwelling, additional allowance does not have to be made for installed lighting because it is included in the 3 VA per square foot for illumination unless installed lighting is actually greater than provided for in Table 220.12.

(E) Heavy-Duty Lampholders provides that such equipment is to be calculated at 600 VA.

(F) Sign and Outline Lighting states that these outlets are to be calculated at 1,200 VA for each branch circuit, as specified in Section 600.5(A), Required Branch Circuit (for Electric Signs and Outline Lighting).

(G) Show Windows allows a choice of either of two alternatives for these loads:

- 180 VA for each receptacle required by Section 210.62, which calls for one receptacle for each 12 linear feet

- 200 VA for each linear foot of show window

(H) Fixed Multioutlet Assemblies provides that these items in nondwellings or guest rooms or guest suites of hotels or motels are calculated as follows:

- If appliances will not be used simultaneously, each 5 feet or fraction is 180 VA.

- If appliances are to be used simultaneously, each 1 foot or fraction is 180 VA.

(I) Receptacle Outlets states that except in dwellings and banks, receptacle outlets are to be calculated at 180 VA for each single or multiple receptacle on a single yoke. Four or more receptacles for a single piece of equipment are calculated at 90 VA per receptacle. These receptacle outlets are not the lighting outlets installed for general illumination, nor are they small-appliance branch circuits, but they are in addition to these items. The receptacle load for general illumination in dwellings and related is included in the general lighting load given in Table 220.12. In other words, in dwellings, you don't have to count individual receptacles to figure the load; they are already figured in with the lighting. Recall back in Article 210 for nondwellings that you don't have to provide receptacles in accordance with wall-spacing specifications; you just put in what is needed for the particular application. (This all makes sense.) In nondwellings and related, to repeat, receptacles have to be counted outside general illumination.

(J) Dwelling Occupancies provides what we just said in (I), and it can't be overemphasized that an error in this area can be costly in terms of getting the service wrong. In one-family, two-family, and multifamily dwellings and guest rooms or guest suites of hotels and motels, the following outlets are included in the general lighting load calculations of 220.12, and no additional load calculations are required:

(1) All general-use receptacle outlets of 20-ampere rating or less

(2) The receptacle outlets specified in 210.52(E) and (G) (outdoor outlets and basement, garage, and accessory building outlets)

(3) The lighting outlets specified in 210.70 (A) and (B) (dwellings and guest rooms and guest suites)

(K) Banks and Office Buildings states that in these occupancies, the receptacle load (not the lighting load) is to be the larger of the following:

- The calculated load from 220.14(I)

- 1 VA per square foot

(L) Other Outlets provides that other outlets not covered above are to be calculated at 180 VA per outlet.

Section 220.16, Loads for Additions to Existing Installations, provides that the preceding specifications also apply for structural additions to existing buildings or for new or extended circuits in previously wired buildings. This applies to dwellings and nondwellings, as you would expect.

Section 220.18, Maximum Loads, states that the total load must not exceed the rating of the branch circuit and also not exceed the following:

(A) Motor-Operated and Combination Loads provides that Article 430 applies to exclusive motor loads and that Article 440 applies to exclusive air-conditioning, refrigeration, or both. For fastened-in-place utilization equipment with a motor larger than 1/8 horsepower in combination with other loads, the total calculated load is to be based on 125 percent of the largest motor load plus the sum of the other loads.

(B) Inductive and LED Lighting Loads states that calculations for these loads are to be based on total ampere rating of these units and not on the watt rating of the lamps.

(C) Range Loads permits these loads to be sized with demand factors in accordance with Table 220.55, including Note 4. We will be looking closely at Table 220.55 with its associated notes because it provides commonly used demand factors for cooking equipment in dwellings, where there is great diversity of load.

Part III, Feeder and Service Load Calculations, opens with:

Section 220.40, General, which states that the calculated load of a feeder or service is not less than the sum of the loads on the branch circuits supplied, as determined by Part II, which we just finished examining. This is after demand factors permitted by the rest of the present chapter have been applied.

Section 220.42, General Lighting, authorizes use of Table 220.42, Lighting Load Demand Factors, for various occupancies. Various types of occupancies have differing demand factors according to size, except for the general category titled All Others, which allows no reduction regardless of size. Notice that hospitals and hotels and motels, including apartment houses with no provision for cooking, allow reduction right from the start, whereas other occupancies, such as dwellings, have demand factors that don't kick in below a specified size. Notice also that for hospitals and hotels and motels, the demand factors do not apply in areas where the entire lighting is likely to be simultaneous.

Section 220.43, Show-Window and Track Lighting, provides the following:

(A) Show Windows: A load of 200 VA per linear foot is to be used for lighting.

(B) Track Lighting: A load of 150 VA for every 2 linear feet or fraction thereof is to be used for lighting.

Section 220.44, Receptacle Loads—Other Than Dwelling Units, provides that such loads are subject to a demand factor in Table 220.44, Demand Factors for Non-Dwelling Receptacle Loads. This table has two lines: A demand factor of 100 percent, i.e., no reduction, applies to the first 10 kVA. For the remainder, we are allowed to apply a 50 percent demand factor. An alternative is available if we recall Table 220.42, Lighting Load Demand Factors. The receptacle loads are added to the lighting load, and the demand factor in Table 220.42 is applied.

Section 220.50, Motors, reminds us that motor loads are a different matter altogether and are to be calculated in accordance with 430.24, Several Motors or a Motor(s) and Other Load(s), 430.25, Multimotor and Combination-Load Equipment, 430.26, Feeder Demand Factor, and 440.6, Ampacity and Rating (for Air-Conditioning and Refrigeration Equipment).

Section 220.51, Fixed Electric Space Heating, provides that this type of equipment is not subject to demand factors. The AHJ may allow reduction in feeder and service ampacity if units operate on duty cycle, intermittently, or not simultaneously.

Section 220.52, Small-Appliance and Laundry Loads—Dwelling Unit, tells you in (A) that for each small-appliance load, you need to allow 1,500 VA load. This has to be added for every circuit you provide, even if you provide more than the two required. The refrigerator circuit may be excluded. These loads may be added to the general lighting load and are subject to the demand factors in Table 220.42.

(B) Laundry Circuit Load, provides that load also must be figured at 1,500 VA and that it also may be added to the general lighting load so that it becomes subject to the demand factors permitted.

Section 220.53, Appliance Load—Dwelling Units, provides that in dwelling units, only a very large demand factor, 75 percent, i.e., small reduction, may be applied to the nameplate rating of four or more appliances fastened in place, other than electric ranges, clothes dryers, space-heating equipment, or air-conditioning equipment served by the same feeder in a one-family, two-family, or multifamily dwelling. The types of appliances that are excluded are treated elsewhere. What you do, therefore, is make a list of eligible appliances such as water pump, dishwasher, water heater, furnace fan, and so on. Total the nameplate ratings, and apply the 75 percent demand factor as the first step in determining feeder and service conductor size. This simple calculation is often neglected on exams and in the field, resulting in an incorrect service, especially in a large building project.

Section 220.54, Electric Clothes Dryers—Dwelling Units, provides that in dwelling units, each electric clothes dryer is figured at 5,000 VA or the nameplate rating, whichever is larger. The use of Table 220.54, Demand Factors for Household Electric Clothes Dryers, then is permitted. These demand factors become very generous as the numbers grow larger, which makes sense because a household dryer is on for a limited period of time, and this may occur at various times during the day. It does not peak out just before the evening meal like electric cooking equipment does. If two or more single-phase dryers are supplied by a three-phase, four-wire feeder or service, the total load is to be calculated on the basis of twice the number connected between any two phases. (In dwellings, three-phase services are not too common unless it is a very large apartment building. Nevertheless, three-phase questions often appear on exams. Notice that the 1.73 factor does not come into play here because these are single-phase loads connected to two phases of an overall three-phase system.)

Section 220.55, Electric Ranges and Other Cooking Appliances—Dwelling Units, is somewhat more difficult than what we have seen earlier. But there shouldn't be a problem as long as we take it one step at a time. This section applies to household electric ranges, wall-mounted ovens, counter-mounted cooking units, and other household cooking units individually rated in excess of 1¾ kW. These loads may be calculated in accordance with Table 220.55, Demand Factors and Loads for Household Electric Ranges, Wall-Mounted Ovens, Counter-Mounted Cooking Units, and Other Household Cooking Appliances over 1¾ kW Rating. This table is similar to the demand-factor tables we have been using with a few differences. The left-hand column lists number of appliances, which includes 30 categories ranging from 1 to 61 and over. The next two columns give demand factors. Column A is for the above-named appliances with less than a 3½-kW rating. Column B is for the above-named appliances with a 3½- through 8¾-kW rating. Column C gives values for maximum demand in kilowatts, not demand factor expressed as a percentage. This is an odd type of table because we are used to seeing consistent units throughout a given table. Notice that the numbers in Column C are the final kilowatt values, not demand factors. Notice also that Column C is for not over 12-kW rating. So Column C overlaps Columns A and B but also has more (units over 8¾-kW up to and including 12-kW rating). If the foregoing is comprehended, you can figure loading for any job. The only task that remains is going through the notes.

Note 1 provides that for over 12- through 27-kW ranges all of the same rating, the maximum demand in Column C is to be increased by 5 percent for each kilowatt or major fraction exceeding 12 kW.

Note 2 gives the method for calculating the total load for ranges individually rated 8¾ through 27 kW. An average is figured by adding the rating of all ranges and dividing this sum by the number of ranges. Use 12 kW for any range rated less than 12 kW. Then the maximum demand in Column C is increased by 5 percent for each kilowatt or major fraction exceeding 12 kW.

Note 3 provides an optional alternate method of calculating the load of all household electric appliances rated more than 1¾ kW but not more than 8¾ kW.

Note 4, Branch-Circuit Load, states that it is permissible to calculate the branch-circuit load for one range in accordance with Table 220.55. The branch-circuit load for one wall-mounted oven or one counter-mounted cooking unit is the nameplate rating of that appliance. The branch-circuit load for a counter-mounted cooking unit and not more than two wall-mounted ovens, all supplied by a single branch circuit and located in the same room, is calculated by adding the nameplate ratings of the individual appliances and treating the total as equivalent to one range.

Note 5 states that the table is applicable to household cooking appliances rated over 1¾ kW and used in instructional programs.

The final now-familiar point is made that where two or more single-phase ranges are supplied by a three-phase, four-wire feeder or service, the total load is calculated on the basis of twice the maximum number connected between any two phases.

Section 220.56, Kitchen Equipment—Other Than Dwelling Unit(s), provides that it is permissible to calculate the load for commercial electric cooking equipment, dishwasher booster heaters, water heaters, and other kitchen equipment in accordance with Table 220.56, Demand Factors for Kitchen Equipment—Other Than Dwelling Unit(s). The demand factors do not apply to space-heating, ventilating, or air-conditioning equipment. When you examine the table in your Code book, notice that the figures are less generous than those for dwellings. This is so because during a busy season, all such equipment may run full blast for long periods of time. Likewise, in cold weather, all heating equipment may run simultaneously for a long time.

Section 220.60, Noncoincident Loads, states that when it is unlikely that two or more such loads will be used simultaneously, it is permissible to use only the largest load(s) that will be used at one time for calculating total feeder or service load. The classic examples of noncoincident loads are heating and air-conditioning.

Section 220.61, Feeder or Service Neutral Load, is divided into three parts:

(A) Basic Calculation defines neutral load as maximum unbalance of the load. It is the maximum net calculated load between the neutral conductor and any one ungrounded conductor. An exception states that for three-wire, two-phase or five-wire two-phase systems, the preceding resultant has to be multiplied by 140 percent. Two-phase systems are quite rare and not to be confused with single-phase systems, which are said to have two phases.

(B) Permitted Reductions provides that an additional 70 percent demand factor may be applied for

(1) A feeder or service supplying household electric ranges, wall-mounted ovens, counter-mounted cooking units, and electric dryers, where the maximum unbalanced load has been determined in accordance with Table 220.55 for ranges and Table 220.54 for dryers.

(2) That portion of the unbalanced load in excess of 200 amperes where the feeder is supplied from a three-wire dc or single-phase ac system or a four-wire, three-phase or a three-wire, two-phase system or a five-wire, two-phase system.

(C) Prohibited Reductions states that there is to be no neutral or grounded conductor reduction in the following:

(1) Any portion of a three-wire circuit consisting of two ungrounded conductors and the neutral conductor of a four-wire, three-phase, wye-connected system.

(2) That portion consisting of nonlinear loads supplied from a four-wire, three-phase, wye-connected system.

Part IV, Optional Feeder and Service Load Calculations, provides an alternate method of calculating loads, and it can be used in lieu of the standard method in Part III. However, whereas the standard method is applicable to all installations, the optional method can be used only for certain environments, although these are quite common. Specifically, Part IV is for dwellings, schools, and new restaurants. In the case of dwellings, application of this method is limited to those having the total load served by a single 120/240-volt or 208Y/120-volt set of three-wire service or feeder conductors with an ampacity of 100 amperes or more.

Section 220.82, Dwelling Unit, states in (A) that for the preceding service types, the neutral load is determined by 220.61, Feeder or Service Neutral Load, which we discussed earlier, and the calculated load is the result of adding the following:

(B) General loads, made up of 100 percent of the first 10 kVA plus 40 percent of

(1) 3 VA per square foot for general lighting and general-use receptacles. As always, floor areas are figured on the basis of outside building dimensions. Porches, garages, and unused or unfinished spaces not adaptable for future use are not counted.

(2) 1,500 VA for each two-wire, 20-ampere small-appliance branch circuit and each laundry branch circuit.

(3) The nameplate rating of

    a. All appliances that are fastened in place, permanently connected, or located to be on a specific circuit.

    b. Ranges, wall-mounted ovens, and counter-mounted cooking units.

    c. Clothes dryers that are not connected to the laundry branch circuit.

    d. Water heaters.

(4) The nameplate ampere or kVA rating of all permanently connected motors not included in (3).

(C) Heating and air-conditioning load, made up of the largest one of the following:

(1) 100 percent of the nameplate rating(s) of air-conditioning and cooling.

(2) 100 percent of the nameplate rating(s) of a heat pump when used without supplemental electric heating.

(3) 100 percent of the nameplate rating(s) of the heat pump compressor and 65 percent of the supplemental electric heating for central electric space-heating systems. If the heat pump and supplementary heat cannot operate concurrently, the heat pump does not have to be added to the supplementary heat.

(4) 65 percent of the nameplate rating(s) of electric space heating if there are fewer than four separately controlled units.

(5) 40 percent of the nameplate rating(s) of electric space heating if there are four or more separately controlled units.

(6) 100 percent of the nameplate ratings of electric thermal storage and other heating systems where the usual load is expected to be continuous at the full nameplate value. Systems qualifying under this selection are not to be calculated under any other selection in Section 220.82(C).

That's all there is to it for calculating a dwelling under the optional method. Notice that the principle of noncoincidental loading plays a large role. These straightforward calculations take the place of all the demand-factor tables, notes, exceptions, and associated language in Part III. To answer exam questions based on these optional calculations, it is only necessary to go down the list and choose each applicable item and then total them. Do not forget, however, the voltage and system restrictions, and also remember that this is not applicable to nondwellings except in limited situations (i.e., schools and new restaurants).

We shall not treat separately the material on existing dwelling units, multifamily dwellings, schools, and new restaurants because it is substantially the same, with different numbers. The calculations are simple, the material is organized under separate headings, and it is all in Part IV. If you get an exam question on optional feeder and service load calculations, consider yourself fortunate.

Part V, Farm Load Calculations, is a short, simple part. Why are farms in a category by themselves? The answer is in the nature of the venue. Farms are unique in that they usually combine a dwelling and a large enterprise, both of which use lots of electricity. The dwelling includes a great amount of laundry and cooking, often for hired workers, with heat and lighting for a large house. Additionally, the barn may have motors, pumps, large water heaters, electric grain drying equipment, and other loads. The service may be organized around a central yard pole. Proper grounding and adequate safety provisions are essential in this environment.

Section 220.100, General, states that farm loads *shall* be calculated in accordance with Part V. Keep in mind that this is not part of the optional calculation method of Part IV. For farms, it is necessary to use Part V.

Section 220.102, Farm Loads—Buildings and Other Loads, states that a farm dwelling load is to be calculated in accordance with Part III (standard) or Part IV (optional) of Article 220. If the dwelling has electric heat and the farm has an electric grain drying system, the optional method is not to be used where the dwelling and farm loads are supplied by a common service. It states further that if a feeder or service supplies a farm building having two or more branch circuits, loads are to be calculated using Table 220.102, Method for Calculating Farm Loads for Other Than Dwelling Unit. Thus the first step is to figure each qualifying building using these demand factors.

Section 220.103, Farm Loads—Total, provides for the second and final step. Table 220.103, Method for Calculating Total Farm Load, gives the demand factors. Each building is added to make a subtotal, which, in turn, is added to the dwelling load. That's all there is to it.

Article 225

Article 225, Outside Branch Circuits and Feeders, opens with 225.1, Scope. It states that this article covers branch circuits on or between buildings, structures, or poles on the premises.

Frequently, this type of wiring is outside NEC jurisdiction because each building has its own service, and such wiring is upstream of the service point and so under utility control. Some installations, though, such as college and commercial or industrial facilities under single ownership, have transformers, substations, and high-voltage distribution systems. They are under NEC jurisdiction. These installations are covered in Article 225, as are installations at the other end of the spectrum—outside feeders or branch circuits in a residential setting, such as where a detached garage or outbuilding is supplied by a dwelling entrance panel.

Part I, General, contains a number of basic provisions:

Section 225.4, Conductor Covering, states that within 10 feet of any building or structure excluding supporting poles or towers, open individual aerial conductors are to be covered or insulated. Outside this 10-foot limit, we often see bare aluminum wires of various voltages, as in utility power lines. An exception, of course, permits bare equipment-grounding conductors and bare grounded-circuit conductors as permitted elsewhere in the Code.

Section 225.5, Size of Conductors 600 Volts, Nominal, or Less, refers us once again to Section 310.15, Ampacities for Conductors Rated 0–2,000 Volts, which, in turn, refers us to the allowable ampacity tables, to be considered when we get to Chapter 3.

Section 225.6, Conductor Size and Support, takes up a different sizing mode, which is in addition to the ampacity tables. The concern is strength to support conductors and withstand the ravages of wind, ice, and other environmental pressures.

(A) Overhead Spans states that these are to be no smaller than

(1) 10 AWG copper or 8 AWG aluminum for spans up to 50 feet in length and 8 AWG copper or 6 AWG aluminum for a longer span unless supported by messenger wire

(2) For over 600 volts, nominal, 6 AWG copper or 4 AWG aluminum where open individual conductors and 8 AWG copper or 6 AWG aluminum where in cable

(B) Festoon Lighting was defined in Article 100, Definitions. It is a string of outdoor lights that is suspended between two points. It is provided that festoon lighting overhead conductors are not to be smaller than 12 AWG unless the conductors are supported by messenger wires. Keep in mind that these minimum sizes may be superseded by the minimum sizes required in the ampacity tables in Chapter 3. It is further provided that in all spans greater than 40 feet, the conductors are to be supported by messenger wire. Additionally, conductors or messenger wires are not to be attached to fire escapes, downspouts, or plumbing equipment. These items could become energized if not grounded adequately and chafing occurred.

Section 225.7, Lighting Equipment Installed Outdoors, provides

(A) General states that branch circuits are to comply with Article 210, Branch Circuits, and also with the following:

(B) Common Neutral states that the ampacity of the neutral conductor must be not less than the maximum calculated load current between the neutral conductor and all ungrounded conductors connected to any one phase of the circuit. The circuit configuration can be single phase or three phase as long as the neutral can handle the maximum current found in any one phase.

(C) 277 Volts to Ground permits these circuits (>120 volts, nominal, between conductors and not exceeding 277 volts, nominal, to ground) for outdoor lighting in industrial establishments, office buildings, schools, stores, and other commercial or public buildings. This means that these circuits are not permitted for dwellings.

(D) 600 Volts Between Conductors provides that circuits over 277 volts, nominal, to ground but not over 600 volts, nominal, between conductors are permitted to supply the auxiliary equipment of electric-discharge lamps in accordance with Section 210.6(D)(1). That section contains minimum height requirements. These higher voltages require greater heights to protect persons who might be in the area. *Auxiliary equipment of electric-discharge lamps* is a recurring Code phrase, and it means equipment that transforms voltage to a higher level so that it is able to ionize gases and produce light in nonincandescent lights. An example is the common ballast within the enclosure of a fluorescent light fixture.

Section 225.8, Calculation of Loads 600 Volts, Nominal, or Less, refers back to Section 220.10 for branch circuits and Part III of Article 220 for outdoor feeders.

Section 225.10, Wiring on Buildings, allows wiring for circuits not over 600 volts, nominal, to be affixed to the outside of buildings. The section lists permitted wiring methods such as multiconductor cable and various types of raceways. Notice that type NM (trade name Romex) is not among the list of permitted wiring types to be mounted on the outside of a building, but type UF (underground feeder) is permitted.

Section 225.11, Circuit Exits and Entrances, provides that where outside branch circuits or feeders enter or leave a building, they must comply with 230.52, Individual Conductors Entering Buildings or Other Structures, and 230.54, Overhead Service Locations. These sections have a number of requirements regarding bushings, goosenecks, drip loops, and the like. It is interesting to note that some outside branch circuits and feeder requirements are found in Article 230, Services. In the context of a licensing exam, it is useful to know about these jumps so that you don't have to search.

Section 225.12, Open-Conductor Supports, mandates that this wiring method be supported on glass or porcelain knobs, racks, brackets, or strain insulators.

Section 225.14, Open Conductor Spacings, provides the following:

(A) 600 Volts, Nominal, or Less states that these conductors are to have spacings as provided in Table 230.51(C), which is found in Article 230, Services. This table gives minimum distances between supports, between conductors, and from the surface for various voltage levels, also depending on exposure to weather.

(B) Over 600 Volts, Nominal, references spacings for these higher voltages, found in Chapters 1 and 4.

(C) Separation from Other Circuits provides for a minimum separation of 4 inches from open conductors of other circuits or systems.

(D) Conductors on Poles provides for a minimum separation of 1 foot where not on racks or brackets. Horizontal climbing space is also required: 30 inches where power conductors are below communications conductors, 24 inches where power conductors are alone or above communications conductors for 300 volts or less, and 30 inches for over 300 volts. Where communication conductors are below power conductors, the requirement is the same as for power conductors. Where communication conductors are alone, there is no requirement. These provisions are to ensure that workers can climb over or through conductors.

Section 225.15, Supports over Buildings, looks ahead to Section 230.29, which contains requirements for services passing over a roof. It says simply that these conductors must be securely supported by substantial structures, and where practicable, the supports are to be independent of the building.

Section 225.16, Attachment to Buildings, also references material in the article on services.

Section 225.17, Masts as Supports, requires that where a mast is used for the support of final spans of feeders or branch circuits, it is to be of adequate strength or be supported by braces or guys to withstand safely the strain imposed by the overhead drop. For raceway masts, all fittings must be identified for use with masts. Only the feeder or branch circuits specified may be attached to the mast. This means that it is improper to use a mast to support other wiring such as for a telephone or a satellite dish, a common violation. These masthead rules are the same as rules for service masts in Article 230.

Section 225.18, Clearance for Overhead Conductors and Cables, provides height clearances for various conditions:

(1) 10 feet above finished grade, sidewalks, or platforms accessible to pedestrians only if voltage does not exceed 150 volts to ground

(2) 12 feet over residential property and driveways and commercial areas not subject to truck traffic where voltage does not exceed 300 volts to ground

(3) 15 feet for all areas listed in (2) where voltage exceeds 300 volts to ground

(4) 18 feet over public streets, alleys, roads, and parking areas subject to truck traffic, driveways on other than residential property, and other land traversed by vehicles, such as cultivated, grazing, forests, and orchards

(5) 24.5 feet over railroad tracks

It is not reasonable to think that most practicing electricians or exam candidates would have this sort of information committed to memory, but it is necessary for them to know where to find it instantly within the Code without resorting to Code hopping or searching the Index. Especially in Chapter 2, it is recommended that you know the article and section arrangements.

Section 225.19, Clearances from Buildings for Conductors of Not Over 600 Volts, Nominal, provides a number of clearances that are likely to be the subject of exams questions:

(A) Above Roofs states that overhead spans of open conductors and open multiconductor cable must have a vertical clearance of 8 feet minimum above the roof surface. This distance has to be maintained at least 3 feet in all directions from the edge of the roof. This is the basic rule, but it is modified by four exceptions, which should be observed to avoid exam or job-site errors:

Number 1: If subject to pedestrian or vehicular traffic, refer to the preceding section.

Number 2: A reduction in clearance to 3 feet is permitted if the voltage between conductors is less that 300 volts and if the roof slopes 4 inches per foot or greater.

Number 3: A reduction in clearance to 18 inches is permitted for the roof overhang if not more than 6 feet of the conductors (4 feet horizontally) pass over the roof overhang, if the voltage between conductors is not more than 300 volts, and if the conductors terminate at a through-the-roof raceway or approved support.

Number 4: Where the conductors are attached to the side of the building, the 3-foot clearance from the edge of the roof does not apply to the final conductor span.

(B) From Nonbuilding or Nonbridge Structures provides that a 3-foot clearance must be maintained from signs, chimneys, radio and television antennas, tanks, and other nonbuilding or nonbridge structures.

(C) Horizontal Clearances states that such clearances must not be less than 3 feet.

(D) Final spans of feeders or branch circuits must comply with the following:

(1) Clearance from Windows: Final spans to the building must be 3 feet from windows that open, doors, porches, balconies, ladders, stairs, fire escapes, or similar structures. An exception removes the required 3-foot clearance for the area above the top of a window.

(2) Vertical Clearance: Final spans must maintain a vertical clearance of 3 feet horizontal from platforms, projections, or surfaces where the conductors may be reached.

(3) Building Openings: These conductors must not be installed beneath openings through which materials may be moved, such as on farm and industrial buildings.

(E) Zone for Fire Ladders: Where buildings exceed three stories or 50 feet in height, overhead lines are to be arranged where practicable to leave a clear space or zone at least 6 feet to facilitate raising of ladders when necessary for firefighting.

Section 225.20, Mechanical Protection of Conductors, refers, once again, to a section pertaining to services, Section 230.50, Protection Against Physical Damage, which we will be examining.

Section 225.21, Multiconductor Cables on Exterior Surfaces of Buildings, likewise refers to a section on services, Section 230.51, Mounting Supports.

Section 225.22, Raceways on Exterior Surfaces of Buildings or Other Structures, provides that these are to be arranged to drain and must be suitable for use in wet locations. A common Code violation is the use of electrical metallic tubing (EMT) outside with the interior set-screw fittings.

Section 225.24, Outdoor Lampholders, indicates that if these items are attached as pendants, the connections to the circuit wires are to be staggered, and if they have terminals that puncture the insulation to make electrical contact, the wire is to be stranded. This puncturing will not work on solid conductors. The pin will deflect and make only partial contact, causing a series arc fault.

Section 225.25, Location of Outdoor Lamps, states that these are to be below all energized conductors, transformers, or electric utilization equipment, unless

(1) Clearances or other safeguards are provided for relamping operations or

(2) A disconnecting means that can be locked in the open position is provided.

Section 225.26, Vegetation as Support, provides that trees and other vegetation are not to be used as support. Notwithstanding, outdoor luminaires may be supported by trees. They have to be fed by an underground wiring method, not aerial, to comply with this section.

Section 225.27, Raceway Seal, specifies that when a raceway enters a building from an underground distribution system, it is to be sealed, as are any spare or unused raceways.

Part II, Buildings or Other Structures Supplied by a Feeder or Branch Circuit, contains provisions for outside branch circuits and feeders on multibuilding facilities under single ownership or management. The feeder may emanate from a building that contains the initial service, or the service may be outdoor switchgear or a substation that is privately owned, i.e., not utility owned. Often we see high-voltage power lines on poles or underground that resemble utility wiring, but if they are downstream from the service point or power is owner-generated, they fall under NEC jurisdiction.

Section 225.30, Number of Supplies, frequently is the focus of an exam question. The basic rule is that only one feeder or branch circuit (or service, as we shall see later) is permitted to supply a single building or structure, but there are a number of exceptions:

(A) Special Conditions states that additional feeders or branch circuits may supply

(1) Fire pumps

(2) Emergency systems

(3) Legally required standby systems

(4) Optional standby systems

(5) Parallel power-production systems

(6) Systems designed for connection to multiple sources of supply for the purpose of enhanced reliability

(B) Special Occupancies allows by special permission, i.e., from the AHJ, additional feeders or branch circuits for

(1) Multioccupancy buildings where there is insufficient space for supply equipment accessible to all occupants

(2) A single building or structure sufficiently large to make two or more supplies necessary

(C) Capacity Requirements permits additional branch circuits or feeders where the capacity requirements are in excess of 2,000 amperes at a supply voltage of 600 volts or less.

(D) Different Characteristics allows additional feeders or branch circuits for different voltages, frequencies, or phases or for different uses, such as control of outside lighting from multiple locations.

(E) Documented Switching Procedures permits additional branch circuits or feeders to supply installations under single management where documented safe switching procedures are maintained for disconnection.

Section 225.31, Disconnecting Means, requires a disconnect for all ungrounded connectors that supply or pass through a building or structure.

Section 225.32, Location, states that the disconnect may be either inside or outside the building. It must be at a readily accessible location nearest the point of entrance. Again, reference is made to a section that appears in the article on services, namely, Section 230.6, Conductors Considered Outside the Building, which states that conductors are considered to be outside a building if under 2 inches of concrete beneath the building, within a building in a raceway encased in 2 inches of concrete, within a vault as defined in Article 450, Part III, Transformer Vaults, where installed in conduit under 18 inches of earth beneath a building, or where installed in overhead service masts on the outside surface of a building, going through the eave. Four exceptions apply to very limited types of installations.

Section 225.33, Maximum Number of Disconnects, allows not more than six disconnects for each supply. These may be switches or circuit breakers. They may be in a single enclosure, in a group of separate enclosures, or in or on a switchboard. In case of an emergency, it may become necessary for responders to cut the power to a building, so it is important to limit the number of disconnects. Single-pole breakers connected by listed handle ties to control multiwire branch circuits count as one disconnect.

Section 225.34, Grouping of Disconnects, emphasizes that the two to six disconnects permitted above are to be grouped. Each disconnect is to be marked to indicate the load served by it. An exception to this grouping permits the disconnect for a water pump, if also intended for fire protection, to be located in a remote location. The purpose is to prevent firefighters from shutting down the water supply when cutting power to a building.

Section 225.35, Access to Occupants, provides that in a multi-occupancy building, each occupant is to have access to the occupant's supply-disconnecting means.

Section 225.36, Suitable for Service Equipment, states that the disconnecting means is to be suitable for service equipment. An exception allows a snap switch or set of three- or four-way snap switches to suffice for garages and outbuildings on residential property.

Section 225.37, Identification, provides that if a building has a combination of feeders, branch circuits, or services passing through or supplying it, a permanent plaque or directory must be posted at each disconnect identifying all such services, feeders, or branch circuits. This is important for firefighters if they decide to power down a building. It is also important for electricians so that they don't work on energized equipment inadvertently.

Section 225.38, Disconnect Construction, states that the disconnect is to meet these requirements (except for residential garages and outbuildings, as mentioned earlier):

(A) Manually or Power Operable provides for either of these alternatives. If power operable, the disconnect must be capable of being opened by hand in the event of power failure.

(B) Simultaneous Opening of Poles states that the disconnecting means must open all ungrounded supply conductors simultaneously. This is to prevent disabled equipment from remaining hot at one pole.

(C) Disconnection of Grounded Conductor states that such a disconnection is also required, but an actual switch is not necessary. The Code allows pressure connectors at a terminal or bus to which all ungrounded conductors are attached to be considered the disconnecting means.

(D) Indicating states that the disconnecting means must plainly indicate whether it is open or closed.

Section 225.39, Rating of Disconnect, provides that the disconnecting means must have a rating not less than the calculated load to be supplied, determined as we have seen from the various parts of Article 220. Minimum ratings are

(A) One-Circuit Installation: If only a limited load of a single branch circuit is supplied, the branch circuit disconnecting means must have a rating of not less than 15 amperes.

(B) Two-Circuit Installations: If not more than two two-wire branch circuits are installed, the disconnecting means must have a rating of not less than 30 amperes.

(C) One-Family Dwelling: For a one-family dwelling, the feeder disconnecting means must have a rating of not less than 100 amperes, three-wire.

(D) All Others: The feeder or branch-circuit disconnecting means is to have a rating of not less than 60 amperes.

Section 225.40, Access to Overcurrent Protective Devices, provides that if a feeder overcurrent device is not readily accessible, branch-circuit overcurrent devices are to be installed on the load side, must be mounted in a readily accessible location, and are to be of a lower ampere rating than the feeder overcurrent device.

Part III, Over 600 Volts, contains these provisions for outside branch circuits and feeders:

Section 225.50, Sizing of Conductors, states that conductors over 600 volts are to be sized in accordance with 210.19(B), Branch Circuits Over 600 Volts, and 215.2(B), Feeders Over 600 Volts. Branch circuits over 600 volts must have conductors whose ampacity is 125 percent of the designed potential load of utilization equipment that will be operated simultaneously. Feeders over 600 volts that supply transformers only must have an ampacity not less than the sum of the transformer nameplate ratings. Feeders over 600 volts that supply transformers and utilization equipment must have ampacity not less than the sum of the transformer nameplate ratings and 125 percent of the designed potential load of the utilization equipment that will be operated simultaneously.

Section 225.52, Disconnecting Means provides

(A) Location: A building or structure disconnecting means must be located in accordance with 225.32, which gives the requirements for disconnects rated not more than 600 volts. Alternately, these devices may be electrically operated by a similarly located remote-control device.

(B) Type: These disconnects for over 600 volts must disconnect all ungrounded conductors simultaneously and must have a fault-closing rating not less than the maximum short-circuit current available at its terminals. It may be necessary to consult utility representatives in order to ascertain this value, also taking into account premises conductor impedances.

(C) Locking: Disconnecting means must be capable of being locked in the open position. The lockout capability must remain in place with or without the lock.

(D) Indicating: The disconnect must show whether it is off or on.

(E) Uniform Position: If handles are operated vertically, up must be on. An exception exempts double-throw and similar switches.

(F) Identification: A permanent plaque is required for multiple feeders, services, or branch circuits in a single building.

Section 225.56, Pre-Energization and Operating Tests, states that such tests are required for over 600-volt systems to the satisfaction of the AHJ.

Section 225.60, Clearance Over Roadways, Walkways, Rail, Water and Open Land, provides a table of heights with greater values than clearances for not over 600 volts.

Section 225.61, Clearances Over Buildings and Other Structures, provides clearances of over 600-volt conductors. These are greater than the lower voltage level. They mirror similar requirements in the *National Electrical Safety Code* (NESC).

Section 225.70, Substations, contains requirements for indoor and outdoor substations. Questions on this topic do not appear on most electricians' licensing exams, and the work is highly specialized. If you need to find the information, you know where it is.

Article 230

Article 230, Services, is a lengthy article with lots of detailed information. Requirements are somewhat more stringent than for branch circuits and feeders because of the larger available fault current and the potential for taking out an entire dwelling or facility in the event of damage and because there is no overcurrent protection except as provided by the utility. An arcing series fault at an entrance panel is a very serious matter. For these and other reasons, design and installation of a service have to be done right, and the NEC should be followed with great precision.

In one respect, services are easy to learn. For one thing, you can drive around a suburban area and view a vast number of configurations, at least the outside portion. There are multicable and conduit, three-phase, multiple-occupancy, high-capacity with paralleled service conductors, outdoor switch-gear services and much more right in plain view. Also, utilities have spec sheets on the myriad varieties of services, and they provide them free of charge to area electricians. It's all right there in NEC Article 230 as well, so it should be easy to get everything right.

Section 230.1, Scope, states that the article covers service equipment for control and protection of services and their installation requirements. An Informational Note draws attention to Figure 230.1, Services, which is a generalized schematic of a service from source to branch circuits with all components labeled and referenced to Code articles and parts.

Part I, General, has a number of basic requirements. Many of them are similar to material we covered in Article 225, Outside Branch Circuits and Feeders, but there are some key differences.

Section 230.2, Number of Services, states that a building or other structure is to be supplied by only one service unless permitted in (A) through (D):

(A) Special Conditions provides that additional services are allowed for

(1) Fire pumps

(2) Emergency systems

(3) Legally required standby systems

(4) Optional standby systems

(5) Parallel power-production systems

(6) Systems designed for connection to multiple sources of supply

(B) Special Occupancies provides that by special permission, additional services are permitted for

(1) Multiple-occupancy buildings where there is no available space for service equipment available to all occupants

(2) A single building or other structure sufficiently large to make two or more services necessary

It must be emphasized that these two loopholes are available only with the consent of the AHJ, who should be consulted early in the planning stages of any such undertaking.

(C) Capacity Requirements indicates that additional services are permitted

(1) Where the capacity requirements are in excess of 2,000 amperes at a supply voltage of 600 volts or less

(2) Where the load requirements of a single-phase installation are greater than the utility normally supplies through one service

(3) By special permission, generally meaning written permission given by the AHJ

(D) Different Characteristics permits additional services when there are different voltages, frequencies, or phases or if there are different uses. An example would be more than one rate schedule. A hot-water heater may be metered at a different rate if it is controlled by a timer that permits it to run only during off-peak hours.

(E) Identification states that a permanent plaque is required if the building or structure is supplied by more than one service, feeder, or branch circuit.

Section 230.3, One Building or Other Structure Not to Be Supplied Through Another, provides that service conductors supplying one building may not pass through the interior of another building. It is permitted, however, for service conductors to be installed along the exterior wall of one building to supply another building. In this connection, refer to the next section.

Section 230.6, Conductors Considered Outside the Building, indicates that a conductor is considered outside a building

(1) If it is under at least 2 inches of concrete beneath a building

(2) If it is inside the building in a raceway encased in 2 inches of brick or concrete

(3) If it is in a vault

(4) If it is in conduit and under at least 18 inches of earth beneath the building

(5) If it is installed in an overhead service mast on the outside of the building traveling through the eave

Section 230.7, Other Conductors in Raceway or Cable, provides that other conductors may not be in the service raceway or cable. Generally, conductors of different voltages may be in the same conduit, provided that all are insulated for the highest voltage present. A service, however, is a more sensitive matter for reasons discussed earlier and needs a dedicated conduit. (For underground, same trench, different conduit.) There are two exceptions to the dedicated service raceway rule:

Number 1: Grounding conductors and bonding jumpers

Number 2: Load management control conductors having overcurrent protection

Section 230.8, Raceway Seal, states that where a service raceway enters a building or structure from an underground distribution system, it is to be sealed. Spare raceways are also to be sealed. Sealants are to be identified for the application. The purpose is to prevent condensation (owing to temperature differences) from entering the building, main disconnect enclosure, or entrance panel.

Section 230.9, Clearances on Buildings, provides that service conductors installed as open conductors or multiconductor cable without an overall outer jacket must have a clearance of not less than 3 feet from windows that open, doors, porches, balconies, ladders, stairs, fire escapes, or similar locations. As in 225.20, there is an exception for the top level of a window.

(B) Vertical Clearance states that within 3 feet horizontally of platforms, projections, or surfaces, a clearance that varies with voltage must be maintained for service conductors.

(C) Building Openings provides that service conductors are not to be run under such openings if materials will be moved through them. We have seen this requirement before, in Article 225.

Section 230.10, Vegetation as Support, states that vegetation such as trees is not to be used to support service conductors. Who would do this?

Part II, Overhead Service Conductors, provides the following:

Section 230.22, Insulation or Covering, requires individual service conductors to be insulated or covered. An exception allows the grounded conductor of a multiconductor cable to be bare. This bare, grounded conductor is used to support overhead service cable, which under icy conditions may become very heavy and exert a strong tensile force tending to pull the cable off the house.

Section 230.23, Size and Rating, requires that service conductors have sufficient ampacity to carry current for the load calculated in Article 220 and also have adequate mechanical strength.

(B) Minimum Size states that service conductors must be no smaller than 8 AWG copper or 6 AWG aluminum. An exception permits service conductors for limited loads of a single branch circuit to be not smaller than 12 AWG copper.

(C) Grounded Conductors refers ahead for minimum size to Section 250.24(C), Grounded Conductor Brought to Service Equipment, which has several sizing requirements. We shall have a look at these when we consider Article 250, Grounding and Bonding. Code exams frequently focus on service sizing and grounding, and because more than one Code location is involved, a well-organized approach to the material is required. For now, we shall stay with Article 230.

Section 230.24, Clearances, provides that overhead service conductors are not to be readily accessible. Moreover, these rules apply as follows:

(A) Above Roofs states that service conductors are to have a vertical clearance of 8 feet above the roof surface. The modifications we saw in Article 225 apply here as well.

(B) Vertical Clearance for Overhead Service Conductors provides that these conductors, where not in excess of 600 volts, are to have these clearances from final grade:

(1) 10 feet at the service entrance to buildings and at the lowest point of the drip loop, as well as above areas of sidewalks accessible only to pedestrians, only for service-drop cables supported on and cabled together with a grounded bare messenger wire where the voltage does not exceed 150 volts to ground

(2) 13 feet over residential property and driveways and commercial areas not subject to truck traffic where the voltage does not exceed 300 volts to ground

(3) 15 feet for those areas in (2) where the voltage exceeds 300 volts to ground

(4) 18 feet over public streets, alleys, roads, parking areas subject to truck traffic, driveways on other than residential property, and other land such as cultivated, grazings, forests, and orchards

(C) Clearance from Building Openings refers back to 230.9.

(D) Clearance from Swimming Pools refers ahead to 680.8, Overhead Conductor Clearances, and associated Table 680.8. These categories are a little complex owing to the varying voltage levels and swimming pool structures, which go together to determine clearances.

(E) Clearance from Communications Wires and Cables, refers ahead to 800.44(A)(4), which provides for a minimum separation of 12 inches anywhere along the span and a minimum separation of 40 inches at the pole. Clearance between services and communications wires in overhead spans is of great importance because wind can cause chafing if the separation is not maintained, and higher service voltages could come into the building within communication wiring. (This refers to telephone, CATV, broadband, and similar wiring.)

Section 230.26, Point of Attachment, gives point of attachment for service conductors to a building surface. The intent is that the point of attachment of the service-drop conductors must be sufficiently high as to maintain clearance from finished grade along the entire span. This point of attachment must be at least 10 feet above finished grade.

Section 230.27, Means of Attachment, states that multiconductor cable used for overhead service conductors must be attached to the building by fittings identified for such use. Open conductors have to be attached to fittings identified for use with service conductors or to noncombustible, nonabsorbent insulators securely attached to the building.

Section 230.28, Service Masts as Supports, makes three points:

- Where a service mast is used to support service drop conductors, it must be of adequate strength or be supported by braces or guys to withstand the strain imposed by the service drop. Wind and ice add to this load, so care must be taken to make the installation strong enough to endure for years to come.

- Where raceway-type service masts are used, all raceway fittings must be identified for the application.

- Only power service-drop conductors are to be attached to the mast. If you drive around a suburban neighborhood, you likely will see violations of this mandate.

Section 230.29, Supports Over Buildings, provides that service conductors passing over a roof must be securely supported by substantial structures, preferably independent of the building.

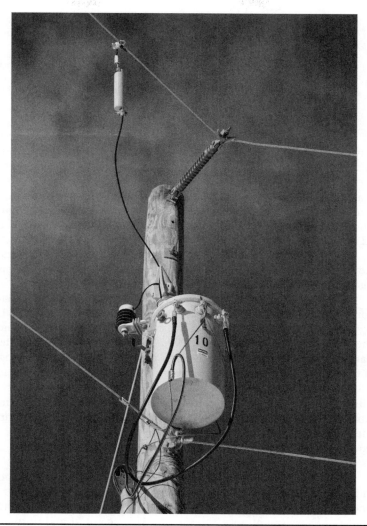

Figure 2-6   This utility-owned transformer supplies a dwelling through an underground service lateral. The point of connection is within the meter socket enclosure because there is no weatherhead at the building, as in an overhead service.

Part III, Underground Service Conductors, states that service-lateral conductors are to be insulated for the applied voltage (Figure 2-6). There are four exceptions for grounded conductors:

Number 1: Bare copper in raceway

Number 2: Bare copper for direct burial where judged suitable for soil conditions

Number 3: Bare copper for direct burial without regard to soil conditions where part of a cable assembly is identified for underground use

Number 4: Aluminum or copper-clad aluminum without individual insulation or covering where part of a cable assembly is identified for underground use or direct burial

Section 230.31, Size and Rating, provides the following:

(A) General states that underground service conductors must have sufficient ampacity to carry the current as calculated under Article 220. They also must have adequate mechanical strength.

(B) Minimum Size states that the conductors must be at least 8 AWG copper or 6 AWG aluminum or copper-clad aluminum. An exception is the same as for overhead service conductors and allows for small loads of a single branch circuit to be supplied by 12 AWG copper.

(C) Grounded Conductors refers ahead to 250.24(C), Grounded Conductor Brought to Service Equipment. There are numerous provisions that we shall discuss shortly.

Section 230.32, Protection Against Damage, requires that service conductors entering a building be installed in accordance with 230.6, which, as we have seen, enumerates conductors considered outside the building, primarily isolated by 2 inches of concrete. An alternative, protection by raceway, will be covered in 230.43, which lists 16 wiring methods applicable to services.

Section 230.33, Spliced Conductors, notes that splices are permitted in service conductors in accordance with 110.14, Electrical Connections, 300.5(E), Splices and Taps, 300.13, Mechanical and Electrical Continuity—Conductors, and 300.15, Boxes, Conduit Bodies, or Fittings—Where Required.

Part IV, Service-Entrance Conductors, specifies the following:

Section 230.40, Number of Service-Entrance Conductor Sets, provides that each service drop, set of overhead service conductors, set of underground service conductors, or service lateral may supply only one set of service-entrance conductors. This seems straightforward, until you see that there are five exceptions, any of which is likely to appear on a licensing exam:

Number 1: A building with more than one occupancy.

Number 2: Where two to six service-disconnecting means in separate enclosures are grouped at one location and supply separate loads from one service drop, set of overhead service conductors, set of underground service conductors, or service lateral, one set of service-entrance conductors is permitted to supply each or several such service-equipment enclosures.

Number 3: A single-family dwelling and its accessory structures is permitted to have one set of service-entrance conductors run to each from a single service drop, set of overhead service conductors, set of underground service conductors, or service lateral.

Number 4: Two-family dwellings, multifamily dwellings, and multiple-occupancy buildings are permitted to have one set of service-entrance conductors installed to supply the circuits covered in 210.25, Branch Circuits in Buildings with More Than One Occupancy, which deals with common areas.

Number 5: One set of service-entrance conductors connected to the supply side of the normal service-disconnecting means is permitted to supply each or several systems covered by 230.82(5) or 230.82(6). These sections deal with equipment permitted to be connected to the supply side of the service disconnect. The sections cited list taps used only to supply load-management devices, circuits for standby power systems, fire pump equipment, and fire and sprinkler alarms, as well as solar photovoltaic systems, fuel-cell systems, and interconnected electric power-production sources.

Section 230.41, Insulation of Service-Entrance Conductors, states that all such conductors entering or on the exterior of buildings or other structures are to be insulated, with these exceptions:

Number 1: Bare copper used in a raceway or part of a service cable assembly

Number 2: Bare copper for direct burial where bare copper is judged to be suitable for the soil conditions

Number 3: Bare copper for direct burial without regard to soil conditions where part of a cable assembly that is identified for underground use

Number 4: Aluminum or copper-clad aluminum without individual insulation or covering where part of a cable assembly or identified for underground use in a raceway or for direct burial

Number 5: Bare conductors used in an auxiliary gutter

Exceptions 1 through 4 are the same as direct burial. Auxiliary gutters are used often on the outsides of buildings to facilitate wiring large services.

Section 230.42, Minimum Size and Rating, provides in (A) that the ampacity of service-entrance conductors before the application of adjustment or correction factors is to be not less than

(1) The sum of the noncontinuous loads plus 125 percent of the continuous loads (An exception permits grounded conductors that are not connected to an overcurrent device to be sized at 100 percent of the continuous and noncontinuous load.)

(2) The sum of the noncontinuous load plus the continuous load if the service-entrance conductors terminate in an overcurrent device where both the overcurrent device and its assembly are listed for operation at 100 percent of their rating

This is a familiar formulation that is repeated in numerous Code locations.

(B) Specific Installations states that in addition to the preceding requirements, the minimum ampacity for ungrounded conductors for specific installations is to be not less than the rating of the service-disconnecting means specified in Section 230.79(A) through (D). These minimums are in addition to conductor ampacity requirements in the Chapter 3 tables.

(C) Grounded Conductor states that such conductor size is as required by Section 250.24(C), Grounded Conductor Brought to Service Equipment, which has sizing requirements for single raceways, parallel conductors in two or more raceways, delta-connected service, and high-impedance grounded neutral systems. Grounding conductor sizes are also considered.

Section 230.43, Wiring Methods for 600 Volts, Nominal, or Less, restricts service-entrance conductor wiring methods to 19 varieties, which are enumerated. Certain types of wiring, such as type NM, are not appropriate for services. See also Uses Permitted and Uses Not Permitted in Articles 320 through 399, which discuss individual wiring and raceway types.

Section 230.44, Cable Trays, permits these systems to support service-entrance conductors. Cable trays used for this purpose are permitted to contain only service-entrance cables unless a barrier is installed. Only these types of cable may be used in this way:

(1) Type SE cable
(2) Type MC cable
(3) Type MI cable
(4) Type IGS cable
(5) Single thermoplastic-insulated conductors 1/0 and larger with CT rating

These cable trays are to be identified with permanently affixed labels.

Section 230.46, Spliced Conductors, mirrors Section 230.33, which has the same title but appears in Part III, Underground Service Conductors. Splices must be in an enclosure or use an underground splice kit, which has to be listed.

Section 230.50, Protection Against Physical Damage, states that where subject to physical damage, service-entrance conductors are to be protected by one of these:

- Rigid metal conduit
- Intermediate metal conduit
- Schedule 80 polyvinyl chloride (PVC) conduit
- Electrical metallic tubing
- Reinforced thermosetting resin conduit
- Other approved means

Section 230.51, Mounting Supports, provides that service-entrance cable is to be supported by straps or other approved means within 12 inches of every service head, gooseneck, or connection to a raceway or enclosure and at intervals not exceeding 30 inches.

Section 230.52, Individual Conductors Entering Buildings or Other Structures, provides that where individual open conductors enter a building or other structure, they must enter through roof bushings or through the wall in an upward slant through individual, noncombustible, nonabsorbent insulating tubes. Drip loops are to be formed on the conductors before they enter the tubes.

Section 230.53, Raceways to Drain, provides that where exposed to the weather, raceways enclosing service-entrance conductors are to be suitable for wet locations and arranged to drain.

Section 230.54, Overhead Service Locations, provides the following:

(A) Service raceways are to be equipped with a service head at the point of connection to service-drop or overhead-service conductors. The service head must be listed for wet locations.

(B) Service-entrance cables are to be equipped with a service head or gooseneck. The service head is to be listed for wet locations. An exception allows type SE cable to be formed in a gooseneck and taped with a self-sealing weather-resistant thermoplastic.

(C) Service heads and goosenecks are to be above the service-drop or overhead-service attachment.

(D) Service-entrance cables are to be held securely in place.

(E) Service heads must have conductors of different potential brought out through separately bushed openings.

(F) Drip loops are to be formed on individual conductors. To prevent the entrance of moisture, service-entrance and overhead-service conductors are to be arranged so that water will not enter the service raceway or equipment.

Section 230.56, Service Conductor with the Higher Voltage to Ground, states, as we have seen before, that on a four-wire, delta-connected service where the midpoint of one phase winding is grounded, the service conductor having the higher phase voltage to ground is to be durably and permanently marked by an outer finish that is orange in color or by other effective means at each termination or junction point. This marking must be at both ends of the service conductor.

Part V, Service Equipment—General, provides the following:

Section 230.62, Service Equipment—Enclosed or Guarded, provides that energized parts of service equipment are to be enclosed or guarded as specified:

(A) Enclosed: Energized parts are to be enclosed so that they will not be exposed to accidental contact or guarded.

(B) Guarded: Energized parts that are not enclosed are to be installed on a switchboard, panelboard, or control board and guarded. Where energized parts are guarded, a means for locking or sealing doors providing access to the energized parts is to be provided.

Section 230.66, Marking, states that service equipment rated at 600 volts or less is to be marked to identify it as being suitable for use as service equipment. All service equipment is to be listed. Individual meter socket enclosures are not considered service equipment and do not have to be listed under this section.

Part VI, Service Equipment—Disconnecting Means, provides the following:

Section 230.70, General, states that means are to be provided to disconnect all conductors in a building or other structure from the service-entrance conductors.

(A) Location states that the disconnecting means is to be installed in accordance with the following:

(1) The service-disconnecting means is to be installed at a readily accessible location either outside of a building or inside the nearest the point of entrance of the service conductors.

(2) The service-disconnecting means may not be installed in bathrooms.

(3) If a remote-control device is used to actuate the service-disconnecting means, the service-disconnecting means is to be located as provided in (1).

The reason for locating the disconnecting means close to where the service-entrance conductors enter the building is that these conductors are not protected by any reasonable degree of overcurrent protection, so they have to receive extra protection from damage. If there is a remote-control device, its location is not important in this regard. It is the main disconnect location that determines exposure to damage of the service-entrance conductors. Often it is possible to have the main disconnect within the entrance panel, but not always. In all cases, the conductors from the main disconnect to the entrance panel are a feeder and as such require a fourth equipment-grounding conductor.

(B) Marking provides that each service disconnect must be permanently marked to identify it as a service disconnect.

(C) Suitable for Use states that each service disconnect must be suitable for prevailing conditions. Examples include outdoors and hazardous (classified) locations.

Section 230.71, Maximum Number of Disconnects, provides the following:

(A) General states that the service disconnect is to consist of not more than six switches or sets of circuit breakers mounted in a single enclosure, in a group of separate enclosures, or in or on a switchboard. There may be no more than six sets of disconnects per service and they must (with exceptions) be grouped in one location.

Section 230.72, Grouping of Disconnects, states that these two to six disconnects are to be grouped. An exception allows a disconnect for a water pump also intended for fire protection to be installed in a remote location. If this is done, a plaque must be posted at the grouped disconnects indicating the location of the remote disconnect.

(B) Additional Service-Disconnecting Means provides that the one or more additional service-disconnecting means for fire pumps, emergency systems, and legally required standby or optional standby services are to be installed remote from the service-disconnecting means to preclude interruption of these vital supplies.

(C) Access to Occupants states that in a multiple-occupancy building, each occupant is to have access to the occupant's service-disconnecting means. This provision parallels the requirement in 240.24(B) for access to service, feeder, and branch-circuit overcurrent protective devices.

Section 230.74, Simultaneous Opening of Poles, provides that each service disconnect is to disconnect all ungrounded service conductors from the premises wiring system simultaneously. This is a familiar mandate, and it applies to many levels of wiring in all occupancies.

Section 230.75, Disconnection of Grounded Conductor, says that where the service-disconnecting means does not disconnect the grounded conductor from the premises wiring, other means are to be provided for this purpose in the service equipment. A terminal or bus to which all grounded conductors can be attached by means of pressure connectors will suffice.

Section 230.76, Manually or Power Operable, states that the service-disconnecting means for ungrounded service conductors is to consist of either

(1) A manually operable switch or circuit breaker equipped with a handle or other suitable operating means or

(2) A power-operated switch or circuit breaker, provided that the switch or circuit breaker can be opened by hand in the event of a power-supply failure.

Note that similar language appeared previously, in 225.38, Disconnect Construction, in Article 225, Outside Branch Circuits and Feeders.

Section 230.77, Indicating, provides that the service-disconnecting means is to plainly indicate whether it is on or off.

Section 230.79, Rating of Service-Disconnecting Means, states that the service-disconnecting means is to have a rating not less than the calculated load to be carried, determined in accordance with Article 220, as we have seen. In no case is the rating to be lower than the following:

(A) One-Circuit Installations states that for installations to supply only limited loads of a single branch circuit, the service-disconnecting means must have a rating of not less than 15 amperes.

(B) Two-Circuit Installations states that for installations consisting of not more than two two-wire branch circuits, the service disconnecting means must have a rating of not less than 30 amperes.

(C) One-Family Dwellings states that for a one-family dwelling, the service-disconnecting means must have a rating of not less than 100 amperes, three-wire.

(D) All Others states that for all other installations, the service-disconnecting means must have a rating of not less than 60 amperes.

Section 230.80, Combined Rating of Disconnects, provides that where the service-disconnecting means consists of more than one switch or circuit breaker, the combined ratings of all the switches or circuit breakers used is to be not less than the rating required by 230.79. This refers to up to six individual switches or circuit breakers grouped together.

Section 230.81, Connection to Terminals, says that the service conductors are to be connected to the service-disconnecting means by pressure connectors, clamps, or other approved means. Connections that depend on solder are prohibited.

Section 230.82, Equipment Connected to the Supply Side of Service Disconnect, provides an exclusive list of equipment that is permitted to be connected to the supply side of the service-disconnecting means:

(1) Cable limiters or other current-limiting devices

(2) Meters and meter sockets nominally rated not less than 600 volts, provided that all metal housings and service enclosures are grounded and bonded

(3) Meter disconnect switches nominally rated not in excess of 600 volts that have a short-circuit current rating equal to or greater than the available short-circuit current (These also must be grounded and bonded.)

(4) Instrument transformers (current and voltage), impedance shunts, load-management devices, surge arresters, and type 1 surge-protective devices

(5) Taps used only to supply load-management devices, circuits for standby power systems, fire pump equipment, and fire and sprinkler alarms, if provided with service equipment and installed in accordance with requirements for service-entrance conductors

(6) Solar photovoltaic systems, fuel-cell systems, or interconnected electric power-production sources

(7) Control circuits for power-operable service-disconnecting means, if suitable overcurrent protection and disconnecting means are provided

(8) Ground-fault protection systems or type 2 surge-protective devices, where installed as part of listed equipment, if suitable overcurrent protection and disconnecting means are provided

(9) Connections used only to supply listed communications equipment under the exclusive control of the serving electric utility, if suitable overcurrent protection and disconnecting means are provided (For installations of equipment by the serving electric utility, a disconnecting means is not required if the supply is installed as part of a meter socket such that access can be gained only with the meter removed.)

Part VII, Service Equipment—Overcurrent Protection, provides the following:

Section 230.90, Where Required, provides that each ungrounded service conductor is to have overload protection. It is known that the service conductors are unfused, or if they have overcurrent protection, it is furnished by the utility at a high level, well beyond their ampacity. So how is this possible?

Overload protection is different from short-circuit and ground-fault protection. Service conductors do not have short-circuit and ground-fault protection in the form of an overcurrent device. The protection they have is by virtue of their location, length limitation on the indoor segment, conductor type, and any raceway that may be in place. Overload protection is accomplished by the overcurrent device that is at the downstream end of the conductors that are protected, i.e., the main breaker in the entrance panel or in the main disconnect enclosure. To summarize, short-circuit and ground-fault protection has to be at the upstream end of the protected conductor. Overload protection may be at the downstream end. The same concept comes into play with taps. Taps typically are protected from overload at their downstream end, and as a tradeoff, the length of the tap is limited, and enhanced raceway protection is required.

(A) Ungrounded Conductor provides that the overload protection just mentioned must be installed in series with each ungrounded service conductor. Since this section does not specify the location of that overload protection, the main breaker(s) or fuses at the downstream end meet the requirement. There are five exceptions:

Number 1 permits motor-starting current to be figured on the basis of Article 430, Motors, Motor Circuits, and Controllers.

Number 2 permits fuses and circuit breakers that comply with 240.4(B), Overcurrent Devices Rated 800 Amperes or Less, 240.4(C), Overcurrent Devices Rated Over 800 Amperes, and 240.6, Standard Ampere Ratings. These sections provide standard ampere ratings and provide for rounding off under 800 amperes, but not over 800 amperes.

Number 3 states that two to six circuit breakers or sets of fuses are permitted to provide the overload protection. The sum of the ratings of the circuit breakers or fuses is permitted to exceed the ampacity of the service conductors, provided that the calculated load does not exceed the ampacity of the service conductors.

Number 4 provides that overload protection for fire pump supply conductors is to comply with 695.4(B)(2)(a). This section states that the overcurrent protective device for a fire pump is to be rated to carry indefinitely the sum of the locked-rotor current of the fire pump motor and the pressure-maintenance pump. The principle here is that a fire pump must not be shut down by a protective device while it is pumping water necessary to fight a fire. It's better to save the building than save the pump and lose the building.

Number 5 states that overload protection for 120/240-volt, three-wire, single-phase dwelling services is permitted in accordance with the requirements of 310.15(B)(7), which refers to a table. This table gives sizes for residential services, and the sizes are less than required by the general ampacity tables. Notice that these sizes cannot be used for nondwelling services.

(B) Not in Grounded Conductor provides that no overcurrent device is to be inserted in a grounded service conductor except a circuit breaker that simultaneously opens all conductors of the circuit. This is to preclude the possibility that the grounded conductor would lose continuity while the circuit remained energized.

Section 230.91, Location, states that the service overcurrent device is to be an integral part of the service-disconnecting means or is to be located immediately adjacent.

Section 230.92, Locked Service, says that if service overcurrent devices are locked or sealed or are not readily accessible to the occupant, branch-circuit or feeder overcurrent devices must be installed on the load side, must be mounted in a readily accessible location, and must be of lower ampere rating than the service overcurrent device.

Section 230.93, Protection of Specific Circuits, provides that where necessary to prevent tampering, an automatic overcurrent device that protects service conductors supplying only a specific load, such as a water heater, may be locked or sealed where located so as to be accessible.

Section 230.94, Relative Location of Overcurrent Device and Other Service Equipment, states that the overcurrent device is to protect all circuits and devices. This basic principle, however, is amended by six exceptions:

Number 1: The service switch is permitted on the supply side.

Number 2: High-impedance shunt circuits, surge arrestors, type 1 surge-protective devices, surge-protective capacitors, and instrument transformers (current and voltage) are permitted to be connected and installed on the supply side of the service-disconnecting means.

Number 3: Circuits for load-management devices are permitted to be connected on the supply side of the service overcurrent device where separately provided with overcurrent protection.

Number 4: Circuits used only for the operation of fire alarm, other protective signaling systems, or the supply to fire pump equipment are permitted to be connected on the supply side of the service overcurrent device where separately provided with overcurrent protection.

Number 5: Meters nominally rated not in excess of 600 volts are permitted, provided that all metal housings and service enclosures are grounded.

Number 6: Where service equipment is power operable, the control circuit is permitted to be connected ahead of the service equipment if suitable overcurrent protection and disconnecting means are provided.

Section 230.95, Ground-Fault Protection of Equipment, applies to solidly grounded wye electric services of more than 150 volts to ground but not exceeding 600 volts phase to phase where the service disconnect is rated 1,000 amperes or more. It states that ground-fault protection is to be installed for such equipment, with the grounded conductor connected directly to ground through a grounding electrode system with no resistor or impedance device. The rating of the service disconnect is the rating of the largest fuse that can be installed or the highest continuous-current trip setting for which the actual overcurrent device installed in a circuit breaker is rated or can be adjusted.

We shall be considering impedance grounding with a grounding resister inserted between the grounded conductor and the grounding electrode (ground rod or other grounding means) when we get to Article 250, Grounding and Bonding. An impedance grounding system is in contrast to a solidly grounded system.

(A) Setting provides that the ground-fault protection system is to operate to cause the service disconnect to open all ungrounded conductors of the faulted circuit. The maximum setting of the ground-fault protection is to be 1,200 amperes, and the maximum delay is to be 1 second for ground-fault currents equal to or greater than 3,000 amperes.

Keep in mind that the foregoing applies only for the larger wye electric services described earlier.

(B) Fuses states that if a fuse and switch combination is used, the fuses must be capable of interrupting any current higher than the interrupting capacity of the switch during a time that the ground-fault protective system will not cause the switch to open.

(C) Performance Testing mandates testing requirements for the preceding equipment. A written record of the test must be made and must be available to the AHJ.

Part VIII, Services Exceeding 600 Volts, Nominal. Most of us do not normally install services of over 600 volts. (The 600-volt service is not in this category because it is not over 600 volts.) Licensing exams usually do not focus on the over 600-volt material. Nevertheless, it is useful to know about these matters if for no other reason than to gain perspective. Some states require special licensing to work on higher voltages.

An Informational Note references the *National Electrical Safety Code* for clearances of conductors over 600 volts.

Section 230.202, Service-Entrance Conductors, has two fundamental mandates for over 600-volt services:

(A) Conductor Size states that these service-entrance conductors are not to be smaller than 6 AWG unless in multiconductor cable. Multiconductor cable conductors must be at least 8 AWG.

(B) Wiring Methods states that service-entrance conductors are to be installed by one of the wiring methods covered in 300.37, Above-Ground Wiring Methods (for Over 600 Volts), and 300.50, Underground Installations (for Over 600 Volts).

Section 230.204, Isolating Switches, requires isolating switches for over 600-volt disconnecting means and all associated service equipment, where the service-disconnecting means consists of oil switches or air, oil, vacuum, or sulfur hexafluoride circuit breakers. The isolating switch must have visible break contacts. It is to be installed on the supply side of the disconnecting means and all associated service equipment. The isolating switch is to be equipped with a means for readily connecting the load-side conductors to a grounding electrode system, equipment-ground busbar, or grounded steel structure when disconnected from the source of supply.

Section 230.205, Disconnecting Means, provides that it may be in a location that is not readily accessible if it can be operated by mechanical linkage from a readily accessible point. Electronic means to accomplish the same purpose is also permitted.

Section 230.206, Overcurrent Devices as Disconnecting Means, permits the circuit breaker to constitute the service-disconnecting means provided that it complies with 230.205.

Other sections give protection requirements, surge-arrester information, notes on metal-enclosed switchgear, and additional requirements for services over 35,000 volts.

While exam questions may not focus on over 600-volt services, it is important to know where the information can be accessed, at the end of Article 230, Services.

Article 240

Article 240, Overcurrent Protection, is one of the major Code articles. Without overcurrent protection, electrical fires would be far more common. The idea is very simple: A device is inserted on the supply side in series with the conductor, device, or equipment to be protected. It is designed to open the circuit when the current reaches a certain level, as determined by the downstream load. The rating of that device, circuit breaker, or fuse is chosen in order to achieve the desired level of protection. The trip or burn-out value may be set lower than the ampacity of the conductor. In this way, the current flowing through the conductor is interrupted before the conductor heats up to a point where the insulation is damaged. Heat damage to insulation is proportional to the temperature reached and length of time the insulation is exposed to excessive heat. Repeated overheating events also take a toll.

Section 240.1, Scope, notes that Parts I through VII provide requirements for overcurrent protection and associated devices for voltages not over 600 volts, nominal. Part VII pertains to portions of supervised industrial installations operating at not more than 600 volts, nominal. Part IX covers overcurrent protection over 600 volts, nominal.

Section 240.2, Definitions, adds three new definitions that do not appear in Article 100, Definitions, because they are specific to Article 240, Overcurrent Protection:

A current-limiting overcurrent protective device is one of the really important Code concepts, especially in industrial work. It is a device that is capable of reducing current flowing in a faulted circuit to a level much lower than if it were not in the circuit. It functions to reduce arc-flash hazard. A current-limiting overcurrent protective device is capable of interrupting fault current in less than one-half cycle, thereby preventing that current from reaching full available fault level.

Supervised industrial installations are portions of a facility where

(1) Conditions of maintenance and engineering supervision ensure that only qualified persons monitor and service the system. This is a recurring Code formulation. *Qualified person* is defined in Article 100, Definitions, as one who has skills and knowledge related to the construction and operation of the electrical equipment and installations and has received safety training to recognize and avoid the hazards involved.

(2) The premises wiring system has 2,500 kVA or greater of load used in industrial processes, manufacturing activities, or both, as calculated in accordance with Article 220.

(3) The premises has at least one service or feeder that is more than 150 volts to ground and more than 300 volts phase to phase.

If all these conditions are met, the facility or part thereof is considered a supervised industrial installation. As such, it qualifies for various relaxed requirements.

Tap conductors are conductors, other than service conductors, that have overcurrent protection ahead of their point of supply exceeding the value permitted for similar conductors that are protected as described in Section 240.4, Protection of Conductors. General tap rules are covered in 240.21(B), Feeder Taps.

Section 240.3, Other Articles, lists equipment with relevant article numbers that contain specific overcurrent requirements.

Section 240.4, Protection of Conductors, is a short but intense section. It is the heart of the article. It provides that conductors, other than flexible cords, flexible cables, and fixture wires, which are covered separately, are to be protected against overcurrent in accordance with their ampacities specified in 310.15 and associated tables unless otherwise permitted or required as follows:

(A) Power Loss Hazard states that conductor overload protection is not required where the interruption of the circuit would create a hazard, such as in a materials-handling circuit or fire pump circuit. Nevertheless, short-circuit protection is to be provided. Obviously, if power to a materials-handling magnet circuit were to be interrupted owing to overload, the load would drop, endangering workers below. Similarly, a fire pump should continue running despite any perceived overload so that fire suppression is not compromised. Short-circuit protection is to be provided on the theory that if it comes to that, the equipment would not function anyway.

(B) Overcurrent Devices Rated 800 Amperes or Less provides that the next-higher standard overcurrent device (above what is required for the conductor) may be installed provided that

(1) The conductors being protected are not part of a branch circuit supplying more than one receptacle for cord-and-plug-connected portable loads.

(2) The ampacity of the conductors does not correspond with the standard ampere rating of a fuse or circuit breaker without overload trip adjustments above its rating.

(3) The next-higher standard rating selected does not exceed 800 amperes.

This and the next section could affect an exam or job calculation. Beware of the following:

(C) Overcurrent Devices Rated Over 800 Amperes provides that where the overcurrent device is rated over 800 amperes, the ampacity of the conductors must be greater than the rating of the associated overcurrent device. To summarize, you can go to the next-higher size (with exceptions) if the overcurrent device is 800 amperes or less, but not if it is over 800 amperes.

(D) Small Conductors provides that after correction factors for ambient temperature and number of conductors (see Chapter 3) have been applied, overcurrent protection is not to exceed that required in 1 through 7 below:

(1) 18 AWG Copper: 7 amperes provided that

- Continuous loads do not exceed 5.6 amperes.

- Overcurrent protection is provided by branch-circuit-rated circuit breakers listed and marked for use with 18 AWG copper wire, branch-circuit fuses listed and marked for use with 18 AWG copper wire, and Class CC, Class J, or Class T fuses.

(2) 16 AWG Copper: 10 amperes provided that

- Continuous loads do not exceed 8 amperes.

- Overcurrent protection is provided by branch-circuit-rated circuit breakers listed and marked for use with 16 AWG copper wire, branch-circuit-rated fuses listed and marked for use with 16 AWG copper wire, or Class CC, Class J, or Class T fuses.

(3) 14 AWG Copper: 15 amperes

(4) 12 AWG Aluminum and Copper-Clad Aluminum: 15 amperes

(5) 12 AWG Copper: 20 amperes

(6) 10 AWG Aluminum and Copper-Clad Aluminum: 25 amperes

(7) 10 AWG Copper: 30 amperes

(E) Tap Conductors provides that tap conductors are permitted to be protected against overcurrent in accordance with the following:

(1) 210.19(A)(3) and (4), Household Ranges and Cooking Appliances and Other Loads

(2) 240.5(B)(2), Fixture Wire

(3) 240.21, Location in Circuit

(4) 368.17(B), Reduction in Ampacity Size of Busway

(5) 368.17(C), Feeder or Branch Circuits (busway taps)

(6) 430.53(D), Single Motor Taps

This section provides a table of contents for overcurrent rules regarding taps. Though far from a complete treatment, it is a good introduction to a subject that, as mentioned previously, is difficult because of the fact that tap rules occur in many different Code locations. Exam questions may be difficult for this reason.

(F) Transformer Secondary Conductors states that certain transformer secondaries may be considered protected by overcurrent devices in the primary circuits. The basic requirement in 240.4, Protection of Conductors, is that protection is to be provided at the point where a conductor is supplied. However, this section allows the following secondary circuits to be protected by overcurrent protection in the primaries:

- Secondary of a transformer with a two-wire primary and a two-wire secondary, provided that the primary is protected

- A three-wire, single-voltage secondary with a delta-delta-connected transformer, provided that the primary is protected

These special cases are allowed only when the value of protection in the primary does not exceed the value determined by multiplying the secondary conductor ampacity by the secondary-to-primary transformer voltage ratio.

Section 240.5, Protection of Flexible Cords, Flexible Cables, and Fixture Wires, refers ahead in (A) to special ampacity tables for these items, Table 400.5(A)(1), Table 400.5(A)(2), and Table 402.5.

(B) (1) and (2) contain additional supply-cord and fixture-wire ampacity details.

(3) Extension Cord Sets states that flexible cord used in listed extension-cord sets is considered adequately protected when used within listing requirements.

(4) Field-assembled extension-cord sets by their nature are not listed. If, however, the individual components, plug, cord, and connector are listed, 16 AWG or larger cord may be supplied by a 20-ampere or smaller circuit. If the field-assembled cord is supplied by a circuit higher than 20 amperes, standard ampacity rules apply.

Section 240.6, Standard Ampere Ratings, lists these values for fuses and inverse-time circuit breakers: 15, 20, 25, 30, 35, 40, 45, 50, 60, 70, 80, 90, 100, 110, 125, 150, 175, 200, 225, 250, 300, 350, 400, 450, 500, 600, 700, 800, 1,000, 1,200, 1,600, 2,000, 2,500, 3,000, 4,000, 5,000, and 6,000 amperes. Notice that the intervals become greater for higher values. Values up to 200 amperes should be memorized, and higher values can be accessed in this section. The use of fuses with nonstandard ratings is permitted. Nonstandard overcurrent devices are available from manufacturers.

The next-higher value of standard rated overcurrent devices may be substituted, as we have seen, subject to certain conditions.

(B) Adjustable-Trip Circuit Breakers provides that where the conditions of (C) below are not met, the rating of adjustable-trip circuit breakers is the maximum setting possible.

(C) Restricted-Access Adjustable-Trip Circuit Breakers permits the rating to be equal to the adjusted value provided that access is restricted as follows:

(1) Removable and sealable covers over the adjusting means

(2) Bolted equipment enclosure doors

(3) Locked doors accessible only to qualified personnel

Section 240.8, Fuses or Circuit Breakers in Parallel, provides that fuses and circuit breakers may be connected in parallel only if factory assembled in parallel and listed as a unit.

Section 240.9, Thermal Devices, states that these are not permitted as short-circuit or ground-fault protection but may be used to protect motor branch-circuit conductors from overload.

Section 240.10, Supplementary Overcurrent Protection, states that it may not be used as a substitute for required branch-circuit protection. It is not required to be readily accessible. This type of overcurrent protection may take many forms. It is sometimes embedded inside factory-made equipment such as low-ampere glass fuses mounted on a circuit board within a television.

Section 240.12, Electrical System Coordination, permits a system of coordination where an orderly shutdown is required to minimize hazards to personnel and equipment. It is to be based on

- Coordinated short-circuit protection (A faulty circuit will actuate only the closest overcurrent device so that other loads are not affected.)

- Overload indication based on monitoring systems or devices

An Informational Note states that a monitoring system may cause the condition to go to alarm, allowing corrective action to be taken.

Notice that an alarm is allowed to substitute for standard overload protection but not for short-circuit protection. In the latter case, coordinated short-circuit protection will have to do.

Section 240.13, Ground-Fault Protection of Equipment, provides that for solidly grounded wye systems of more than 150 volts to ground but not exceeding 600 volts phase to phase, each individual device used as a building or structure main disconnect must include ground-fault protection. This mandate does not apply

- For continuous industrial processes where a nonorderly shutdown will introduce additional or increased hazards

- For installations where ground-fault protection is provided by other requirements for services or feeders

- For fire pumps

Section 240.15, Ungrounded Conductors, provides the following:

(A) Overcurrent Device Required: A fuse or circuit breaker is to be connected in series with each ungrounded conductor. A current transformer and overcurrent relay is to be considered equivalent to an overcurrent trip unit.

(B) Circuit Breaker as Overcurrent Device: A circuit breaker is to open all ungrounded conductors of the circuit both automatically and manually unless otherwise permitted as follows:

(1) Multiwire Branch Circuits: Individual single-pole circuit breakers, with identified handle ties, are permitted as the protection for each ungrounded conductor of multiwire branch circuits that serve only single-phase line-to-neutral loads.

(2) Grounded Single-Phase AC Circuits: Individual single-pole circuit breakers, rated 120/240 volts ac, with identified handle ties, are permitted as the protection for each ungrounded conductor for line-to-line connected loads for single-phase circuits.

(3) Three-Phase and Two-Phase Systems: For line-to-line loads in four-wire, three-phase systems or five-wire, two-phase systems, individual single-pole circuit breakers rated 120/240 volts ac with identified handle ties are permitted as the protection for each ungrounded conductor if the systems have a grounded neutral point and the voltage to ground does not exceed 120 volts.

(4) Three-Wire DC Circuits: Individual single-pole circuit breakers rated 125/250 volts dc with identified handle ties are permitted as the protection for each ungrounded conductor for line-to-line connected loads for three-wire dc circuits supplied from a system with a grounded neutral where the voltage to ground does not exceed 125 volts.

It is important to realize that two or three single-pole breakers with handle ties may be used only in certain limited applications. It is always safe to use two- or three-pole breakers. In no case is it permissible to use improvised handle ties made with nails or wire. Ties must be identified.

Part II, Location, provides the following:

Section 240.21, Location in Circuit, contains many of the tap rules. It is surprising to find branch-circuit tap rules in Article 210 and feeder tap rules here in Article 240, Overcurrent Protection. Then there is the matter of transformer secondaries, which are not taps but share many of the same definitions and mandates and which also appear in this article.

(A) Branch-Circuit Conductors states that branch-circuit tap conductors meeting the requirements of Section 210.19, Conductors Minimum Ampacity and Size, are permitted to have overcurrent protection specified in Section 210.20, Overcurrent Protection.

(B) Feeder Taps states that conductors are permitted to be tapped, without overcurrent protection at the tap, from a feeder as specified in (1) through (5) below. Section 240.4(B), Overcurrent Devices Rated 800 Amps or Less, does not apply. This is the provision that says that you can use the next-higher-rated standard overcurrent device under certain conditions. And what is being stated here is that such substitution is prohibited for feeder taps. This is a subtle distinction, but it could become important in a real-world installation, and it might make a good exam question.

(1) Taps Not Over 10 Feet Long provides the taps that comply with these four provisions:

   a. The ampacity of the tap conductor is

      • Not less than the combined loads on the circuits supplied by the tap conductors and

      • Not less than the rating of the device supplied by the tap conductors or not less than the rating of the overcurrent protective device at the termination of the tap conductors.

   b. The tap conductors do not extend beyond the switchboard, panelboard, disconnecting means, or control devices they supply.

   c. Except at the point of connection to the feeder, the tap conductors are enclosed in a raceway, which must extend from the tap to the enclosure of an enclosed switchboard, panelboard, or control devices or to the back of an open switchboard.

   d. For field installations, if the tap conductors leave the enclosure or vault in which the tap is made, the ampacity of the tap conductors is not less than one-tenth the rating of the overcurrent device protecting the feeder conductors.

(2) Taps Not Over 25 Feet Long states that such taps must comply with these three provisions:

   a. The ampacity of the tap conductors is not less than one-third the rating of the overcurrent device protecting the feeder conductors.

b. The tap conductors terminate in a single circuit breaker or a single set of fuses that limit the load to the ampacity of the tap conductors. This device is permitted to supply any number of additional overcurrent devices on its load side.

c. The tap conductors are protected from physical damage by being enclosed in an approved raceway or by other approved means.

(3) Taps Supplying a Transformer (Primary Plus Secondary Not Over 25 Feet Long) states that such taps must comply with these five provisions:

a. The conductors supplying the primary of a transformer have an ampacity at least one-third the rating of the overcurrent device protecting the feeder conductors.

b. The conductors supplied by the secondary of the transformer have an ampacity that is not less than the value of the primary-to-secondary voltage ratio multiplied by one-third the rating of the overcurrent device protecting the feeder conductors.

c. The total length of one primary plus one secondary conductor, excluding any portion of the primary conductor that is protected at its ampacity, is not over 25 feet.

d. The primary and secondary conductors are protected from physical damage by being enclosed in an approved raceway or by other approved means.

e. The secondary conductors terminate in a single circuit breaker or set of fuses that limit the load current to not more than the conductor ampacity that is permitted by Section 310.15, Ampacities for Conductors Rated 0–2,000 Volts.

(4) Taps Over 25 Feet Long states that where the feeder is in a high-bay manufacturing building over 35 feet high at the walls, such taps must comply with these nine provisions:

a. Conditions of maintenance and supervision ensure that only qualified persons service the systems.

b. The tap conductors are not over 25 feet long horizontally and not over 100 feet total length.

c. The ampacity of the tap conductors is not less than one-third the rating of the overcurrent device protecting the feeder conductors.

d. The tap conductors terminate at a single circuit breaker or a single set of fuses that limit the load to the ampacity of the tap conductors. This single overcurrent device may supply any number of additional overcurrent devices on its load side.

  e. The tap conductors are protected from physical damage by being enclosed in an approved raceway or by other approved means.

  f. The tap conductors are continuous from end to end and contain no splices.

  g. The tap conductors are sized 6 AWG copper or 4 AWG aluminum or larger.

  h. The tap conductors do not penetrate walls, floors, or ceilings.

  i. The tap is made no less than 30 feet from the floor.

(5) Outside Taps of Unlimited Length states that where the conductors are located outdoors, except at the point of load termination, such taps must comply with these four provisions:

  a. The conductors are protected from physical damage in an approved manner.

  b. The conductors terminate at a single circuit breaker or a single set of fuses that limit the load to the ampacity of the conductors. This single overcurrent device may supply any number of additional overcurrent devices on its load side.

  c. The overcurrent device for the conductors is an integral part of a disconnecting means or is located immediately adjacent.

  d. The disconnecting means for the conductors is installed at a readily accessible location complying with one of the following:

   • Outside of a building or structure

   • Inside, nearest the point of entrance of the conductors

   • Where installed in accordance with Section 230.6, Conductors Considered Outside the Building, nearest the point of entrance of the conductors (Section 230.6, you will recall, is the section that talks about 2 inches of concrete, a vault, in conduit and at least 18 inches under a building, or in overhead service mast. Conductors installed in any of these manners may be considered outside a building.)

(C) Transformer Secondary Conductors states that a set of conductors feeding a single load or each set of conductors feeding separate loads may be connected to a transformer secondary without overcurrent protection at the secondary as specified in (1) through (6) below. The provisions of Section 240.4(B), which permits, under certain circumstances, substituting the next-higher-rated overcurrent device for 800 amperes or less, is not applicable to transformer secondary conductors.

(1) Protection by Primary Overcurrent Device permits two-wire secondaries and three-wire secondaries from a delta-delta-connected transformer to be protected by the overcurrent protection in the primaries, provided that the protection complies with 450.3, Overcurrent Protection (of transformers), and does not exceed the value determined by multiplying the secondary-conductor ampacity by the secondary-to-primary transformer voltage ratio.

(2) Transformer Secondary Conductors Not over 10 Feet Long states that the length of the secondary conductor must not exceed 10 feet and mandates compliance with these four provisions:

a. The ampacity of the secondary conductors is

- Not less than the combined calculated loads on the circuits supplied by the secondary conductors and

- Not less than the rating of the device supplied by the secondary conductors or not less than the rating of the overcurrent protective device at the termination of the secondary conductors.

b. The secondary conductors do not extend beyond the switchboard, panelboard, disconnecting means, or control devices they supply.

c. The secondary conductors are enclosed in a raceway, which must extend from the transformer to the enclosure of an enclosed switchboard, panelboard, or control device or to the back of an open switchboard.

d. For field installations where the secondary conductors leave the enclosure or vault in which the supply connection is made, the rating of the overcurrent device protecting the primary of the transformer, multiplied by the primary-to-secondary transformer voltage ratio, is not to exceed 10 times the ampacity of the secondary conductor.

(3) Industrial Installation—Secondary Conductors Not Over 25 Feet Long addresses industrial installations only, where the length of the secondary conductors does not exceed 25 feet and complies with these four provisions:

a. Conditions of maintenance and supervision ensure that only qualified persons service the systems.

b. The ampacity of the secondary conductors is not less than the secondary current rating of the transformer, and the sum of the ratings of the overcurrent devices does not exceed the ampacity of the secondary conductors.

c. All overcurrent devices are grouped.

    d. The secondary conductors are protected from physical damage by being enclosed in an approved raceway or other approved means.

(4) Outside Secondary Conductors states that where the conductors are located outdoors of a building or structure, except at the point of load termination, they must comply with all four of these provisions:

    a. The conductors are protected from physical damage in an approved manner.

    b. The conductors terminate at a single circuit breaker or a single set of fuses that limit the load to the ampacity of the conductors. This single overcurrent device may supply any number of additional overcurrent devices on its load side.

    c. The overcurrent device for the conductors is an integral part of a disconnecting means or is to be located immediately adjacent.

    d. The disconnecting means for the conductors is installed at a readily accessible location complying with one of the following:

- Outside of a building or structure
- Inside, nearest the point of entrance of the conductors
- Where installed in accordance with 240.6, Conductors Considered Outside the Building, nearest the point of entrance of the conductors

(5) Secondary Conductors from a Feeder Tapped Transformer states that transformer secondary conductors installed in accordance with Section 240.21(B)(3), Taps Supplying a Transformer (above), are permitted to have overcurrent protection as specified in that section.

(6) Secondary Conductors Not Over 25 Feet Long states that such conductors must comply with these three provisions:

    a. The secondary conductors must have an ampacity that is not less than the value of the primary-to-secondary voltage ratio multiplied by one-third the rating of the overcurrent device protecting the primary of the transformer.

    b. The secondary conductors terminate in a single circuit breaker or set of fuses that limit the load current to not more than the conductor ampacity that is permitted by Section 310.15.

    c. The secondary conductors are protected from physical damage by being enclosed in an approved raceway or other approved means.

These are some of the tap rules, grouped together in Article 240, Overcurrent Protection. There are no difficult calculations involved. It's just a set of mandates. One of the underlying concepts is that the longer the tap, the greater are the construction requirements. Transformer secondaries are treated similarly to taps, but they are not taps. However, taps may supply a transformer, and that is a different matter altogether, a possible source of confusion. Additionally, there are tap rules for household ranges and cooking appliances that appear in an exception to Section 210.19(A)(3).

Section 240.21(D), Service Conductors, refers back to 230.91, Location (of service-equipment overcurrent protection), which states that the service overcurrent device is to be an integral part of the service-disconnecting means or is to be located immediately adjacent.

Section 240.21(E), Busway Taps, refers ahead to 368.17, Overcurrent Protection (for Busways), which provides in detail overcurrent rules for this wiring method. (Articles 320 through 398 treat various wiring methods, including cables and raceways, with a separate article on each.)

Section 240.21(F), Motor Circuit Taps, states that motor feeder and branch-circuit conductors may be protected against overcurrent in accordance with Section 430.28, Feeder Taps (for Motors), and Section 430.53, Several Motors or Loads on One Branch Circuit, respectively. As we shall see when we look at Article 430, motor mandates differ markedly from mandates for other types of equipment.

Section 240.21(G), Conductors from Generator Terminals, shifts away from the discussion of taps, which do not recur in this chapter. This section provides that conductors from generator terminals that meet the size requirement in Section 445.13, Ampacity of Conductors (for Generators), may be protected against overload by the generator overload protective device(s) required by Section 445.12, Overcurrent Protection (for Generators). This section enumerates several methods of overcurrent protection for generators and includes in the list *by inherent design*, which is fairly open-ended.

Section 240.21(H), Battery Conductors, provides that overcurrent protection for battery conductors may be installed as close as practicable to the terminals. The section notes that (for this reason) it is permitted to install an overcurrent device that protects conductors from battery terminals within a hazardous (classified) location. Enclosure requirements are given in Chapter 5, Special Occupancies.

Section 240.22, Grounded Conductor, notes that no overcurrent device is to be placed in series with an intentionally grounded conductor unless

(1) The overcurrent device opens all conductors, including the grounded conductor, and is designed so that no pole can operate independently. If it were permitted to interrupt only the grounded conductor, it would be possible to shut down a piece of equipment

such as a hand tool and still have the hot side energized—definitely a hazardous situation.

(2) Where required for motor overload protection in Section 430.36, Fuses—In Which Conductor, or Section 430.37, Devices Other Than Fuses—In Which Conductor.

Section 240.23, Change in Size of Grounded Conductor, provides that where a change is made in the size of an ungrounded conductor, a similar change is to be made in the size of the grounded conductor. It is still necessary to observe the minimum size for ampacity, subject to feeder tap allowances. This situation occurs when conductor size is increased to combat excessive voltage drop.

Section 240.24, Location In or On Premises, provides in (A) that overcurrent devices are to be readily accessible. It further states that the center of the grip of the operating handle of the switch or circuit breaker, when in its highest position, must not to be higher than 6 feet, 7 inches above the floor or platform unless one of these four conditions applies:

(1) For busways, reference is made to 368.17(C), Feeder or Branch Circuits (for Busways), which states that these overcurrent devices may be mounted out of reach if suitable means such as ropes, chains, or sticks are provided for operating the disconnecting means from the floor.

(2) For supplementary overcurrent protection, as described in 240.10. This type of overcurrent protection does not have to be readily accessible.

(3) For overcurrent devices as described in 225.40, Access to Over-current Protective Devices, and 230.92, Locked Service Overcurrent Devices. These sections depict scenarios in which the overcurrent devices are not readily accessible or available to tenants. Downstream overcurrent devices of lower ampere rating that are readily accessible are required to mitigate that situation.

(4) For overcurrent devices that are adjacent to utilization equipment that they supply, access is permitted to be by portable means. An example of such utilization equipment would be a unit heater with an electric fan mounted high on a wall, and an example of portable means of access would be a ladder. This situation is seen frequently in industrial settings.

(B) Occupancy states that occupants must have ready access to all overcurrent devices protecting conductors supplying that occupancy unless otherwise permitted in the following two paragraphs:

(1) Service and Feeder Overcurrent Devices makes exception where electric service and maintenance are provided by building management under continuous supervision for multiple-occupancy buildings and guest rooms or guest suites.

(2) Branch-Circuit Overcurrent Devices makes a similar exception for branch-circuit overcurrent devices for guest rooms and guest suites that do not have permanent provisions for cooking.

(C) Not Exposed to Physical Damage states that overcurrent devices are not to be located where they will be exposed to physical damage. Temporary services sometimes are sited where damage may occur—a definite Code violation. Indoor entrance panels and load centers must not be placed where subject to water damage from leaking roofs or pipes.

(D) Not in Vicinity of Easily Ignitable Material emphasizes that overcurrent devices are not to be in clothes closets—another Code violation. If tagged by an inspector, an entrance panel in a clothes closet would have to be relocated, meaning reworking all branch circuits and feeders and possibly redoing the outdoor portion of the service. An entrance panel in a clothes closet, besides being close to ignitable material, probably would not comply with working-space rules in Article 110, Requirements for Electrical Installation.

(E) Not Located in Bathrooms makes the point that overcurrent devices may not be located in bathrooms. This requirement pertains to dwelling units, dormitories, guest rooms, and guest suites. The requirement does not apply to supplementary overcurrent protection.

(F) Not Located Over Steps is another mandate that pertains to overcurrent devices. This would be a hazard and difficult to service as well.

Part III, Enclosures, provides basic information concerning overcurrent protection.

Section 240.30, General, contains these two provisions:

(A) Protection from Physical Damage mandates that this be provided by either

(1) Installation in enclosures, cabinets, cutout boxes, or equipment assemblies or

(2) Mounting on open-type switchboards, panelboards, or control boards that are in rooms or enclosures free from dampness and easily ignitable material and accessible only to qualified personnel.

(B) Operating Handle permits the operating handle to be accessible without opening the door or cover of an enclosure. In this regard, the common breaker box as used as an entrance panel combines excellent protection and ease of operation. Fuse boxes, while Code compliant, for the most part have been replaced by breaker boxes, although fuses are used commonly for over 600-volt applications.

Section 240.32, Damp or Wet Locations, provides that overcurrent device enclosures must comply with Section 312.2, Damp and Wet Locations (for Cabinets, Cutout Boxes, and Meter Socket Enclosures), which contains regulations for mounting boxes and installing raceways and cables. Enclosures installed in wet locations must be weatherproof.

Section 240.33, Vertical Position, states that with very few exceptions, enclosures for overcurrent devices must be mounted in vertical position, never with the back attached to the floor or ceiling.

Part IV, Disconnecting and Guarding, contains safety-related provisions for fuses and circuit breakers, especially pertaining to large switchgear.

Section 240.40, Disconnecting Means for Fuses, provides that cartridge fuses of circuits of any voltage, where accessible to other than qualified persons, and all fuses in circuits over 150 volts to ground must have disconnecting means on their supply side so that each circuit containing fuses can be disconnected independently from the source of power. Current-limiting devices on the supply side of a service-disconnecting means are permitted by Section 230.82(1), Equipment Connected to the Supply Side of Service Disconnect. No disconnecting means is required on the supply sides of these devices.

Section 240.41, Arcing or Suddenly Moving Parts, pertains to fuses and circuit breakers. It states that they are to be located so that persons will not be burned or injured by their operation. Handles or levers of circuit breakers that may cause injury by sudden motion are to be guarded or isolated.

Part V, Plug Fuses, Fuseholders and Adapters, permits various types of plug fuses for different voltages, ampere ratings, and applications.

Section 240.50, General, gives some basic provisions:

(A) Maximum Voltage states that plug fuses may be used in two types of circuits:

(1) Circuits not exceeding 125 volts between conductors

(2) Circuits supplied by a system having a grounded neutral point where the line-to-neutral voltage does not exceed 150 volts

The two systems that meet these specifications are 120/240-volt, single-phase, three-wire systems and 208Y/120-volt, three-phase, four-wire systems.

(B) Marking states that each fuse, fuseholder, and adapter is to be marked with its ampere rating.

(C) Hexagonal Configuration provides that plug fuses rated 15 amperes or lower are to have a hexagonal configuration of the window, cap, or elsewhere to distinguish them from higher-rated fuses. The purpose of this provision is to help prevent overfusing, which can lead to uncontrolled heating of conductors in the event of overload.

(D) No Energized Parts states that plug fuses, fuseholders, and adapters are to have no exposed energized parts after installation. This is important because an occupant could be feeling around for a fuse in the dark during an outage.

(E) Screw Shell mandates that the screw shell of a plug-type fuseholder is to be connected to the load side of the circuit. Thus, when the fuse is removed, the outer shell is not energized, similar to the outer shell of a lamp socket being connected to the grounded neutral.

Section 240.51, Edison-Base Fuses, provides that

(A) These devices are not to be classified at over 125 volts or 30 amperes.

(B) Edison-base plug fuses are to be used for replacement only in existing installations where there is no evidence of overfusing or tampering.

Section 240.53, Type S Fuses, states that these are to be of the plug type and

(A) They are to be classified at not over 125 volts and 0 to 15 amperes, 16 to 20 amperes, and 21 to 30 amperes.

(B) They are not to be interchangeable with lower-ampere fuses. They are to be designed so that they cannot be used in any fuseholder other than Type S or a standard fuseholder with Type S adapter inserted. The purpose of this section is to prevent overfusing. Trunk slammers may repair chronic overloads by overfusing, and a Type S fuseholder makes this strategy more difficult.

Section 240.54, Type S Fuses, Adapters and Fuseholders, gives further details on how these devices work to prevent overfusing and tampering and further provides that they are to be standardized to permit interchangeability among manufacturers.

Part VI, Cartridge Fuses and Fuseholders, provides the following:

Section 240.60, General, states

(A) Maximum Voltage—300-Volt Type indicates that these cartridge fuses are permitted in circuits not over 300 volts between conductors and in single-phase line-to-neutral circuits supplied from a three-phase, four-wire, solidly grounded neutral source where line-to-neutral voltage does not exceed 300 volts. Recall that *solidly grounded* means that there is a direct electrical connection to ground as opposed to connection through electromagnetic induction.

(B) Non-Interchangeable—0–6,000-Ampere Cartridge Fuseholders provides that these fuseholders are to be designed so that it will be difficult to put a fuse of any given class into a lower-rated fuseholder in terms of either voltage or current. Moreover, fuseholders for current-limiting fuses are not to permit insertion of non-current-limiting fuses.

Additional paragraphs contain provisions on marking (must have interrupting rating if other than 10,000 amperes) and state that Class H renewal cartridge fuses may be used only for replacement in existing installations where there is no evidence of tampering or overfusing.

Part VII, Circuit Breakers, states that such breakers are overcurrent protective devices that have several advantages over fuses and have largely replaced them except for higher voltages and ampere levels, where circuit breakers become extremely expensive. One advantage is that on performing their task of interrupting overload, ground faults, or line-to-line faults, circuit breakers do not have to be replaced—they are merely reset. Moreover, a modern breaker panel is much easier to work on than a fusebox. In troubleshooting, a faulted circuit is easier to find when dealing with circuit breakers. It may be possible to identify a blown fuse by visual examination, but such is not always the case, whereas a tripped breaker moves to the middle position. Square D breakers are particularly convenient because a red indicator appears when they trip out. Also, typically, breaker boxes provide more space for initial wiring and maintenance.

Answers to exam questions on circuit breakers, in an open-book setting, are easy to locate because the relevant material is in the major article on overcurrent protection.

Section 240.80, Method of Operation, provides that circuit breakers are to be trip-free and capable of being closed and opened by manual operation. Their normal method of operation by other than manual means, such as electrical or pneumatic, is permitted if means for manual operation are also provided.

Section 240.81, Indicating, says that circuit breakers are to clearly indicate whether they are on or off. If arranged vertically, up is always "on." Many entrance panels have a main breaker in vertical configuration at the top, and branch-circuit and feeder breakers in two columns, horizontal configuration, with "on" toward the middle and "off" toward the outside.

Section 240.82, Nontamperable, provides that any alteration of calibration or time required for operation requires dismantling the breaker or breaking a seal for other than intended adjustments.

Low-ampere, single-pole breakers are quite inexpensive. They are a molded case and, as the name implies, cannot be taken apart or repaired. Some larger breakers cost thousands of dollars. They are rebuildable, but the job takes specialized know-how. If you come across a cast-off, take it apart just to see what is involved.

Section 240.83, Marking, states the following:

(A) Circuit breakers are to be marked with ampere rating that is durable and visible after installation, with removal of a trim or cover.

(B) Circuit breakers of 100 amperes or less and 600 volts or less are to have the ampere rating on their handle or escutcheon.

(C) If the interrupting rating is other than 5,000 amperes, that rating is to be marked on the circuit breaker. This rule does not apply to circuit breakers used for supplementary protection.

(D) Circuit breakers used as switches in 120- and 277-volt fluorescent lighting circuits are to be listed and marked "SWD" or "HID." Where used as switches in high-intensity discharge lighting circuits, they are to be marked "HID."

(E) Circuit breakers are to be marked with a voltage rating not less than the nominal system voltage that is indicative of their capability to interrupt fault currents between phases or phase to ground.

Section 240.85, Applications, provides that a circuit breaker with a straight voltage rating, such as 240 or 480 volts, may be used in a circuit in which the nominal voltage between any two conductors does not exceed the circuit breaker's rating. A two-pole circuit breaker must not be used to protect a three-phase, corner-grounded delta circuit unless the circuit breaker is marked "one phase, three phase" to indicate such suitability.

A circuit breaker with a slash rating, such as the 120/240- or 480Y/277-volt breaker, may be applied in a solidly grounded circuit where the nominal voltage of any conductor to ground does not exceed the lower of the two values of the circuit breaker's voltage rating and the nominal voltage between any two conductors does not exceed the higher value of the circuit breaker's voltage rating. A slash rating marked on a breaker means that the higher voltage is between phases and the lower voltage is between one phase and ground.

Section 240.86, Series Ratings, is a strategy for dealing with high available fault current. When this value is higher than the rating marked on a circuit breaker, it is permissible to insert a properly rated overcurrent device, either fuse or circuit breaker, upstream to protect the underrated breaker from potential fault-current damage. Such a series-rated ensemble must be designed carefully if it is to be reliable. The requirements are as follows:

(A) Selected Under Engineering Supervision in Existing Installations states that a licensed professional engineer engaged primarily in the design or maintenance of electrical installations is needed to select the series-rated combination devices contemplated in this section. The selection is to be documented and stamped by the engineer, and the documentation is to be available to those who design, install, inspect, maintain, and operate the system. Also, the end-use equipment is to be field marked with the combination rating, including identification of the upstream device. The engineer must ensure that the downstream circuit breakers remain passive during the interruption period. Note that this provision applies to existing installations. It is applicable in situations where a change in utility or premises wiring increases available fault current. Inserting a breaker or fuse in an upstream location achieves the desired level of protection without replacing an entire installation.

(B) Tested Combinations provides that the combination is to be tested and marked on the end-use equipment. The marking is of great importance because otherwise alterations could be made to the electrical system in the future that would compromise the protective function of the original series-rated equipment.

(C) Motor Contribution states that series ratings are not to be used where motors are connected between the devices and where the sum of the motor full-load currents exceeds 1 percent of the interrupting rating of the lower-rated circuit breaker.

Section 240.87, Noninstantaneous Trip, outlines important requirements for when a noninstantaneous trip device is in place. Documentation must be available to those who design, install, operate, or inspect the installation. This documentation must include the location of the noninstantaneous device. Furthermore, one of the following must be provided:

- Zone-selective interlocking
- Differential relaying
- Energy-reducing maintenance switching with local status indicator

An energy-reducing maintenance switch, according to an Informational Note, allows workers to set a circuit breaker trip unit to "No intentional delay." The purpose is to limit arc-flash duration and hence intensity in case of a fault during maintenance, when a worker might be within an arc-flash boundary. This is defined in National Fire Protection Association® (NFPA®) 70E-2009, Standard for Electrical Safety in the Workplace. On completion of the maintenance operation, the trip unit is to be set back to normal operation, with the delay once again in place.

Part VIII, Supervised Industrial Installations, notes that provisions in this part apply only to areas of the supervised industrial installation that are used for manufacturing or process-control activities. This part provides alternatives to the general overcurrent protection rules as promulgated in Section 240.21. Part VIII modifies these requirements, provided that only qualified personnel monitor and maintain the installation.

Section 240.91, Protection of Conductors, indicates that conductors are to be protected in accordance with (A) or (B):

(A) General: Conductors are to be protected in accordance with Section 240.4, Protection of Conductors, which, as we have seen, contains the basic overall rules for conductor protection in accordance with Chapter 3 ampacities, with exceptions for power-loss hazard, tap conductors, and transformer secondary conductors.

(B) Devices Rated Over 800 Amperes: Where the overcurrent device is rated over 800 amperes, the ampacity of the conductors it protects is to be equal to or greater than 95 percent of the overcurrent device. This section is applicable, however, only if the conductors are protected within recognized time versus current limits for short-circuit currents and all equipment in which the conductors terminate is listed and marked for the application.

And so we see that a reduction is allowed only under limited conditions, foremost of which is that it is a supervised industrial installation.

Section 240.92, Location in Circuit, provides reduced requirements to the general overcurrent rules in Section 240.21, here again, only within a supervised industrial facility. The material in this and Part IX, Overcurrent Protection Over 600 Volts, Nominal, is not likely to appear on licensing exams, and it is applicable only to highly specialized electrical work. However, it is important to know where to find it if need arises. Look for this information at the end of Article 240, Overcurrent Protection.

Article 250

Article 250, Grounding and Bonding, is one of the really major Code articles. Licensing exams focus intently on this important topic, and it is worthy of close attention. Electrical inspectors frequently take a hard, close look at this aspect of any electrical installation, and it is the location of numerous deficiencies. Mistakes in grounding and bonding can result in electric shock to persons and building fires, sometimes with tragic consequences. In terms of electricians' licensing exams, design work at the drawing board, and field installations, your knowledge of grounding and bonding should be complete and accurate.

One major source of error is the mistaken assumption that *good grounding* is the whole story. It is beneficial to have adequate and strong grounding, i.e., low-resistance connection of the proper conductors to earth, but this is just the beginning. Not to say that grounding is not essential, but the more fundamental issue is low-impedance bonding back to the system neutral, which in a 120/240-volt, single-phase system translates into a direct connection to the center tap of the transformer on the pole or pad.

Another error that occurs, especially in residential applications, is nonbonding or inadequate bonding to premises water piping, leading to voltage where it shouldn't be, unprotected by an overcurrent device.

Another area of concern is grounding and bonding of communications systems such as satellite TV and Internet access, cable television (CATV), and telephone.

To prevent these and other grounding and bonding errors and score high on a licensing exam, the place to start is NEC Article 250, Grounding and Bonding. Throughout, our attention should focus on the difference between grounding and bonding, two separate, albeit related, concepts that serve different, though overlapping, purposes.

Part I, General, contains provisions relating to the fundamental concepts of grounding and bonding. Grounded versus ungrounded systems is discussed, as well as objectionable current and connection of grounding and bonding jumpers.

Section 250.1, Scope, presents an overview of topics covered in Article 250. Besides general requirements for grounding and bonding of electrical installations, topics covered include

1. Systems, circuits, and equipment required, permitted, or not permitted to be grounded
2. Circuit conductor to be grounded on grounded systems

3. Location of grounding connections

4. Types and sizes of grounding and bonding conductors and electrodes

5. Methods of grounding and bonding

6. Conditions under which guards, isolation, or insulation may be substituted for grounding

Section 250.2, Definitions, contains three definitions that are specific to this article:

Bonding jumper, supply side, is a conductor installed on the supply side of a service or within a service-equipment enclosure(s), or for a separately derived system, that ensures the required electrical conductivity between metal parts to be connected electrically. These are connecting on the upstream side of the service-disconnecting means. As we shall see, they are required to ensure electrical conductivity between the service-disconnecting means and metal raceways or an enclosure. In contrast, bonding jumpers installed on the load side or downstream from a service, feeder, or branch-circuit overcurrent protective device enclosure are referred to as *equipment-bonding jumpers.*

Effective ground-fault current path is an intentionally constructed low-impedance electrically conductive path designed and intended to carry current under ground-fault conditions from the point of a ground fault on a wiring system to the electrical supply source and that facilitates operation of the overcurrent protective device or ground-fault detectors on high-impedance grounded systems. (High-impedance grounded systems are highly specialized installations used in industrial facilities for three-phase systems of 480 to 1,000 volts and higher. We'll take a look at these when we get to Section 250.36, High-Impedance Grounded Neutral Systems.)

Ground-fault current path is an electrically conductive path from the point of a ground fault on a wiring system through normally non-current-carrying conductors, equipment, or the earth to the electrical supply source.

For definitions of other Article 250–related terms, refer to Article 100, Definitions. Defined terms include *bonded, grounded, equipment-grounding conductor, grounding electrode,* and *grounding-electrode conductor.*

Section 250.3, Application of Other Articles, refers to Table 250.3, Additional Grounding and Bonding Requirements, which lists conductors and equipment with Code references that pertain to grounding and bonding. These sections contain requirements that add to or modify those in Article 250.

Section 250.4, General Requirements for Grounding and Bonding, outlines performance requirements in regard to grounding and bonding, whereas the remainder of the article contains prescriptive methods that are to be followed or permitted to be used to accomplish these objectives.

(A) Grounded Systems must meet these performance requirements:

(1) Electrical System Grounding states that grounded electrical systems must be connected to earth in a manner that will limit the voltage imposed by lightning, line surges, or unintentional contact with higher-voltage lines and that will stabilize the voltage to earth during normal operation.

An Informational Note stresses the importance of routing bonding and grounding electrode conductors. They should be no longer than necessary. Bends and loops are to be avoided.

It is sometimes said that lightning is like a fast railroad train traveling along a track. A sharp bend in the track will cause it to fly off—it can't make the curve. The actual reason is that a bend in a conductor is like part of the winding of a coil. The very fast rise time of a lightning surge makes it behave like a high-frequency waveform, and the bend in the conductor, at that frequency, has enough inductive reactance to oppose the flow of electrons so that they take the easier route of going straight through the air. In the presence of combustible material, a dangerous situation exists, and it is due to a sharp bend in the grounding electrode conductor.

Sometimes a water analogy is useful. A conductive object, when it becomes unintentionally energized, is like a bucket of water. If the object is grounded, it is as if the bucket had a hole in it. The better it is grounded, the larger is the hole. A very solidly grounded object, i.e., one that has a low-impedance connection to ground, is like a bucket with no bottom. It is impossible to fill it with water. A well-grounded metal object is difficult to energize, which is an advantage in an electrical storm. Notice how lightning tears through a tree, producing lots of heat. This is so because the tree is not conductive enough to be well grounded.

(2) Grounding of Electrical Equipment states that normally non-current-carrying conductive materials enclosing conductors or equipment or forming part of such equipment are to be connected to earth so as to limit the voltage to ground on these materials. Keep this statement in mind as you read the following statement, and notice that these are two altogether different concepts. These two performance requirements are the basis of NEC Article 250.

(3) Bonding of Electrical Equipment states that normally non-current-carrying conductive materials enclosing electrical conductors or equipment or forming part of such equipment are to be connected together and to the electrical supply source in a manner

that establishes an effective ground-fault current path. The following statement is also related:

(4) Bonding of Electrically Conductive Materials and Other Equipment states that normally non-current-carrying electrically conductive materials that are likely to become energized are to be connected together and to the electrical supply source in a manner that establishes an effective ground-fault current path.

(5) Effective Ground-Fault Current Path states that electrical equipment and wiring and other electrically conductive material likely to become energized is to be installed in a manner that creates a low-impedance circuit facilitating operation of the overcurrent device (or ground detector for high-impedance grounded systems). It is to be capable of safely carrying the maximum ground-fault current likely to be imposed on it from any point on the wiring system where a ground fault may occur to the electrical supply source. The earth is not to be considered an effective ground-fault current path.

There are two separate hookups involved: One is the connection to ground, which will bleed out any high-voltage charge imposed by lightning and stabilize system voltage with respect to earth, and the other is the connection back to the grounded point at the service, the purpose of which is to facilitate the protective operation of the overcurrent device. This connection is to be low impedance.

(B) Ungrounded Systems states that such systems must meet these performance requirements:

(1) Grounding Electrical Equipment states that non-current-carrying conductive materials enclosing electrical conductors or equipment or forming part of such equipment are to be connected to earth in a manner that will limit the voltage imposed by lightning or unintentional contact with higher-voltage lines and limit the voltage to ground on these materials. This section is telling us that for an ungrounded system, the equipment-grounding system must be in place just as for a grounded system. What is not grounded is one of the current-carrying conductors—the system is ungrounded only in that sense.

(2) Bonding of Electrical Equipment states that non-current-carrying conductive materials enclosing electrical conductors or equipment or forming part of such equipment are to be connected together and to the supply-system grounded equipment in a manner that creates a low-impedance path for ground-fault current that is capable of carrying the maximum fault current likely to be imposed on it.

(3) Bonding of Electrically Conductive Materials and Other Equipment states that electrically conductive materials that are likely to become energized are to be connected together and to the supply-system grounded equipment in a manner that creates a low-impedance path for ground-fault current that is capable of carrying the maximum fault current likely to be imposed on it.

(4) Path for Fault Current states that electrical equipment, wiring, and other electrically conductive material likely to become energized are to be installed in a manner that creates a low-impedance circuit from any point on the wiring system to the electrical supply source to facilitate the operation of overcurrent devices should a second ground fault from a different phase occur on the wiring system. The earth is not to be considered an effective fault-current path.

Grounding requirements in these four paragraphs for ungrounded systems are essentially the same as those for grounded systems. The differences in wording reflects changed circuit realities since there is not a grounded current-carrying conductor.

Section 250.6, Objectionable Current, is highly problematic in office settings where there is heavy use of computers and in some industrial settings with computerized controls and Programmable Logic Controllers (PLCs). These types of equipment are sensitive to objectionable current, which consists of circulating currents on equipment-grounding conductors. This section provides the following:

(A) Arrangement to Prevent Objectionable Current: Grounding of electrical systems and components is to be accomplished in a manner that will prevent objectionable current.

(B) Alterations to Stop Objectionable Current: If the use of multiple grounding connections results in objectionable current, one or more of the following alterations may be made, provided that the requirements of Section 250.4(A)(5) or (B)(4) are met:

1. Discontinue one or more but not all such grounding connections.

2. Change the locations of the grounding connections.

3. Interrupt the continuity of the conductor or conductive path causing the objectionable current.

4. Take other suitable remedial and approved action.

It is to be emphasized that these steps may be taken only when they do not conflict with the fundamental mandates of 250.4(A)(5), Effective Ground-Fault Current Path, and 250.4(B)(4), Path for Fault Current. These two sections, the first for grounded systems and the second for ungrounded systems, lay out the principle that electrical equipment, wiring, and other electrically conductive materials are to be bonded back to the electrical supply source (not the grounding electrode) to facilitate the operation of overcurrent protection. This means that the preceding actions may be taken only when they do not interrupt that required continuity. Trunk slammers correctly analyze the problem but fix it by cutting off the bonding means required by the preceding sections. The protective function of the overcurrent devices is compromised, and shock hazard may be present.

(C) Temporary Currents Not Classified as Objectionable Currents makes the point that these currents resulting from abnormal conditions such as ground faults do not constitute objectionable currents for the purposes of (A) and (B) above.

(D) Limitations to Permissible Alterations emphasizes that proper equipment grounding must be maintained when steps are taken to eliminate objectionable current.

(E) Isolation of Objectionable Direct-Current Ground Currents recognizes use of a listed ac-coupling/dc-isolating device when cathodic protection systems are in place. These systems prevent corrosion in metal piping systems by imposing a constant low-level dc voltage on them.

Section 250.8, Connection of Grounding and Bonding Equipment, continues the discussion of performance-based requirements by listing means for connecting equipment-grounding conductors, grounding-electrode conductors, and bonding jumpers.

(A) Permitted Methods allows the following:

(1) Listed pressure connectors

(2) Terminal bars

(3) Pressure connectors listed as grounding and bonding equipment

(4) Exothermic welding processes

(5) Machine screw–type fasteners that engage not less than two threads or are secured by a nut

(6) Thread-forming machine screws that engage not less than two threads in the enclosure

(7) Connections that are part of a listed assembly

(8) Other listed means

Sheet-metal screws, wood screws, and improvised hookups may not be used to make the connections addressed in this section. Twist-on wire connectors (wire nuts) are permissible for these ground connections. The color green for wire nuts connecting equipment grounding conductors is useful for circuit tracing but not required by the Code.

(B) Methods Not Permitted prohibits grounding connections that depend solely on solder.

Section 250.10, Protection of Ground Clamps and Fittings, provides that these must be approved for general use without protection, or they must be protected from physical damage by being installed in a setting where they are not likely to be damaged, or they must be enclosed in metal, wood, or equivalent.

Section 250.12, Clean Surfaces, states that nonconductive coatings such as paint on equipment to be grounded are to be removed from contact surfaces, or the connection is to be made by means of fittings designed so as to make such removal unnecessary. Care must be taken if removal of paint is contemplated for outdoor equipment, where the coating is necessary for corrosion protection.

Part II, System Grounding, is the heart of Article 250. It covers system grounding, by which is meant grounding of one of the current-carrying conductors as opposed to the equipment-grounding conductor, which keeps enclosures, raceways, and other normally non-current-carrying conductive materials at ground potential. Key concepts in this part are systems required to be grounded, permitted to be grounded, and not permitted to be grounded. Also covered are main bonding jumpers and system bonding jumpers. Of great importance is where to connect the grounding conductor to the grounded conductor and where not to connect them. If they are connected at more than one point, objectionable circulating current may result. In a typical single-phase, three-wire service, this connection takes place through the main bonding jumper, after which the two current paths diverge, never to rejoin. Even in an appliance or portable tool such as an electric drill, the neutral is not to connect to the equipment-grounding conductor or metal case.

Section 250.20, Alternating-Current Systems to Be Grounded, provides that the following ac systems are required to be grounded:

(A) Alternating-Current Systems of Less Than 50 Volts, where

(1) Supplied by transformers, if the transformer supply system exceeds 150 volts to ground

(2) Supplied by transformers, if the transformer supply system is ungrounded

(3) Installed outside as overhead conductors

Grounding of a primary circuit does not carry over to the secondary of a transformer via electromagnetic induction. If the secondary is to be grounded, a new ground connection has to be established.

(B) Alternating-Current Systems of 50 Volts to Less Than 1,000 Volts provides that these systems are to be grounded under any of the following conditions:

(1) The system can be grounded so that the maximum voltage to ground on the ungrounded conductors does not exceed 150 volts.

(2) The system is three-phase, four-wire, wye connected in which the neutral conductor is used as a circuit conductor.

(3) The system is three-phase, four-wire, delta connected in which the midpoint of one phase winding is used as a circuit conductor.

(C) Alternating-Current Systems of 1 kV and Over states that these systems, if they supply portable or mobile equipment, are to be grounded. If they supply other than portable or mobile equipment, they are permitted to be grounded.

(D) Impedance Grounded Neutral Systems states that such systems are to be grounded in accordance with 250.36 or 250.186. These are highly specialized systems used in industrial settings and will be discussed later.

Section 250.21, Alternating-Current Systems of 50 Volts to Less Than 1,000 Volts Not Required to Be Grounded, provides the following:

(A) General states that the following ac systems of 50 volts to less than 1,000 volts are permitted to be grounded but not required to be grounded:

(1) Electrical systems used exclusively to supply industrial furnaces for melting, refining, tempering, and the like

(2) Separately derived systems used exclusively for rectifiers that supply only adjustable-speed industrial drives

(3) Separately derived systems supplied by transformers that have a primary voltage rating of less than 1,000 volts, provided that all the following conditions are met:

    a. The system is used exclusively for control circuits.

    b. Conditions of maintenance and supervision ensure that only qualified persons service the installation.

    c. Continuity of control power is required.

(4) Other systems that are not required to be grounded in accordance with the requirements of Section 250.20(B)

The purpose of installing an ungrounded system where permitted is to ensure continuity of operation because a line-to-ground fault would cause an outage. It must be decided whether this consideration should be given priority in view of the fact that ground-fault protection would not be provided.

(B) Ground Detectors states that such detectors are to be installed as follows:

(1) Ungrounded ac systems operating at not less than 120 volts and not exceeding 1,000 volts must have ground detectors.

(2) The ground-detection sensing equipment is to be installed as close as practicable to where the system receives its supply.

(C) Marking states that ungrounded systems are to be marked at the source or first disconnecting means.

Section 250.22, Circuits Not to Be Grounded, lists these circuits:

(1) Circuits for electric cranes operating over combustible fibers in Class III locations, as provided in Section 503.155, Electric Cranes, Hoists, and Similar Equipment—Class III, Divisions 1 and 2

(2) Circuits in health care facilities, as provided in Section 517.61, Wiring and Equipment (for Health Care Facilities), and Section 517.160, Isolated Power Systems (for Health Care Facilities)

(3) Circuits for equipment within an electrolytic-cell working zone, as provided in Article 668, Electrolytic Cells

(4) Secondary circuits of lighting systems, as provided in Section 411.5(A), Grounding (for Secondary Circuits of Lighting Systems Operating at 30 Volts or Less)

(5) Secondary circuits of lighting systems, as provided in Section 680.23(A)(2), Transformers and Power Supplies (for Underwater Luminaires)

Section 250.24, Grounding Service-Supplied Alternating-Current Systems, contains information of great importance. The section should be memorized.

(A) System Grounding Connections states that a premises wiring system supplied by a grounded ac service is to have a grounding-electrode conductor connected to the grounded service conductor, at each service, in accordance with (1) through (5) below:

(1) General provides that the grounding electrode conductor connection is to be made at any accessible point from the load end of the service drop or service lateral to and including the terminal or bus to which the grounded service conductor is connected at the service-disconnecting means. The most common connection is to a ground lug provided for that purpose within the meter socket enclosure. In this way, the grounding-electrode conductor does not have to be run through the wall and is usually directly over the first ground rod.

(2) Outdoor Transformer states that where the transformer supplying the service is located outside the building, at least one additional grounding connection is to be made from the grounded service conductor to a grounding electrode, either at the transformer or elsewhere outside the building. An exception excludes high-impedance grounded neutral systems. Most of the time, the outside transformer is utility property and therefore outside NEC jurisdiction. Nevertheless, it will be grounded. When the transformer is customer-owned, the NEC regulation applies.

(3) Dual-Fed Services provides that for these services, a single grounding-electrode connection to the tie point of the grounded conductor(s) from each power source is permitted.

(4) Main Bonding Jumper as Wire or Busbar permits this grounding-electrode conductor to be connected, in the service equipment enclosure, to the equipment-grounding terminal, bar, or bus to which the main bonding jumper is connected.

(5) Load-Side Grounding Connections says that a grounded conductor is not to be connected to normally non-current-carrying metal parts of equipment or to equipment-grounding conductors or to be reconnected to ground on the load side of the service-disconnecting means except as otherwise permitted in Section 250.140, Frames of Ranges and Clothes Dryers, and Section 250.142, Use of Grounded Circuit Conductor for Grounding Equipment, which allow exceptions to this general rule under very limited conditions. For example, the frames of ranges, wall-mounted ovens, counter-mounted cooking units, and clothes dryers may be grounded by the neutral-grounded conductor only for existing installations where an equipment-grounding conductor is not present in the outlet or junction box and

- The supply circuit is 120/240-volt, single-phase, three-wire or 208Y/120-volt derived from a three-phase, four-wire, wye-connected system.

- The grounded conductor is not smaller than 10 AWG copper or 8 AWG aluminum.

- The grounded conductor is insulated, or the grounded conductor is uninsulated and part of a Type SE service-entrance cable, and the branch circuit originates at the service equipment.

- Grounding contacts of receptacles furnished as part of the equipment are bonded to the equipment.

The preferred method would be to run a new branch circuit with an equipment-grounding conductor to the location of the appliance. Be sure to open the appliance and make sure that the equipment-grounding conductor is connected to the frame, including a jumper for any hinged metal doors, and that the grounded conductor is not connected to the frame.

Section 250.26, Conductor to be Grounded—Alternating-Current Systems, specifies the following conductors to be grounded for ac systems:

(1) Single-phase, two-wire—one conductor (For residences, this is a very rare, obsolete type of service.)

(2) Single-phase, three-wire—the neutral conductor

(3) Multiphase systems having one wire common to all phases—the common conductor

(4) Multiphase systems where one phase is grounded—one-phase conductor

(5) Multiphase systems in which one phase is used as in (2)—the neutral conductor

Refer back to Section 250.20(B) to see if a system is required to be grounded; then refer to Section 250.26 to ascertain which conductor is to be grounded.

Section 250.28, Main Bonding Jumper and System Bonding Jumper, brings together several previously unresolved elements. This is a supremely important section. These two jumpers fulfill the same function, and they are hooked up the same. They are different in name only. The main bonding jumper is for a conventional service, whereas the system bonding jumper is for a separately derived system. These two terms appear in Article 100, Definitions, under *bonding jumper, main*, and *bonding jumper, system*.

(A) Material provides that main bonding jumpers and system bonding jumpers are to be copper or other corrosion-resistant material. They may consist of wire, bus, screw, or similar suitable conductor.

(B) Construction states that where a main bonding jumper or a system bonding jumper is a screw only, the screw is to be green and visible with the screw installed.

(C) Attachment refers back to Section 250.8, Connection of Grounding and Bonding Equipment. It provides that main bonding jumpers and system bonding jumpers are to be connected in accordance with that section which, as we have seen, contains specifications for permitted methods and methods not permitted for connecting these jumpers and other devices.

(D) Size says that main bonding jumpers and system bonding jumpers are to be sized in accordance with (1) through (3) below:

(1) General states that main bonding jumpers and system bonding jumpers are not to be smaller than the sizes shown in Table 250.66, Grounding Electrode Conductor for Alternating-Current Systems. Notice that this table displays the required sizes of grounding-electrode conductors as a function of sizes of largest ungrounded service-entrance conductors. Despite the fact that the stated purpose of the table is to size out grounding-electrode conductors, it has other applications as well, as we are seeing in this section on main bonding jumper and system bonding jumper sizes.

It is further provided that where the supply conductors are larger than 1,100 kcmil copper or 1,750 kcmil aluminum, i.e., off the chart, these bonding jumpers are to have an area that is not less than 12½ percent of the area of the largest phase conductor, except that where the phase conductors and bonding jumper are of different materials (copper or aluminum), the minimum size of the bonding jumper is to be based on the assumed use of phase conductors of the same material as the bonding jumper and with an ampacity equivalent to that of the installed phase conductors. An entrance panel intended to be installed as service equipment is supplied by the manufacturer with a bonding jumper, with instructions, to be installed in the field. The reason that it is not preinstalled is that the box may be used as a feeder panel or nonservice load center, in which case the bonding jumper is not to be used because there must be no connection at that downstream location between the grounded circuit conductor and the equipment-grounding conductor.

(2) Main Bonding Jumper for Service with More Than One Enclosure provides that where a service consists of more than a single enclosure, as permitted in Section 230.71, Maximum Number of Disconnects, the main bonding jumper for each enclosure is to be sized in accordance with Section 250.28(D)(1) based on the largest ungrounded service conductor serving that enclosure.

(3) Separately Derived System with More Than One Enclosure provides two alternate methods for sizing the system bonding jumper for this installation. The system bonding jumper for each enclosure is to be sized in accordance with Section 250.28(D)(1), just like a main bonding jumper, based on the largest ungrounded feeder serving that enclosure. Alternately, a single system bonding jumper is to be installed at the source and sized in accordance with the same section based on the equivalent size of the largest supply conductor,

determined by the largest sum of the areas of the corresponding conductors of each set.

Section 250.30, Grounding Separately Derived Alternating-Current Systems, provides that these systems must comply with

- Section 250.30(A), Grounded Systems (below)
- Section 250.30(B), Ungrounded Systems (below)
- Section 250.20, Alternating-Current Systems to Be Grounded (above)
- Section 250.21, Alternating-Current Systems of 50 Volts to Less Than 1,000 Volts Not Required to Be Grounded (above)
- Section 250.22, Circuits Not to Be Grounded (above)
- Section 250.26, Conductor to Be Grounded—Alternating-Current Systems (above)

An Informational Note makes this very important distinction: An alternate ac power source, such as an on-site generator, is not a separately derived system if the grounded conductor is solidly interconnected to a service-supplied system-grounded conductor. An example of such a situation is where alternate-source transfer equipment does not include a switching action in the grounded conductor and allows it to remain solidly connected to the service-supplied grounded conductor when the alternate source is operational and supplying the load served.

If the generator neutral is connected to a neutral bar in the transfer switch, and the service-equipment neutral bar is wired to that same transfer-switch neutral bar, it is not a separately derived service. If, on the other hand, for a 208Y/120-volt, three-phase, four-wire system, the transfer switch has a switch with the grounded conductor connected to a pole in the switch so that it is capable of being interrupted, then the system is separately derived. This distinction must be kept in mind throughout the discussion that follows (Figure 2-7).

(A) Grounded Systems states that except as otherwise permitted in Article 250, a grounded conductor is not to be connected to normally non-current-carrying metal parts of equipment, be connected to equipment-grounding conductors, or be reconnected to ground on the load side of the system bonding jumper. A separately derived ac system is to comply with (1) through (8) below:

(1) System Bonding Jumper provides that an unspliced system bonding jumper must comply with Section 250.28(A) through (D). These, as we have seen, are bonding jumper specifications containing requirements for such parameters as size and material. These are the same for both types of bonding jumpers. It is further provided that this connection is to be made at any single (*single!*) point on the separately derived system from the source to the first system-disconnecting means or overcurrent device, or it is to be made at the source of a separately

Figure 2-7   Back-up power for a dwelling, connected to the entrance panel through a transfer switch. This may or may not be a separately derived system depending on how the grounded conductor is switched.

derived system that has no disconnecting means or overcurrent devices, in accordance with 250.30(A)(1)(a), Installed at the Source, or 250.30(A) (1)(b), Installed at the First Disconnecting Means. The system bonding jumper must remain within the enclosure where it originates. If the source is located outside the building or structure supplied, a system bonding jumper is to be installed at the grounding-electrode connection in compliance with 250.30(C), Outdoor Source (below). Three exceptions modify this section:

Number 1 permits a single system bonding jumper connection to the tie point of the grounded circuit conductors from each power source in systems installed in accordance with Section 450.6, Secondary Ties (for Transformers).

Number 2 permits a system bonding jumper at both the source and the first disconnecting means if doing so does not establish a parallel path for the grounded conductor.

Number 3 provides that the size of the system bonding jumper for a system that supplies a Class 1, Class 2, or Class 3 circuit and is derived from a transformer rated not more than 1,000 volt-amperes is not to be smaller than the derived ungrounded conductors and is not to be smaller than 14 AWG copper or 12 AWG aluminum. What's all this about Class 1, 2, and 3 circuits? We shall see when we get to Article 725, Class 1, Class 2, and Class 3 Remote-Control,

Signaling, and Power-Limited Circuits. For now, we'll say that Article 725 provides alternatives to the normal wiring methods in Chapters 1 through 4 for those types of circuits. This is sometimes called *low-voltage wiring*, but that terminology is not definitive.

(2) Supply-Side Bonding Jumper states that if the source of a separately derived system and the disconnecting means are located in separate enclosures, a supply-side bonding jumper is to be installed with the circuit conductors from the source enclosure to the first disconnecting means. A supply-side bonding jumper is not required to be larger than the derived ungrounded conductors. The supply-side bonding jumper is permitted to be of nonflexible metal raceway type or the wire or bus type as follows:

a. A supply-side bonding jumper of the wire type must comply with 250.102(C), Size—Supply-Side Bonding Jumper, which goes back to Table 250.66 and incorporates the 12½ percent rule.

b. A supply-side bonding jumper of the bus type is to have a cross-sectional area not smaller than the wire type described earlier.

(3) Grounded Conductor provides that if a grounded conductor is installed and the system bonding jumper connection is not located at the source, the following apply:

a. Sizing for a Single Raceway states that the grounded conductor is not to be smaller than the required grounding electrode conductor specified in Table 250.66 but is not required to be larger than the largest derived ungrounded conductor. Also, the 12½ percent rule applies. Here we see another application of Table 250.66 that goes beyond its original title.

b. Parallel Conductors in Two or More Raceways states that if the ungrounded conductors are installed in two or more raceways, the grounded conductor also must be installed in parallel. The size of the grounded conductor in each raceway is to be based on the total circular mil area of the parallel derived ungrounded conductors in the raceway but not smaller than 1/0 AWG.

A more complete treatment of this topic comes later in Section 310.10(H), Conductors in Parallel. Conductors for each phase, polarity, neutral, or grounded circuit may be connected in parallel only in sizes of 1/0 AWG and larger, subject to numerous conditions. This is an optional type of installation that is helpful in reducing the need for huge, unwieldy conductors. Generally, it is a good thing because it enhances heat dissipation owing to the fact that the paralleled conductors are spatially separated. Special twin-lug adapters are available and necessary for the terminations. In each metal race-way, it is essential to run the grounded and ungrounded conductors together to prevent inductive heating.

    c. Delta-Connected System states that the grounded conductor of a three-phase, three-wire delta system is to have an ampacity not less than that of the ungrounded conductors.

    d. Impedance Grounded System states that the grounded conductor must comply with 250.36, High-Impedance Grounded Neutral Systems, or 250.186, Impedance Grounded Neutral Systems (for Over 1 kV).

(4) Grounding Electrode, as applicable to separately derived systems, provides that the grounding electrode is to be as near as practicable to and preferably in the same area as the grounding electrode connection to the system. It is to be the nearest of the following:

    a. Metal water pipe grounding electrode

    b. Structural metal grounding electrode

An exception states that if neither of the preceding is available, any of the other electrodes listed in Section 250.52(A) may be substituted. This section lists seven permitted grounding electrodes, such as ground ring and plate electrodes. (At present, we are considering grounding of separately derived systems. A complete treatment of grounding electrodes appears in Part III, Grounding Electrode System and Grounding Electrode Conductor.)

Another exception states that if a separately derived system originates in listed equipment suitable for use as service equipment, the grounding electrode used for the service or feeder equipment may be used as the grounding electrode for the separately derived system.

(5) Grounding Electrode Conductor, Single Separately Derived System provides that this conductor is to be sized in accordance with Table 250.66 for the derived ungrounded conductors. It is to be used to connect the grounded conductor of the separately derived system to the grounding electrode. This connection is to be made at the same point on the separately derived system where the system bonding jumper is connected. There are three exceptions:

Number 1: If the system bonding jumper is a wire or busbar, it is permitted to connect the grounding-electrode conductor to the equipment-grounding terminal, bar, or bus, provided that it is of sufficient size for the separately derived system.

Number 2: If a separately derived system originates in listed equipment suitable as service equipment, the grounding-electrode conductor from the service or feeder equipment to the grounding electrode is permitted as the grounding-electrode conductor for the separately derived system, provided that the grounding-electrode conductor is of sufficient size for the separately derived system. If the equipment-grounding bus internal to the equipment is not smaller than the required grounding-electrode conductor for the separately derived system, the grounding-electrode connection for the separately derived system is permitted to be made to the bus.

Number 3: A grounding-electrode conductor is not required for a system that supplies a Class 1, Class 2, or Class 3 circuit and is derived from a transformer rated not more than 1,000 volt-amperes, provided that the grounded conductor is bonded to the transformer frame or enclosure by a jumper sized in accordance with Section 250.30(A)(1), Exception Number 3, and the transformer frame or enclosure is grounded by one of the means specified in Section 250.134, Equipment Fastened in Place or Connected by Permanent Wiring Methods (Fixed)—Grounding.

(6) Grounding Electrode Conductor, Multiple Separately Derived Systems provides that a common grounding-electrode conductor for multiple separately derived systems is permitted. If installed, the common grounding-electrode conductor is to be used to connect the grounded conductor of the separately derived systems to the grounding electrode. A grounding-electrode conductor tap then is to be installed from each separately derived system to the common grounding-electrode conductor. Each tap conductor is to connect the grounded conductor of the separately derived system to the common grounding-electrode conductor. This connection is to be made at the same point on the separately derived system where the system bonding jumper is connected. Two exceptions are the same as attached to (5) for a single separately derived system (above).

   a. Common Grounding-Electrode Conductor states that such a conductor is permitted to be one of the following:

   • A conductor of the wire type not smaller than 3/0 AWG copper or 250 kcmil aluminum

   • The metal frame of the building or structure that complies with 250.52(A)(2) or is connected to the grounding-electrode system by a conductor not smaller than 3/0 copper or 250 kcmil aluminum

   b. Tap Conductor Size states that Table 250.66 is to be used. An exception provides that if a separately derived system originates in listed equipment suitable as service equipment, the grounding-electrode conductor from the service or feeder equipment to the grounding electrode is permitted as the grounding-electrode conductor for the separately derived system, provided that it is of sufficient size. If the equipment-ground bus internal to the equipment is not smaller than the required grounding-electrode conductor for the separately derived system, the grounding-electrode connection for the separately derived system may be made to the bus.

c. Connections provides that all tap connections to the common grounding-electrode conductor are to be made at an accessible location by one of these methods:

1. A connector listed as grounding and bonding equipment

2. Listed connection to aluminum or copper busbars not smaller than ¼ inch × 2 inches

3. The exothermic welding process

Tap conductors must be connected to the common grounding-electrode conductor in such a manner that the common grounding-electrode conductor remains without a splice or joint.

The designer/installer has the option of using a common grounding-electrode conductor for multiple separately derived systems as opposed to running separate grounding-electrode conductors for each.

(7) Installation is to comply with 250.64(A), (B), (C), and (E), Grounding Electrode Conductor Installation, which contains details not specific to separately derived systems but applicable here as well.

(8) Bonding states that structural steel and metal piping are to be connected to the grounded conductor of a separately derived system in accordance with Section 250.104(D), Separately Derived Systems (for Bonding of Piping Systems and Exposed Structural Steel). This section, as we shall see, provides details for bonding these items to the electrical-system ground. Neglecting this item is a frequent error and can lead to a dangerous situation where large amounts of metal can become energized.

(B) Ungrounded Systems states that the equipment of an ungrounded separately derived system is to be grounded and bonded as specified in 250.30(B)(1) through 250.30(B)(3):

(1) Grounding Electrode Conductor states that this conductor, sized in accordance with the ubiquitous Table 250.66, is to be used to connect the metal enclosures of the derived system to the grounding electrode. The connection is to be made at any point on the separately derived system from the source to the first system-disconnecting means.

(2) Grounding Electrode lists provisions that are the same as for grounded systems, given in Section 250.30(A)(4).

(3) Bonding Path and Conductor states that a supply-side bonding jumper is to be installed from the source of a separately derived system to the first disconnecting means in compliance with Section 250.30(A)(2). Here again, the requirement is the same as for grounded systems. The bottom line is that ungrounded systems are only ungrounded insofar as the current-carrying conductors are

all ungrounded. But there is always an equipment-grounding conductor with grounding electrode and grounding-electrode conductor.

(4) Outdoor Source provides that if the source of the separately derived system is located outside, a grounding-electrode connection is to be made at the source location. The reason for this rule is that an outside separately derived source is vulnerable to lightning damage and falling power lines. A grounding-electrode connection at this point reduces these hazards.

Section 250.32, Buildings or Structures Supplied by a Feeder(s) or Branch Circuit(s), provides as follows:

(A) Grounding Electrode states that buildings or structures supplied by feeder(s) or branch circuit(s) must have a grounding electrode or grounding-electrode system. An exception states that a grounding electrode is not required where the building is supplied by not more than one branch circuit. For present purposes, a multiwire branch circuit is considered a single circuit.

(B) Grounded Systems provides the following:

(1) Supplied by a Feeder or Branch Circuit states that an equipment-grounding conductor is to be run with the supply conductors and is to be connected to the building or structure disconnecting means and to the grounding electrode(s). The equipment-grounding conductor is to be used for grounding or bonding of equipment, structures, or frames required to be grounded or bonded. The equipment-grounding conductor is to be sized in accordance with Table 250.122, Minimum Size Equipment Grounding Conductors for Grounding Raceway and Equipment. Any installed grounded conductor must not be connected to the equipment-grounding conductor or to the grounding electrode. An exception for installations made in compliance with previous editions of the Code that permitted such connection permits the grounded conductor that runs with the supply to the building or structure to serve as the ground-fault return path. There are three conditions:

a. An equipment-grounding conductor is not run with the supply to the building or structure.

b. There are no continuous metallic paths bonded to the grounding system in each building or structure served.

c. Ground-fault protection of equipment has not been installed on the supply side of the feeder(s).

(2) Supplied by Separately Derived System provides the following:

a. With Overcurrent Protection states that if overcurrent protection is provided where the conductors originate, the installation is to comply with Section 250.32(B)(1).

b. Without Overcurrent Protection states that if overcurrent protection is not provided where the conductors originate, the installation must comply with Section 250.30(A), Grounded Systems (for Separately Derived Alternating-Current Systems). If installed, the supply-side bonding jumper is to be connected to the building or structure disconnecting means and to the grounding electrodes.

(C) Ungrounded Systems provides the following:

(1) Supplied by a Feeder or Branch Circuit states that an equipment-grounding conductor is to be installed with the supply conductors, and it must be connected to the building or structure disconnecting means and to the grounding electrode(s). The grounding electrode(s) is also to be connected to the building or structure disconnecting means.

This section reminds us that an ungrounded system still requires an equipment-grounding conductor and associated grounding electrode and grounding-electrode conductor. When fed by a feeder or branch circuit, the entrance panel is in another building or outside, and the equipment-grounding conductor is run with the feeder or branch circuit. This is always required. If there is more than a single branch circuit in the second building, it needs a grounding electrode and grounding-electrode conductor as well.

(2) Supplied by a Separately Derived System provides the following:

a. With Overcurrent Protection states that the installation is to comply with (C)(1) above.

b. Without Overcurrent Protection states that where the conductors originate, the installation is to comply with Section 250.30(B), Ungrounded Systems (for Separately Derived Alternating-Current Systems). If installed, the supply-side bonding jumper is to be connected to the building or structure disconnecting means and to the grounding electrode.

(D) Disconnecting Means Located in Separate Building or Structure on the Same Premises states that where one or more disconnecting means supply one or more additional buildings or structures under single management, and where these disconnecting means are located remote from those buildings or structures, all the following conditions must be met:

(1) The connection of the grounded conductor to the grounding electrode, to normally non-current-carrying metal parts of equipment, or to the equipment grounding conductor at a separate building or structure is not to be made.

(2) Where an equipment-grounding conductor for grounding and bonding any normally non-current-carrying metal parts of equipment, interior metal piping systems, and building or structural metal frames

is run with the circuit conductors to a separate building or structure and connected to existing grounding electrode(s) required in Part III of Article 250 or where there are no existing electrodes, the grounding electrode(s) required in Part III are to be installed where a separate building or structure is supplied by more than one branch circuit.

(3) The connection between the equipment-grounding conductor and the grounding electrode at a separate building or structure is to be made in a junction box, panelboard, or similar enclosure located immediately inside or outside the separate building or structure.

(E) Grounding Electrode Conductor refers for sizing of this conductor to Table 250.66, based on the largest ungrounded supply conductor. It is the primary application for this table. The entire installation must comply with Part III of Article 250.

Section 250.34, Portable and Vehicle-Mounted Generators, states that

(A) The frame of a portable generator is not required to be connected to a grounding electrode under the following conditions:

(1) The generator supplies only equipment mounted on the generator, cord-and-plug-connected equipment through receptacles mounted on the generator, or both, and

(2) The normally non-current-carrying metal parts of equipment and the equipment-grounding conductor terminals of the receptacles are connected to the generator frame.

(B) Vehicle-Mounted Generators states that the frame of a vehicle is not required to be connected to a grounding electrode for a system supplied by a generator located on this vehicle under the following conditions:

(1) The frame of the generator is bonded to the vehicle frame, and

(2) The generator supplies only equipment located on the vehicle or cord-and-plug-connected equipment through receptacles mounted on the vehicle or both equipment located on the vehicle and cord-and-plug-connected equipment through receptacles mounted on the vehicle or on the generator, and

(3) The normally non-current-carrying metal parts of equipment and the equipment-grounding conductor terminals of the receptacles are connected to the generator frame.

(C) Grounded Conductor Bonding provides that a system conductor that is required to be grounded by Section 250.26, Conductor To Be Grounded—Alternating-Current Systems, is to be connected to the generator frame where the generator is a component of a separately derived system.

Section 250.35, Permanently Installed Generators, provides that a conductor that provides an effective ground-fault current path is to be installed with the supply conductors from a permanently installed generator(s) to the first disconnecting means in accordance with (A) or (B):

(A) Separately Derived System states that if the generator is installed as a separately derived system, Section 250.30, Grounding Separately Derived Alternating-Current Systems, applies. Keep in mind that the definition of a separately derived system is based on whether or not the grounded conductor is switched at the disconnect. If the grounded conductor runs straight through, it is not a separately derived system.

(B) Non-Separately Derived System states that if the generator is installed as a non-separately derived system and overcurrent protection is not integral with the generator assembly, a supply-side bonding jumper must be installed between the generator equipment-grounding terminal and the equipment-grounding terminal, bar, or bus of the disconnecting means. It is to be sized in accordance with Section 250.102(C), Size—Supply-Side Bonding Jumper, based on the size of the conductors supplied by the generator.

Section 250.36, High-Impedance Grounded Neutral Systems, begins with a brief description and limitations on the use of such systems. It says that high-impedance grounded neutral systems in which a grounding impedance, usually a resistor, limits the ground-fault current to a low value are permitted for three-phase ac systems of 480 to 1,000 volts only if the following three conditions are met:

(1) Conditions of maintenance and supervision ensure that only qualified persons service the installation.

(2) Ground detectors are installed on the system.

(3) Line-to-neutral loads are not served.

Systems over 1,000 volts are also permitted. They must comply with Section 250.186, Impedance Grounded Neutral Systems (for Grounding of Systems and Circuits of Over 1 kV).

High-impedance grounded neutral systems are found in industrial settings where continuity of the electrical supply is essential. In the event of a ground fault, an alarm will sound rather than conventional overcurrent protection by means of a circuit breaker. An orderly shutdown rather than a sudden outage will prevent additional hazards.

High-impedance grounded neutral systems must comply with the following provisions:

(A) Grounding Impedance Location states that the grounding impedance is to be installed between the grounding-electrode conductor and the system neutral point. If a neutral point is not available, the grounding impedance is to be installed between the grounding-electrode conductor and the neutral point derived from a grounding transformer.

(B) Grounded System Conductor states that the grounded system conductor from the neutral point of the transformer or generator to its connection point to the grounding impedance is to be fully insulated. This is so because the grounding impedance is likely to cause a voltage to appear on the segment of the grounding-electrode conductor that is between the impedance and the neutral point of the transformer. The segment between the resister and the grounding electrode does not have to be insulated because it is at ground potential at all times.

The grounded system conductor must have an ampacity of not less than the maximum current rating of the grounding impedance, but in no case is the grounded system conductor to be smaller than 8 AWG copper or 6 AWG aluminum or copper-clad aluminum.

(C) System Grounding Connection states that the system is not to be connected to ground except through the grounding impedance.

(D) Neutral Point to Grounding Impedance Conductor Routing provides that the conductor connecting the neutral point of the transformer or generator to the grounding impedance is permitted to be installed in a separate raceway from the ungrounded conductors. It is not required to run this conductor with the phase conductors to the first system-disconnecting means or overcurrent device.

(E) Equipment Bonding Jumper states that the equipment bonding jumper (the connection between the equipment-grounding conductors and the grounding impedance) is to be an unspliced conductor run from the first system-disconnecting means or overcurrent device to the grounded side of the grounding impedance.

(F) Grounding Electrode Conductor Location provides that the grounding-electrode conductor is to be connected at any point from the grounded side of the grounding impedance to the equipment-grounding connection at the service equipment or first system-disconnecting means.

(G) Equipment Bonding Jumper Size states that the equipment bonding jumper is to be sized in accordance with either of the following:

(1) If the grounding-electrode conductor connection is made at the grounding impedance, the equipment bonding jumper is to be sized in accordance with Table 250.66 based on the size of the service-entrance conductors for a service or the derived phase conductors for a separately derived system, or

(2) If the grounding-electrode conductor is connected at the first system-disconnecting means or overcurrent device, the equipment bonding jumper is to be sized the same as the neutral conductor in Section 250.36(B).

That's all there is to NEC's mandates for high-impedance grounded neutral systems up to 1,000 volts. If you need to build one of these, grounding resistor manufacturers' literature is helpful. I-Gard (www. ipc-resistors.com) has a comprehensive technical library available online.

Part III, Grounding Electrode System and Grounding Electrode Conductor, contains provisions for this portion of the overall grounding and bonding picture. The requirements are simple and straightforward for the most part, although, as we shall see, there are a few fine points that could be missed on an exam or in the field. Trunk slammers get the idea that a service has to be grounded, but rarely do they get every detail right. As for satellite-dish installations, the technicians may do an impeccable job of mounting and aiming the dish, but lots of installations are only in partial compliance when it comes to grounding and bonding. Later on (when we get to Chapter 8, Communications Systems) we'll see how these are to be installed in order to be in total compliance. For now, we continue our grounding and bonding journey that started with the service and will end below ground.

Section 250.50, Grounding Electrode System, makes the point that grounding is a complete system. It provides that all grounding electrodes as described in Section 250.52(A)(1) through (7) that are present at each building or structure are to be bonded together to form the grounding-electrode system. Where none of these grounding electrodes exist, one or more of the grounding electrodes specified in Section 250.52(A)(4) through (8) are to be installed and used.

These are the grounding electrodes that, if present, are to be bonded together to form the grounding-electrode system:

- *Metal underground water pipe.* This must be in direct contact with the earth for 10 feet or more to qualify. The 10-foot figure includes metal well casing if bonded to the pipe. In these times of plastic pipe, it may be difficult to verify whether there is 10 feet of metal pipe outside the building. If not, but if there is metal pipe inside, it does not qualify as a grounding electrode but nevertheless needs to be bonded to the grounding-electrode system. If there are any plastic pipe segments or items that may be removed for servicing such as a water filter housing or water meter, these have to be bypassed with a bonding conductor that is the same size as the grounding-electrode conductor.

- *Metal frame of a building or structure.* This must consist of at least one structural member that is in direct contact with the earth, with or without concrete encasement, or hold-down bolts securing the structural steel column that are connected to a concrete-encased electrode that complies with the next item and is located in the support footing or foundation. The hold-down bolts are to be connected to the concrete-encased electrode by welding, exothermic welding, the usual steel tie wires, or other approved means.

- *Concrete-encased electrode.* This must consist of at least 20 feet of either one or more bare or zinc galvanized or other electrically conductive coated-steel reinforcing bars or rods of not less than ½ inch in diameter installed in one continuous 20-foot length or, if in multiple pieces connected by the usual steel tie wires, exothermic welding, welding, or other effective means to create a 20-foot length, or bare copper conductor not smaller than 4 AWG. Metallic components are to be encased by at least 2 inches of concrete and are to be located horizontally within that portion of a concrete foundation or footing that is in direct contact with the earth. If multiple concrete-encased electrodes are present at a building or structure, it is permissible to bond only one into the grounding electrode system. (An Informational Note makes the point that concrete installed with insulation, vapor barriers, films, or similar items separating the concrete from the earth is not considered to be in *direct contact* with the earth.) This structure is probably the most effective grounding electrode among the many that are permitted. A concrete-encased metal grounding electrode is called a *ufer*. It was invented by Herbert Ufer in Arizona in the 1940s. The dry soils in Arizona present difficulties in achieving a good ground because their resistivity is high. To cope with this problem, Ufer found that by connecting a ground wire to rebar in a concrete foundation, an acceptably low-impedance ground connection could be created. The reason it works is that concrete is more conductive than soil. The only problem is that it may be difficult to verify conditions within the concrete if the foundation has been poured and backfilled prior to the electrician's arrival at the site.

- *Ground ring.* This must encircle the building or structure in direct contact with the earth and must consist of at least 20 feet of bare copper conductor not smaller than 2 AWG.

- *Rod and pipe electrodes.* These must be not less than 8 feet in length. There are two alternatives: (A) Grounding electrodes of pipe or conduit are to be not smaller than ¾ inches (trade size) and, where of steel, must have the outer surface galvanized or otherwise metal coated for corrosion protection, or (B) rod-type grounding electrodes of stainless steel or copper- or zinc-coated steel must be at least 5/8 inch in diameter unless listed.

- *Plate electrodes.* Each plate must expose not less than 2 square feet of surface to exterior soil. Electrodes of bare or conductively coated iron or steel plates are to be at least ¼ inch thick. Solid, uncoated electrodes of nonferrous metal are to be 0.06 inch thick.

- *Other listed electrodes.*

These electrodes, if present, are to be bonded together to form the grounding-electrode system. There is no obligation to create any or all of these. For example, a building such as a garage may not have plumbing and so no metal water pipes. Even a residence may have all plastic pipes. Similarly, there may be no suitable concrete-encased electrode, or its exact nature may be unknown. In no case is it ever necessary to chip concrete to make rebar accessible for grounding purposes.

If none of the preceding are present, one of these is to be installed and used:

- Ground ring
- Rod and pipe electrodes
- Other listed electrodes
- Plate electrodes
- Other local metal underground systems or structures—piping systems, underground tanks, and underground well casings that are not bonded to a metal water pipe

(B) Not Permitted for Use as Grounding Electrodes states that metal underground gas piping systems and aluminum are not to be used as grounding electrodes. Even though gas piping may not constitute a grounding electrode, still it is to be bonded to the service equipment enclosure, the grounded conductor at the service, the grounding-electrode conductor if of sufficient size, or to one or more grounding electrodes used. The bonding conductor is sized in accordance with Table 250.122, Minimum Size Equipment Grounding Conductors for Grounding Raceway and Equipment.

Section 250.53, Grounding Electrode System Installation, addresses various types of grounding electrodes.

(A) Rod, Pipe and Plate Electrodes states that these items must be installed as follows:

(1) Below Permanent Moisture Level states that if practicable, rod, pipe, and plate electrodes are to be embedded below permanent moisture level. They must be free from nonconductive coatings such as paint or enamel. To acquire a moist surrounding, a good practice is to drive ground rods along the drip line from the roof.

(2) Supplemental Electrode Required states that a single rod, pipe, or plate electrode is to be supplemented by an additional electrode of any of the permitted types listed earlier. The additional supplemental electrode may be bonded to the primary grounding electrode, a grounding-electrode conductor, the grounded service-entrance conductor, a nonflexible grounded service raceway, or any grounded service enclosure. An exception states that if a single rod, pipe, or plate electrode has a resistance to earth of 25 ohms or less, the supplemental electrode is not required.

Resistance to earth cannot be measured by means of a simple ohmmeter. To do so would require a known perfect reference ground. Since the equipment for making a ground resistance test is expensive and involves driving two reference ground rods, most electricians simply drive a second rod. When a second, supplemental electrode is used, it is not necessary to have the 25-ohm minimum resistance to earth.

(B) Electrode Spacing states that where two electrodes are used, they must be not less than 6 feet apart.

(C) Bonding Jumper states that the bonding jumper used to connect the grounding electrodes together to form the grounding-electrode system is to be installed in accordance with Section 250.64(A), (B), and (E), must be sized in accordance with Section 250.66, and is to be connected in accordance with Section 250.70, Methods of Grounding and Bonding Conductor Connection to Electrodes.

(D) Metal Underground Water Pipe has some exacting requirements to qualify as a grounding electrode:

(1) Continuity of the grounding path or the bonding connection is not to rely on water meters or filtering devices and similar equipment. Wire jumpers are an effective way to meet this requirement.

(2) Supplemental Electrode Required states that a metal underground water pipe is to be supplemented by an additional electrode of a type specified in Section 250.52(A)(2) through (8). The reason for this requirement is that if some time in the future a metal waterline is replaced with plastic, the grounding for the building would be lost without the supplemental grounding electrode.

(E) Supplemental Electrode Bonding Connection Size provides that where the supplemental electrode is a rod, pipe, or plate electrode, that portion of the bonding jumper that is the sole connection to the supplemental grounding electrode is not required to be larger than 6 AWG copper or 4 AWG aluminum wire. This section supersedes Table 250.66, which would seem to require a 4 AWG grounding electrode conductor for a 200-ampere residential service. But it has been calculated that because of the ground impedance of a rod, pipe, or plate electrode, the larger size would not be called for.

(F) Ground Ring provides that a ground ring is to be buried not less than 30 inches below the earth's surface. A ground ring is very effective but far more expensive than other grounding electrodes.

(G) Rod and Pipe Electrodes contains installation requirements. The electrode is to be installed so that at least 8 feet of length is in contact with the soil. It is to be driven to a depth of at least 8 feet. Where rock bottom is encountered, the truck slammer's answer is to cut off the excess that is above grade, but this will result in a substandard installation, i.e., high ground resistance. To be in compliance, the ground rod may be driven at an angle of not more than 45 degrees. If this is not possible, another alternative is to dig a trench and install the rod horizontally at a depth of at least 30 inches. It is further provided that the upper end of the electrode is to be flush with or below ground level unless the above-ground end including the grounding-electrode conductor attachment is protected against physical damage.

(H) Plate Electrode provides that it must be at least 30 inches below the surface of the earth.

Section 250.54, Auxiliary Grounding Electrodes, states that such electrodes are different from supplemental grounding electrodes. The former are not required, whereas the latter are required under certain circumstances, as outlined earlier. Any number of auxiliary grounding electrodes are permitted to be connected at any point along the equipment-grounding conductor. However, they are required if they are specified in manufacturers' installation instructions, which are part of the UL or other organization's listing. Typically, an auxiliary grounding electrode is installed at the site of each parking lot luminaire for lightning protection (Figure 2-8). These definitely cannot be used in lieu of the required equipment-grounding conductor. That is called a *floating ground* and is against the Code. The reason is that the earth is not a sufficiently low-impedance path back to the electrical supply, and overcurrent devices would not function. In case of a ground fault, the circuit would not open, and normally non-current-carrying conductive metal objects would remain energized. Auxiliary grounding electrodes are not required to be bonded to the grounding-electrode system, and they must never be connected to the neutral or ungrounded circuit conductor.

Section 250.58, Common Grounding Electrode, states that where an ac system is connected to a grounding electrode in or at a building or structure, the same electrode is to be used to ground conductor

Figure 2-8 Parking lot luminaire may need its own grounding electrode if required by the manufacturer's instructions. Otherwise, it is optional. In all cases, a separate equipment-grounding conductor is required, tied into any grounding electrode.

enclosures and equipment in or on that building or structure. Where separate services, feeders, or branch circuits supply a building and are required to be connected to a grounding electrode(s), the same grounding electrode(s) is to be used.

Two or more grounding electrodes that are bonded together are to be considered as a single grounding-electrode system in this sense.

Section 250.60, Use of Strike Termination Devices, states that conductors and driven pipes, rods, or plate electrodes used for grounding strike-termination devices are not to be used in lieu of the grounding electrodes required by Section 250.50, Grounding Electrode System, for grounding wiring systems and equipment. This provision does not prohibit the required bonding together of grounding electrodes of different systems.

One of the basic principles of grounding is that all grounding electrodes of all systems are to be bonded together. This includes, as the preceding section provides, lightning-rod grounding electrodes. Also included are ground electrodes for satellite dishes and all the diverse communications systems that come into a modern dwelling or other facility. Homeowners sometimes have a hard time with this because they think that lightning will infiltrate other systems within the house. All you can do is show them your Code book and explain the real danger is when grounding electrode systems are not connected, so voltage potential exists between them.

Section 250.62, Grounding Electrode Conductor Material, provides that the grounding-electrode conductor is to be copper, aluminum, or copper-clad aluminum. The conductors may be solid or stranded, insulated, covered, or bare.

Section 250.64, Grounding Electrode Conductor Installation, states that they are to be installed as follows:

(A) Aluminum or Copper-Clad Aluminum Conductors states that these may be used as grounding-electrode conductors, but their use is restricted owing to corrosion problems at the terminations. They are not to be used where in direct contact with masonry or the earth or where subject to corrosive conditions. If used outside, they are not to be terminated within 18 inches of earth.

(B) Securing and Protection Against Physical Damage states that if a grounding-electrode conductor is cut, the entire premises wiring system becomes compromised, and for this reason, this section contains several mandates that guard against such an unfortunate outcome. Where exposed, a grounding electrode or its enclosure is to be securely fastened to the surface on which it is carried. Grounding-electrode conductors are permitted to be installed on or through framing members. A 4 AWG or larger copper or aluminum grounding-electrode conductor is to be protected if exposed to physical damage. A 6 AWG grounding-electrode conductor that is free from exposure to physical damage is permitted to be run along the surface of a building construction without metal covering or protection if it is securely fastened to the construction; otherwise, it is to be protected in rigid-metal conduit, intermediate-metal conduit, rigid PVC, reinforced thermosetting-resin conduit, electrical metallic tubing, or cable armor. Grounding-electrode conductors smaller than 6 AWG are to be protected in rigid-metal conduit, intermediate-metal conduit, rigid PVC, reinforced thermosetting-resin conduit, electrical metallic tubing, or cable armor. Often, a grounding electrode conductor is run straight down to the ground from a knockout on the bottom of a meter socket enclosure. If metal conduit is used, it must be bonded to ground at both ends (using a grounding bushing). To avoid this unwieldy operation, rigid PVC generally is used with a protective bushing at the bottom end below grade, whereupon the conductor goes first to one ground rod and then to the second.

(C) Continuous states that generally, with exceptions in 250.30(A) (5) and (6), for separately derived systems, 250.30(B)(1) and 250.68(C), Metallic Water Pipe and Structural Metal, grounding-electrode conductors must be one continuous length without splice. Where the preceding exceptions permit a splice or connection, it is to be made as permitted in (1) through (4):

(1) Splicing of the wire-type grounding-electrode conductor is permitted only by irreversible compression-type connectors listed as grounding and bonding equipment or by the exothermic welding process. This is a highly specialized welding process that involves surrounding the pieces to be joined by a disposable form, called a *graphite crucible mold*. It is filled with powder consisting of aluminum and a mixture of copper oxides. The exothermic (heat-releasing) reaction is triggered by means of a flint striker of the type used to light an oxyacetylene torch. The conductors to be joined are in the mold, surrounded by the powder mix. The finished weld is mechanically very strong, corrosion resistant, and possesses excellent electrical properties, including an ability to withstand very high fault current.

(2) Sections of busbar may be connected together to form a grounding-electrode conductor.

(3) Bolted, riveted, or welded connections of structural-metal frames of buildings or structures is permitted.

(4) Threaded, welded, brazed, soldered, or bolted-flange connections of metal water piping is permitted.

(D) Service with Multiple Disconnecting Means Enclosures: The Code provides three alternate methods for grounding a service with two or more disconnecting-means enclosures:

(1) Common Grounding Electrode Conductor and Taps states that this is a widely used method for grounding multiple disconnecting-means enclosures. A common grounding-electrode conductor is run past each of the disconnects. A smaller tap runs into each enclosure, at which location it is bonded. The common grounding-electrode

conductor is sized using Table 250.66 based on the sum of the circular mil area of the largest ungrounded service-entrance conductor(s). The taps are sized in accordance with Table 250.66 based on the largest service-entrance conductor serving the individual enclosure. The taps are connected to the common grounding-electrode conductor in such a manner that the common grounding-electrode conductor remains without a splice or joint. Accepted methods are exothermic welding, connectors listed as grounding and bonding equipment, and busbars.

(2) Individual Grounding Electrode Conductors states that this is another permissible method for grounding multiple enclosures but is used less because of the greater quantity of wire required. Individual conductors are run to each enclosure from the grounding-electrode system. These conductors are sized using Table 250.66 based on the service-entrance conductors supplying each individual service-disconnecting means.

(3) Common Location provides that a grounding-electrode conductor must be connected to the grounded service conductors in a wireway or other accessible enclosure on the supply side of the service-disconnecting means. The connection is to be made by exothermic welding or a connector listed as grounding and bonding equipment. The grounding-electrode conductor is sized in accordance with Table 250.66 based on the service-entrance conductor(s) at the common location where the connection is made.

(E) Enclosures for Grounding Electrode Conductors provides that ferrous metal enclosures for grounding-electrode conductors must be electrically continuous from the point of attachment to cabinets or equipment to the grounding electrode and that they are to be securely fastened to the ground clamp or fitting. The bonding methods, either for service or nonservice locations, must be used at each end and to all intervening ferrous raceways, boxes, and enclosures between cabinets or equipment and the grounding electrode. The bonding jumper is to be the same size as or larger than the enclosed grounding-electrode conductor. If a raceway is used to protect a grounding-electrode conductor, the installation must comply with the requirements of the appropriate raceway article found in Chapter 3.

(F) Installation of Electrode(s) states that grounding-electrode conductors may be run to any convenient grounding electrode provided that all are bonded as previously required. An alternate method is to run grounding-electrode conductors to grounding electrodes individually. A third method is to install a copper or aluminum busbar in an accessible location such as fastened to a wall. Then connections of individual grounding-electrode conductors, upstream from various multiple enclosures and downstream to various grounding electrodes, can be made using a listed connector or the exothermic welding process. The busbar is to be ¼ inch thick by 2 inches wide. The length is not specified. It should be long enough to accommodate all connections with room for future additions if they are contemplated.

Section 250.66, Size of Alternating-Current Grounding Electrode Conductor, with associated Table 250.66, has been referenced repeatedly by us. It has other applications besides what the title denotes. The article provides details. The notes to Table 250.66 are important as well. This is one of the principal Code sections, and we will take a close look at it.

The section opens by referencing Table 250.66, Grounding Electrode Conductor for Alternating-Current Systems. The table is quite simple. On the left are sizes of the largest ungrounded service-entrance conductors or equivalent area for parallel conductors. On the right are sizes of the grounding-electrode conductors. Each of these appears as two columns, one for copper and the other for aluminum or copper-clad aluminum.

Notes to the table state:

Number 1: Where multiple sets of service-entrance conductors are used, the equivalent size of the largest service-entrance conductor is to be determined by the largest sum of the areas of the corresponding conductors of each set.

Number 2: Where there are no service-entrance conductors, the grounding-electrode conductor size is to be determined by the equivalent size of the largest service-entrance conductor required for the load to be served.

It is noted that the table also applies to the derived conductors of separately derived systems. The values in the table are modified by three subsections that place upper limits on the required size of grounding-electrode conductors connected to specific types of grounding electrodes. The thinking behind this is that it would not do any good to use conductors with current-carrying capabilities beyond the ground conductivity of the ground electrodes. The permitted reductions are as follows:

(A) Connections to Rod, Pipe, or Plate Electrodes states that maximum required grounding-electrode size is 6 AWG copper or 4 AWG aluminum.

(B) Connections to Concrete-Encased Electrodes states that maximum required grounding-electrode size is 4 AWG copper.

(C) Connections to Ground Rings states that maximum required grounding-electrode size is not larger than the conductor used for the ground ring. Of course, if you go first to a ground rod and then to a grounding electrode that requires a larger size, you have to use the larger size throughout.

Section 250.68, Grounding Electrode Conductor and Bonding Jumper Connection to Grounding Electrodes, contains three provisions that concern the connection of a grounding-electrode conductor at the service and at each building. Branch circuits, feeders, and separately derived systems are covered:

(A) Accessibility provides that all grounding-electrode conductor terminations including mechanical elements are to be accessible. Recall that this is different from saying that they must be *readily accessible*. Exceptions state that this mandate does not apply to buried or concrete-encased connections and that it does not apply to exothermic or irreversible compression connections or to mechanical means (reversible or not) used to attach the terminations to fireproofed structural metal.

(B) Effective Grounding Path states that a grounding-electrode conductor or bonding-jumper connection must be made in a manner that will ensure an effective grounding path. In a metal piping system used as a grounding electrode or grounding-electrode conductor, it is necessary to provide a jumper to shunt around any insulated joints or equipment likely to be removed for repair. Examples are water meters and water filter housings. The jumpers must be long enough so that the equipment can be removed without disconnecting or interrupting the bonding jumper.

(C) Metallic Water Pipe and Structural Metal permits two locations to serve as extensions for bonding jumpers and grounding-electrode conductors. It is to be emphasized that these are not actual grounding electrodes. These are as follows:

(1) Interior metal water pipe located not more than 5 feet from the point of entry to a building is permitted to be used as a conductor to interconnect electrodes that are part of the grounding-electrode system. An exception allows lengths of interior water pipe of greater than 5 feet from the point of entry to serve this purpose only in industrial, commercial, and institutional buildings or structures if conditions of maintenance and supervision ensure that only qualified persons service the installation. The entire length of the pipe, except for short sections passing perpendicularly through walls, floors, or ceilings, is exposed. This is to ensure that the integrity of the ground path is not broken by having a plastic segment inserted in the course of repairs.

(2) The metal structural frame of a building that is directly connected to a grounding electrode may serve a similar function. If it is verified that a valid grounding electrode such as concrete-encased rebar is in place, the structural metal frame may serve as a grounding-electrode conductor, and it is permissible to connect a grounding conductor to it at any point. If the structural metal building frame is not connected to a grounding electrode, it is still necessary to bond it to the electrical system, but it may not count as part of the grounding-electrode system.

Section 250.70, Methods of Grounding and Bonding Conductor Connection to Electrodes, provides that a grounding or bonding conductor is to be attached to the grounding electrode by exothermic welding, listed lugs, listed pressure connectors, listed clamps, or other listed means. The emphasis on listing denotes that improvised or field-fabricated methods will not do. For most installations, the ground clamp will be buried below grade and so must be listed for direct burial. Clamps that will be encased in concrete must be listed for that application as well. For indoor communications applications, a listed sheet-metal strap-type ground clamp is permitted.

Part IV, Enclosure, Raceway, and Service Cable Connections, provides in Section 250.80 that metal enclosures and raceways for service conductors and equipment are to be connected to the grounded system conductor if the electrical system is grounded or to the grounding-electrode conductor for electrical systems that are not grounded. An exception provides that a metal elbow that is installed in an otherwise nonmetallic underground raceway and is a minimum depth of 18 inches below the earth's surface is not required to be grounded. It is common practice to insert metal sweeps in PVC conduit that is installed underground because PVC is subject to damage when pulling large conductors, and it is a weak link in the underground raceway system.

Section 250.84, Underground Service Cable or Raceway, states that the sheath or armor of an underground service-cable system that is connected to the grounded system conductor on the supply side is not required to be connected to the grounded system conductor at the building or structure. The sheath or armor is permitted to be insulated from the interior raceway or piping. This connection is also not required for a metal service raceway that contains a metal-sheathed or armored cable that is connected to the grounded system conductor.

Section 250.86, Other Conductor Enclosures and Raceways, provides that metal enclosures and raceways for other than service conductors also must be connected to the equipment-grounding conductor. This is a very basic rule, but there are some exceptions. Prominent among these are the exceptions for short sections of metal enclosures or raceways used for support and protection and the metal elbow in a nonmetallic raceway buried 18 inches below the earth's surface, as discussed earlier. Aside from the exceptions, this rule is very important because otherwise, in the event of a fault, the enclosure or raceway could remain hot and be a severe hazard.

Part V, Bonding, contains requirements for bonding raceways and enclosures. Bonding is required for all types of wiring, but the requirements differ based on whether it is for a service, over 250 volts to ground, or within a hazardous (classified) area. A common electrical deficiency noted by inspectors is lack of bonding jumpers where a raceway is attached to service equipment. Exam questions are likely to focus on some of the requirements in this section, and distinctions should be carefully noted. Part V, Bonding, is easy to find. Refer to the Code Index for specific subtopics.

Section 250.90, General, states the basic principle that bonding is to be provided where necessary to ensure electrical continuity and the capacity to conduct safely any fault current likely to be imposed.

Section 250.92, Services, tells us in (A) which service components must be bonded and provides methods for fulfilling this mandate. Normally non-current-carrying metal parts of the following equipment must be bonded together:

- All raceways, cable trays, cable bus frameworks, auxiliary gutters, and service armor or sheaths that enclose, contain, or support service conductors, except for an underground metallic elbow (18 inches below the earth's surface) within a nonmetallic raceway system, as discussed previously

- All enclosures containing service conductors, including meter fittings, boxes, and the like, interposed in the service raceway or armor

(B) Method of Bonding at the Service states that bonding jumpers meeting the requirements of Article 250 are to be used around impaired connections, such as reducing washers or oversized, concentric, or eccentric knockouts. Standard locknuts or bushings are not to be the only means for the bonding required by this section but are permitted to be installed to make a mechanical connection of the raceway(s).

On the line side of the main disconnect, grounding bushings are always required. On the load side, they are required where the knockouts are impaired.

Electrical continuity at service equipment, service raceways, and service enclosures is to be ensured by one of the following methods:

(1) Bonding equipment to the grounded service conductor in a manner provided in Section 250.8, Connection of Grounding and Bonding Equipment (This section enumerates means for connection, including exothermic welding, listed pressure connectors, and others.)

(2) Connections using threaded couplings or threaded hubs on enclosures if made up wrench-tight

(3) Threadless couplings and connectors if made up tight for metal raceways and metal-clad cables

(4) Other listed devices, such as bonding-type locknuts, bushings, or bushings with bonding jumpers

Section 250.94, Bonding for Other Systems, introduces the important, relatively new concept of the intersystem bonding termination. This is a very inexpensive but useful device for allowing other trades such as satellite-dish and telephone installers to tie into the power-system ground. This section provides that an intersystem bonding termination for connecting intersystem bonding conductors required for other systems is to be provided external to enclosures at the service-equipment or metering-equipment enclosure and at the disconnecting means for any additional buildings or structures. It must comply with the following:

(1) It must be accessible for connection and inspection.

(2) It must consist of a set of terminals with the capacity for connection of not less than three intersystem bonding conductors.

(3) It must not interfere with opening the enclosure for a service, building, or structure disconnecting means or metering equipment.

(4) At the service equipment, it must be securely mounted and electrically connected to the service equipment, to the meter enclosure, or to an exposed nonflexible metallic service raceway or be mounted at one of these enclosures and connected to the enclosure or to the grounding-electrode conductor with a minimum 6 AWG copper conductor.

(5) At the disconnecting means for a building or structure, it must be securely mounted and electrically connected to the metallic enclosure for the building or structure disconnecting means or be mounted at the disconnecting means and be connected to the metallic enclosure or to the grounding-electrode conductor with a minimum 6 AWG copper conductor.

(6) The terminals are to be listed as grounding and bonding equipment.

An exception states that for existing buildings or structures where any intersystem bonding electrode conductor is in place, the intersystem bonding termination is not required.

Section 250.96, Bonding Other Enclosures, contains general requirements for bonding non-current-carrying metal parts and also discusses isolated grounding circuits.

(A) General provides that metal raceways, cable trays, cable armor, cable sheath, enclosures, frames, fittings, and other metal non-current-carrying parts that are to serve as equipment-grounding conductors, with or without the use of supplementary equipment-grounding conductors, are to be bonded where necessary to ensure electrical continuity and the capacity to conduct safely any fault current likely to be imposed on them. Any nonconductive paint, enamel, or similar coating is to be removed at threads, contact points, and contact surfaces or be connected by fittings designed to make such removal unnecessary.

An example of the latter is the common locknut placed on the threaded shank of connectors where they extend on the insides of enclosures that have knockouts. The sharp edges dig through the painted surface and contact the metal underneath in order to acquire ground continuity. These locknuts should be driven quite hard in a clockwise direction to make sure that a low-impedance connection is made. Sections of cable tray must be joined in such a way as to ensure ground continuity, and the manufacturer's instructions should be consulted to ensure that this happens.

(B) Isolated Grounding Circuits states that this is a strategy for dealing with electrical noise or electromagnetic interference. This phenomenon is caused by, among other things, fluorescent ballasts, electric motors that may have arcing at the brushes, and any number of nonlinear loads. It causes data loss in computers and all sorts of electronic garbling. It travels along metal raceways and other normally non-current-carrying metal bodies and finds its way into electronic equipment, where it is not wanted. A way to mitigate this problem is by creating isolated grounds. The metal raceway where it enters the enclosure of a piece of sensitive equipment, for example, can be electrically isolated from the equipment by using a listed nonmetallic fitting made for the purpose. The raceway is bonded to the enclosure of the entrance panel or load center where the conductors it encloses receive their supply. An insulated grounding conductor is attached to the equipment-grounding terminal inside the service (or other) panel and run through the raceway to the piece of equipment that has to be protected from electrical noise. At that point, it is terminated by means of a bonding screw to the equipment enclosure. The end result is that both raceway and equipment are grounded properly, and in a good proportion of cases, the electrical noise is eliminated. An Informational Note states that use of an isolated equipment-grounding conductor does not relieve the requirement for grounding the raceway system.

Section 250.97, Bonding for Over 250 Volts, states that for circuits of over 250 volts to ground, the electrical continuity to ground of metal raceways and cables with metal sheaths must be ensured by using any of the methods for bonding raceways that contain service conductors as specified in Section 250.92(B), except for (B)(1). This latter provision refers us to Section 250.8, which lists the eight standard methods for grounding equipment (listed pressure connectors, terminal bars, etc.). An exception, however, states that where oversized, concentric or eccentric knockouts are not encountered or where a box or enclosure with concentric or eccentric knockouts is listed to provide a reliable bonding connection, the following methods shall be permitted:

(1) Threadless couplings and connectors for cables with metal sheathes

(2) Two locknuts on rigid metal conduit or intermediate metal conduit, one inside and one outside the boxes and cabinets

(3) Fittings with shoulders that seat firmly against the box or cabinet, such as electrical metallic tubing connectors, flexible metal conduit connectors, and cable connectors, with one locknut on the inside of boxes and cabinets

(4) Listed fittings

You may have noticed that it is fairly difficult to remove the necessary number of concentric knockouts without distorting the ones that are to remain. This is so because a good amount of metal is necessary where they join to provide a low-impedance ground path. Where there is only one knockout, as in a junction box, it is easy to remove because it is attached at only one point that will not have to carry fault current from an attached raceway.

Section 250.98, Bonding Loosely Jointed Metal Raceways, states that expansion fittings and telescoping sections of metal raceways are to be made electrically continuous by equipment-bonding jumpers or other means. Expansion fittings are not used as often in metal race-way systems as in nonmetallic conduit systems because metal has a much lower expansion coefficient than plastic. (Metal and concrete have almost identical expansion coefficients, which is why rebar is possible. Mounting PVC pipe in a long horizontal run on an outside wall where it is subject to a wide range of temperatures is highly problematic in terms of bowing and buckling.)

Section 250.100, Bonding in Hazardous (Classified) Locations, indicates that any bonding problem in a hazardous location could be disastrous, and therefore, absolute reliability is necessary. A faulty bond in series with a ground fault or static charge could result in a spark sufficient to initiate an explosion in a Class 1, Division 1 area. And a complete break in the ground path could result in energized metal with an equally catastrophic outcome. To ensure electrical con-tinuity, NEC 2011 specifies any of the bonding methods found in Sec-tion 250.92(B)(2) through (4) above. It further states that one or more of these methods are to be used whether or not equipment-grounding conductors of the wire type are installed.

Section 250.102, Bonding Conductors and Jumpers, provides the following:

(A) Material states that bonding jumpers are to be of copper or other corrosion-resistant material. A bonding jumper is to be wire, bus, screw, or other suitable conductor. Is aluminum corrosion-resistant? This would be a question for the AHJ. It is less corrosion-resistant than copper but more corrosion-resistant than iron. The main reason for using aluminum in any application is that it is much cheaper than copper. In short lengths, copper is always the metal of choice, but no one would use copper for 200-ampere service-entrance cable going up the outside of a house. On the other hand, for short pieces going through a conduit stub from meter to entrance panel, copper would be more appropriate. As for bonding jumpers, these usually would fall into the latter category.

(B) Attachment states that bonding jumpers are to be attached in the manner specified by the applicable provisions of Section 250.8, Connection of Grounding and Bonding Equipment, and Section 250.70, Methods of Grounding and Bonding Conductor Connection to Electrodes.

(C) Size—Supply-Side Bonding Jumper provides a different method of sizing upstream of the main disconnect, where we don't have as high a degree of overcurrent protection as for a load-side equipment bonding jumper. The supply-side bonding jumper is not to be smaller than the sizes shown in Table 250.66 for grounding-electrode conductors. As in other applications, the 12½ percent rule kicks in when the ungrounded supply conductors are larger than 1,100 kcmil copper or 1,750 kcmil aluminum.

Where the ungrounded supply conductors are paralleled in two or more raceways or cables and an individual supply-side bonding jumper is used for bonding them, the size of the supply-side bonding jumper for each raceway or cable is to be selected from Table 250.66 based on the size of the ungrounded supply conductors in each raceway or cable. A single supply-side bonding jumper installed for bonding two or more raceways or cables is to be sized in accordance with Section 250.102(C)(1) above. Where the ungrounded supply conductors and the supply-side bonding jumper are of different materials (copper or aluminum), the minimum size of the supply-side bonding jumper is to be based on the assumed use of ungrounded conductors of the same material as the supply-side bonding jumper and with an ampacity equivalent to that of the installed ungrounded supply conductors.

(D) Size—Equipment Bonding Jumper on the Load Side of an Overcurrent Device states that as we move from the supply side to the load side, we shift from Table 250.66 to Table 250.122. Notice that whereas the former is based on the size of the largest ungrounded service-entrance conductor or equivalent area for parallel conductors, the latter is based on the rating or setting of the automatic overcurrent device in the circuit ahead of equipment, conduit, etc. This makes perfect sense because for the supply-side bonding jumper, there is no overcurrent device, and for the load-side bonding jumper, the service-entrance conductors are irrelevant.

(E) Installation permits bonding jumpers or conductors and equipment bonding jumpers to be either inside or outside of a raceway or enclosure. An external bonding jumper, strapped to the outside of flexible metal conduit going to a motor, for example, may look a little strange, but it is actually a very practical solution. For bonding telescoping sections of metal conduit as required previously, there is no workable alternative. The length of an external bonding jumper is limited to 6 feet, but an exception permits greater lengths at outside pole locations for the purpose of bonding or grounding isolated sections of metal raceways or elbows installed in exposed risers and for bonding grounding electrodes. These are not required to be routed with a raceway or enclosure.

Section 250.104, Bonding of Piping Systems and Exposed Structural Steel, requires that a low-impedance fault-current path be established from any piping systems and exposed structural steel that may be present on the premises.

(A) Metal Water Piping outlines bonding requirements for this system. The points of attachment of the bonding jumpers must be accessible.

(1) General lists the four points of attachment for bonding the water piping system to the electrical supply. They are

- The service-equipment enclosure
- The grounded conductor at the service (either end of the main bonding jumper)

- The grounding-electrode conductor where of sufficient size
- One or more grounding electrodes used

The bonding jumpers are to be sized in accordance with Table 250.66. While the metal water piping system may not qualify as a grounding electrode, still it must be bonded to one of the four points of attachment listed earlier (Figure 2-9). Isolated sections of metal piping that are not electrically connected because of intervening plastic pipe do not have to be bonded if they do not extend beyond individual appliances.

(2) Buildings of Multiple Occupancy provides that in these locations, metallic water piping systems that are isolated from one another owing to intervening plastic pipe may be bonded to non-service-grounding terminals within each unit. The bonding jumper is sized using Table 250.122, not Table 250.66. This is so because the connection is downstream from the main disconnecting means within the four-wire region for single-phase 120/240-volt systems.

(3) Multiple Buildings or Structures Supplied by a Feeder(s) or Branch Circuit(s) provides that the metal water piping system(s) installed in or attached to a building or structure is to be bonded to the disconnecting-means enclosure at that location, to the equipment-grounding conductor run with the supply conductors, or to the one

Figure 2-9 It is the electrician's responsibility to make sure that any metallic water piping system is bonded to the electric supply. If the wiring is completed before the plumbing is installed, it is essential that the electrician revisit the premises to make sure that this important step is not omitted, leading to a very hazardous situation.

or more grounding electrodes used. The bonding jumper(s) is to be sized in accordance with Table 250.66 based on the size of the feeder or branch-circuit conductors that supply the building. The bonding jumper is not required to be larger than the largest ungrounded feeder or branch-circuit conductor supplying the building.

(B) Other Metal Piping states that such piping also must be similarly bonded. This section makes the point that gas piping is included. Previously, we saw that underground gas piping may not be used as a grounding electrode. Nevertheless, the gas piping is to be bonded to one of the four attachment points mentioned earlier for water piping systems. Since the requirement says "likely to become energized," it is generally not necessary to separately bond the gas piping if it is attached to one or more appliances that also have electrical or water hookups that create a ground connection.

(C) Structural Metal states that structural metal, if exposed, also must be bonded to the grounding-electrode system by means of one of the points of attachment listed earlier.

(D) Separately Derived Systems provides that metal water piping systems and structural metal must be bonded to separately derived systems. Exceptions provide that a separate bonding jumper is not needed where the items to be bonded are already used as grounding electrodes.

Section 250.106, Lightning-Protection Systems, specifies that the ground terminals for these systems must be bonded to the premises electrical grounding-electrode system. This is consistent with the basic principle that all grounded conductors of all systems in or on a building or structure are to be interconnected.

Part VI, Equipment Grounding and Equipment Grounding Conductors, contains vital Code information that appears frequently on licensing exams and is needed in the field.

Section 250.110, Equipment Fastened in Place (Fixed) or Connected by Permanent Wiring Methods, provides that normally non-current-carrying metal parts of fixed equipment supplied by or enclosing conductors or components that are likely to become energized are to be connected to an equipment-grounding conductor under any of the following conditions:

(1) Where within 8 feet vertically or 5 feet horizontally of ground or grounded metal objects and subject to contact by persons

(2) Where located in a wet or damp location and not isolated

(3) Where in electrical contact with metal

(4) Where in a hazardous (classified) location

(5) Where supplied by a wiring method that provides an equipment-grounding conductor, except for short sections of metal enclosures

(6) Where equipment operates with any terminal at over 150 volts to ground

It would be possible to misinterpret this section. Does (1) mean that if a junction box is mounted on a ceiling over 8 feet vertically from the floor, it does not have to be grounded? The answer is no because Section 250.110 says ". . . under *any* of the following conditions," and such an installation would require grounding because of (5) unless the wiring method did not include an equipment-grounding conductor.

Exceptions allow equipment to be used without an equipment-grounding conductor under certain limited conditions, including where listed equipment is double insulated and so marked.

Section 250.112, Specific Equipment Fastened in Place (Fixed) or Connected by Permanent Wiring Methods, is a natural sequel to the preceding section. It contains a list of 13 specific types of equipment required to be connected to an equipment-grounding conductor regardless of voltage. Most of these are treated in articles that appear in various Code locations:

- Switchboard frames and structures—except for two-wire dc switchboards insulated from ground

- Pipe organs

- Motor frames—see Section 430.242, Stationary Motors

- Enclosures for motor controllers—unless attached to ungrounded portable equipment

- Elevators and cranes

- Garages, theaters, and motion picture studios—except pendant lampholders supplied by circuits not over 150 volts to ground

- Electric signs

- Motion picture projection equipment

- Remote-control, signaling, and fire alarm circuits—except equipment supplied by Class 1 circuits operating at less than 50 volts (A great amount of detail on this difficult subject is provided in Article 725, Class 1, Class 2, and Class 3 Remote-Control, Signaling, and Power-Limited Circuits, and in Article 760, Fire Alarm Systems.)

- Luminaires

- Skid-mounted equipment

- Motor-operated water pumps—including the submersible type

- Metal well casings (If a submersible pump is used in a metal well casing, the well casing is to be connected to the pump circuit equipment-grounding conductor.)

Section 250.114, Equipment Connected by Cord and Plug, provides a comprehensive list by occupancy type of such equipment that is required to be grounded, with exceptions for listed equipment that is double-insulated. Also excepted are tools and portable headlamps that are powered by an isolating transformer with an ungrounded secondary of not over 50 volts.

Section 250.118, Types of Equipment Grounding Conductors, lists those permitted. Generally, a copper or aluminum conductor, solid or stranded, insulated, covered, or bare, will suffice. Also, various types of metal raceways, flexible metal conduit, liquid-tight flexible conduit, flexible metallic tubing, and Type MC cable qualify as equipment-grounding conductors but with various conditions attached to each category. These conditions must be met to ensure reliability.

A major source of shock and fire hazard is a break in the equipment-grounding path. We often see Type EMT (electrical metallic tubing) that has pulled apart owing to poor workmanship or motion of the building. Good electricians "pull a green for everything" so as to provide an additional level of protection.

Section 250.121, Use of Equipment Ground Conductors, states that an equipment-grounding conductor is not to be used as a grounding-electrode conductor. Trunk slammers see that it should work, so why not? Besides violating Section 250.121, sizing requirements would invalidate this approach. Grounding-electrode conductors must be wire or busbar only, so that rules out metal raceways, as permitted for an equipment-grounding conductor.

Section 250.122, Size of Equipment Grounding Conductors, introduces in (A) Table 250.122, which we have used for some previous applications. Notice that unlike Table 250.66, which is based on service-conductor size, this table is based on the rating or setting of the automatic overcurrent device in the circuit ahead of the equipment or conduit that is to be grounded.

(B) Increased in Size makes the point that where ungrounded conductors are increased in size in order to counter voltage drop or for any other reason, the equipment-grounding conductors must be increased in size proportionately according to the circular mil area of the ungrounded conductors.

(C) Multiple Circuits states that if a single equipment-grounding conductor is run with multiple circuits in the same raceway, cable, or cable tray, it is to be sized for the largest overcurrent device protecting those conductors.

(D) Motor Circuits provides that equipment-grounding conductors for motor circuits are to be sized according to Table 250.122. Where the overcurrent device is an instantaneous-trip circuit breaker or a motor short-circuit protector, the equipment-grounding conductor is to be sized not smaller than given by that table using the maximum permitted rating of a dual-element time-delay fuse selected for branch-circuit short-circuit and ground-fault protection.

(E) Flexible Cord and Fixture Wire provides that the equipment-grounding conductor in a flexible cord with the largest circuit conductor 10 AWG or smaller and the equipment-grounding conductor used with fixture wires of any size is not to be smaller than AWG 18 copper and is not to be smaller than the circuit conductors. The equipment-grounding conductor in a flexible cord with a circuit conductor larger than 10 AWG is to be sized in accordance with Table 250.122.

The cutoff size for flexible cords is 10 AWG. Above that, go to the table. Below that, it is the same as the circuit conductors. For fixture wire, there is no cutoff. It is always the same as the circuit conductors.

(F) Conductors in Parallel states that where conductors are installed in parallel in multiple raceways or cables, the equipment-grounding conductors also must be installed in parallel in each raceway or cable. Where conductors are installed in parallel in the same raceway, cable, or cable tray, a single equipment-grounding conductor is permitted. Each equipment-grounding conductor is to be sized in accordance with Table 250.122 based on the rating or setting of the overcurrent device.

Where you could go wrong is to assume that the equipment-grounding conductor size was tied to the individual paralleled-circuit conductor, where in fact each of the individual grounding-electrode conductors has to be full sized based on the overcurrent device.

(G) Feeder Taps provides information for sizing equipment-grounding conductors that serve taps taken from feeders. As we have seen, a feeder has an overcurrent device at each end. This is what makes it a feeder, by definition. The feeder tap equipment-grounding conductor is to be sized by the overcurrent device at the upstream end of the feeder. Even though it goes with the tap, it has to be sized out based on the overall feeder overcurrent device. In no event, however, does it have to be larger than the tap itself.

Section 250.124, Equipment Grounding Conductor Continuity, contains provisions designed to ensure that whenever the current-carrying conductors are energized, there will be ground continuity.

(A) Separable Connections provides that those in such equipment provided in draw-out units or attachment plugs and connectors or receptacles are to be first make, last break of the equipment-grounding conductor. This is not required for interlocked equipment that precludes energization without grounding continuity.

(B) Switches provides that no switch is to be placed in the equipment-grounding conductor unless the opening of that switch disconnects all sources of energy.

Section 250.126, Identification of Wiring Device Terminals, provides that the color green, hexagonal shape, the word *green* or *ground*, the letters *G* or *GR*, or a grounding symbol is to be used to identify equipment-grounding conductor terminals.

Part VII, Methods of Equipment Grounding, contains a number of Code requirements that appear in most licensing exams. Section 250.130, Equipment Grounding Conductor Connections, provides that at the source of separately derived systems, they are to be made in accordance with Section 250.30(A)(1), System Bonding Jumper. Equipment-grounding conductor connections at service equipment are to be made as follows:

(A) For Grounded Systems states that the connection is to be made by bonding the equipment-grounding conductor to the grounded service conductor and the grounding-electrode conductor.

(B) For Ungrounded Systems states that the connection is to be made by bonding the equipment-grounding conductor to the grounding-electrode conductor.

(C) Nongrounding Receptacle Replacement or Branch-Circuit Extensions lists permitted points of connection for the equipment-grounding conductor of a grounding-type receptacle or branch-circuit extension:

(1) Any accessible point on the grounding-electrode system

(2) Any accessible point on the grounding-electrode conductor

(3) The equipment-grounding terminal bar within the enclosure where the branch circuit for the receptacle or branch circuit originates

(4) For grounded systems, the grounded service conductor within the service-equipment enclosure

(5) For ungrounded systems, the grounding terminal bar within the service-equipment enclosure

The intent behind this section is to provide guidance for replacing old nongrounding receptacles (Figure 2-10) or extending nongrounding branch circuits. However, this is only in the context of

FIGURE 2-10 Many old homes, hotels, restaurants, and other occupancies still have these nongrounding receptacles. The Code allows several options for replacing them without running a new branch circuit with an equipment-grounding conductor. If it is too difficult to run new wire back to the panel, the best choice may be to use a GFCI marked, "No Equipment Ground." This device does not require an equipment-grounding conductor to function properly, and such an installation is NEC compliant.

equipment-grounding connections. Other methods for replacing nongrounding receptacles are found in Section 406.4(D)(2), e.g., replacement with another nongrounding receptacle, still sold for that purpose, and replacement with a GFCI, which has to be marked, "No Equipment Ground."

Section 250.132, Short Sections of Raceway, states that where these are required to be grounded, they are to be connected to an equipment-grounding conductor in accordance with the following section.

Section 250.134, Equipment Fastened in Place or Connected by Permanent Wiring Methods (Fixed)—Grounding, states that unless grounded by connection to the grounded circuit conductor as permitted by Sections 250.32, 250.140, and 250.142, non-current-carrying metal parts of equipment, raceways, and other enclosures, if grounded, are to be connected by an equipment-grounding conductor by one of the methods specified in (A) or (B):

(A) Equipment Grounding Conductor Types states that this is done by connecting to any of the equipment-grounding conductors permitted by Section 250.118. Recall that this is the general list of permitted equipment-grounding conductors that includes wire, busbar, and various types of metal conduit with conditions attached to the flexible versions. If you are not close to maximum conduit fill, pull a green.

(B) With Circuit Conductors states that this is done by connecting to an equipment-grounding conductor contained within the same raceway or cable or otherwise run with the circuit conductors.

Section 250.136, Equipment Considered Grounded, specifies two additional conditions under which equipment may be considered grounded:

(A) Equipment Secured to Grounded Metal Supports states that a metal rack or structure that is itself grounded by one of the means listed in Section 250.134 will serve to ground equipment secured to and in electrical contact with it. However, the structural metal frame of a building may not be used in this way. Equipment could be mounted on the structural metal frame of a building, and it would be grounded, but that would not meet the requirement. The equipment also would have to be grounded by one of the means listed in Section 250.134.

(B) Metal Car Frames states that such frames supported by metal hoisting cables attached to or running over metal sheaves or drums of elevator machines that are grounded by one of the means listed in Section 250.134 may be considered grounded and will provide grounding for equipment in or on it.

Section 250.138, Cord-and-Plug-Connected Equipment, contains two means for providing grounding for non-current-carrying metal parts of cord-and-plug-connected equipment, if required:

(A) By Means of an Equipment Grounding Conductor states that such a conductor run with the power-supply conductors in a cable assembly or flexible cord terminated in a grounding-type plug is acceptable.

(B) By Means of a Separate Flexible Wire or Strap states that where part of the equipment, such an arrangement is acceptable.

Section 250.140, Frames of Ranges and Clothes Dryers, indicates that the basic rule is that these appliances, which also include wall-mounted ovens, counter-mounted cooking units, and the outlet or junction boxes connected to them, are to be connected to the equipment-grounding conductor in accordance with Sections 250.134 or 250.138. However, there is an exception that is likely subject matter for a licensing exam question and also important in your field work.

The exception states that for existing branch-circuit installations only where an equipment-grounding conductor is not present in the outlet or junction box, the grounded circuit conductor may be used instead of the usual equipment-grounding conductor only if all the following conditions are met:

(1) The supply circuit is 120/240-volt, single-phase, three-wire or 208Y/120-volt derived from a three-phase, four-wire, wye-connected system.

(2) The grounded conductor is not smaller than 10 AWG copper or 8 AWG aluminum.

(3) The grounded conductor is insulated or the grounded conductor is uninsulated and part of Type SE service-entrance cable and the branch circuit originates at the service equipment.

(4) Grounding contacts of receptacles furnished as part of the equipment are bonded to the equipment.

It must be emphasized that for new installations, a regular three-wire plus equipment grounding conductor must be run. The exception is valid only in instances where there is no available green wire. Also, it is essential that the appliance and circuit are compatible. If a newer appliance is to be powered by one of these grandfathered circuits, the neutral has to be bonded to the frame inside the appliance. If an older appliance is to be connected to a new-style circuit, the equipment-grounding conductor has to be bonded to the frame and the neutral bonding, if present, has to be removed. Under the old rules, prior to NEC 1996, the branch circuit was required to originate at the service equipment to avoid neutral current in the wrong place, so that aspect of working under the preceding exception should also be checked out. You're dealing with a lot of available fault current, so take care!

Section 250.142, Use of Grounded Circuit Conductor for Grounding Equipment, is related to the preceding section. The premises wiring is divided into two zones:

(A) Supply-Side Equipment states that a grounded circuit conductor is permitted to ground non-current-carrying metal parts of equipment, raceways, and other enclosures at any of the following locations:

(1) On the supply side or within the enclosure of the ac service-disconnecting means (An example is the meter socket enclosure,

which is grounded by the grounded circuit conductor. The equipment-grounding conductor splits off later, at the service-disconnecting means enclosure or main disconnecting means.)

(2) On the supply side or within the enclosure of the main disconnecting means for separate buildings, as provided in Section 250.32(B), Grounded Systems [for Buildings or Structures Supplied by a Feeder(s) or Branch Circuit(s)]

(3) On the supply side or within the enclosure of the main disconnecting means or overcurrent devices of a separately derived system where permitted by Section 250.30(A)(1), System Bonding Jumper (for Grounding Separately Derived Alternating-Current Systems)

The basic grounding and bonding rules for a service-supplied system and a separately derived system are similar, but the nomenclature is different. A service-supplied system, for example, has a main bonding jumper, whereas it is called a system bonding jumper in a separately derived system.

(B) Load-Side Equipment states a basic principle: A grounded circuit conductor is not to be used to ground non-current-carrying metal parts on the load side of the service-disconnecting means or on the load side of a separately derived system-disconnecting means or the overcurrent devices for a separately derived system not having a main disconnecting means. However, there are four exceptions to this basic principle:

Number 1: Frames of ranges, etc. for existing installations, as discussed in Section 250.140.

Number 2: It is permissible to ground meter enclosures by connection to the grounded circuit conductor on the load side of the service disconnect, where all the following conditions apply:

1. No service ground-fault protection is installed.

2. All meter enclosures are located immediately adjacent to the service-disconnecting means.

3. The size of the grounded circuit conductor is not smaller than the size specified in Table 250.122 for equipment-grounding conductors.

Of course, the normal location for a meter is upstream of the main disconnect, so exception number 2 does not usually become an issue.

Number 3: Direct-current systems are permitted to be grounded on the load side of the disconnecting means or overcurrent device in accordance with Section 250.164, Point of Connection for Direct-Current Systems.

Number 4: Electrode-type boilers operating at over 600 volts are to be grounded as required in Section 490.72(E)(1), Grounded Neutral Conductor (for Equipment, Over 600 Volts, Nominal) and Section 490.74, Bonding.

Section 250.144, Multiple Circuit Connections, states that where equipment is grounded and is supplied by more than one circuit or grounded premises wiring system, an equipment-grounding conductor termination is to be provided for each connection.

Section 250.146, Connecting Receptacle Grounding Terminal to Box, discusses use of an equipment bonding jumper, which can be factory made with a grounding screw preattached (the so-called greenie) or field fabricated from a piece of green or bare wire. The equipment bonding jumper is sized out in accordance with Table 250.122. This section also discusses situations where an equipment bonding jumper is not required for grounding a receptacle to a box:

(A) Surface Mounted Box states that the equipment bonding jumper may be omitted for a surface-mounted box if at least one of the insulating retaining washers is removed from the receptacle so that there is direct metal-to-metal contact. The reason this solution is not permitted for a flush-mounted box is that the box could be recessed a slight amount so that there would not be metal-to-metal contact. The screw on its own is not sufficient. This provision does not apply to cover-mounted receptacles unless the box and cover combination is listed as providing satisfactory ground continuity between the box and the receptacle.

(B) Contact Devices or Yokes states that such items that are designed and listed as self-grounding are permitted in conjunction with the supporting screws to establish the grounding circuit between the device yoke and flush-type boxes.

(C) Floor Boxes states that such boxes designed and listed as providing satisfactory ground continuity are permitted without an equipment bonding jumper.

(D) Isolated Receptacles states that where installed for the reduction of electrical noise on the grounding circuit, a receptacle in which the grounding terminal is purposely insulated from the receptacle mounting means is permitted. The receptacle grounding terminal is to be connected to an insulated equipment-grounding conductor run with the circuit conductors. The equipment-grounding conductor may pass through boxes, wire ways, or other enclosures without being connected to them.

The purpose of all this is to minimize electrical noise within electronic equipment. When this method is used properly, everything that is required to be grounded—equipment, metal raceways, and enclosures—is indeed grounded so that the performance requirement of Section 250.4(A)(5) has been satisfied.

Section 250.148, Continuity and Attachment of Equipment Grounding Conductors to Boxes, states that with the exception of the equipment-grounding conductor used for isolated receptacles, whenever circuit conductors are spliced or terminated within a box, all equipment-grounding conductor splices and terminations are to comply with (A) through (E) below:

(A) Connections and Splices states that connections and splices are to be made in accordance with Section 110.14(B), Splices (for Electrical Connections), except that insulation is not required. In other words, equipment-grounding conductor connections and splices have to be made to the same standard as if they were current-carrying conductors. This precludes the practice of simply twisting the bare wires together, the so-called Boston wrap.

(B) Grounding Continuity provides that when a luminaire, receptacle, or other device is removed from a box, ground continuity is not interrupted. This may involve the use of a short field-fabricated bare or green jumper with an extra wire nut.

(C) Metal Boxes provides that a connection is to be made between the equipment-grounding conductors and a metal box by means of a grounding screw that is to be used for no other purpose, equipment used for no other purpose, equipment listed for grounding, or a listed grounding device. The practice of using a drywall screw that also mounts the box is out. Such trunk slammer expedients as wrapping the equipment-grounding conductor under a yoke mount or folding it back within a cable connector are not permitted. Use of a grounding clip is permissible.

(D) Nonmetallic Boxes states that one or more equipment-grounding conductors brought into a nonmetallic outlet box are to be arranged so that a connection can be made to any fitting or device in that box requiring grounding. It is to be emphasized that grounding continuity must be preserved even while the fitting or device is removed for replacement.

(E) Solder states that connections that depend solely on solder are not to be used.

Part VIII, Direct-Current Systems, states that such systems are to be grounded in a different manner than the more common alternating-current systems. There are many sources of dc—batteries, thermocouples, photovoltaic cells, wind generators, and engine-generator sets that are made to produce dc.

Section 250.160, General, specifies that dc systems are to comply with Part VIII and other sections of Article 250 not specifically intended for ac systems.

Section 250.162, Direct-Current Circuits and Systems To Be Grounded, provides the following:

(A) Two-wire dc systems supplying premises wiring and operating at greater than 50 volts but not greater than 300 volts must be grounded. There are three exceptions:

Number 1: A system equipped with a ground detector and supplying only industrial equipment in limited areas is not required to be grounded.

Number 2: A rectifier-derived dc system supplied from an ac system complying with Section 250.20, Alternating-Current Systems To Be Grounded, is not required to be grounded.

Number 3: Dc fire alarm circuits having a maximum current of 0.030 ampere as specified in Article 760, Part III, are not required to be grounded.

As in ac systems, a dc system that is not grounded still must have an equipment-grounding conductor connected to a grounding-electrode system. It is just that no current-carrying conductor is grounded. Keep in mind also the distinctions among *required, permitted,* and *not permitted.*

(B) Three-Wire Direct-Current Systems states that the neutral conductor of all three-wire dc systems supplying premises wiring is to be grounded.

Section 250.164, Point of Connection for Direct-Current Systems, shows how dc systems differ markedly from ac systems:

(A) Off-Premises Source states that if a dc system is supplied by an off-premises source, and if it is to be grounded in accordance with the preceding section, the grounding connection is to be made at one or more supply stations, not at individual services or at any point on the premises wiring.

(B) On-Premises Source states that if the dc system source is located on the premises, a grounding connection is to be made at one of the following:

(1) The source

(2) The first system-disconnection means or overcurrent device

(3) By other means that accomplish equivalent system protection and that use equipment listed and identified for use

Section 250.166, Size of the Direct-Current Grounding Electrode Conductor, provides that the size of the grounding-electrode conductor for a dc system is to be as specified in (A) and (B), except as permitted by (C) through (E):

(A) Not Smaller Than the Neutral Conductor states that for a dc system that has a three-wire balancer set or a balancer winding with overcurrent protection as provided in Section 445.12(D), Balancer Sets (for Generators), the grounding-electrode conductor is not to be smaller than the neutral conductor and not smaller than 8 AWG copper or 6 AWG aluminum.

(B) Not Smaller Than the Largest Conductor states that where the dc system is other than (A), the grounding-electrode conductor is not to be smaller than the largest conductor supplied by the system and not smaller than 8 AWG copper or 6 AWG aluminum.

(C) Connected to Rod, Pipe, or Plate Electrodes states that the grounding-electrode conductor is not required to be larger than 6 AWG copper or 4 AWG aluminum.

(D) Connected to a Concrete-Encased Electrode states that the grounding-electrode conductor is not required to be larger than 4 AWG copper. Notice that aluminum is not mentioned here. An aluminum grounding-electrode conductor cannot be used with a concrete-encased electrode.

(E) Connected to a Ground Ring states that the grounding-electrode conductor is not required to be larger than the conductor used for the ground ring.

Section 250.168, Direct-Current System Bonding Jumper, states that for dc systems required to be grounded, an unspliced system bonding jumper is to be used to connect the equipment-grounding conductor to the grounded conductor at the source of the first system-disconnecting means where the system is grounded. The size of the bonding jumper is as specified in Section 250.66.

Section 250.169, Ungrounded Direct-Current Separately Derived Systems, provides that except for portable and vehicle-mounted generators, these systems are to have a grounding-electrode conductor connected to a grounding electrode. It is to be connected to the metal enclosure at any point on the separately derived system from the source to the first system-disconnecting means or overcurrent device or at the source of a separately derived system that has no disconnecting means or overcurrent devices. The size of the grounding-electrode conductor is to be in accordance with Section 250.166.

Part IX, Instruments, Meters, and Relays, contains provisions for grounding these devices. Remember that we are no longer talking about exclusively dc equipment.

Section 250.170, Instrument Transformer Circuits, provides that secondary circuits of current and potential instrument transformers are to be grounded where the primary windings are connected to circuits of 300 volts or more to ground, and where on switchboards, they are to be grounded irrespective of voltage. There are two exceptions:

Number 1: Circuits where the primary windings are connected to circuits of less than 1,000 volts with no live parts or wiring exposed or accessible to other than qualified persons are not required to be grounded.

Number 2: Current transformer secondaries connected in a three-phase delta configuration are not required to be grounded.

Keep in mind that this section is talking about grounding of current-carrying conductors. It has nothing to do with an equipment-grounding conductor.

Section 250.172, Instrument Transformer Cases, states that cases or frames of instrument transformers are to be connected to the equipment-grounding conductor where accessible to other than qualified persons. An exception provides that cases or frames of current transformers with primaries of not over 150 volts to ground that are used exclusively to supply current to meters are not required to be connected to the equipment-grounding conductor.

Section 250.174, Cases of Instruments, Meters, and Relays Operating at Less than 1,000 Volts, are to be connected to the equipment-grounding conductor, except where they are on live-front switchboards and have exposed live parts on the front of panels. These are not to have their cases connected to an equipment-grounding conductor. Where the voltage is greater than 150 volts to ground, mats of insulating rubber or other suitable floor insulation are to be provided for the operator.

Section 250.176, Cases of Instruments, Meters, and Relays—Operating Voltage 1 kV and Over, provides that these are to be isolated by elevation or protected by suitable barriers or grounded metal or insulating covers or guards. Their cases are not to be connected to the equipment-grounding conductor. An exception excludes cases of electrostatic ground detectors where the internal ground segments of the instrument are connected to the instrument case and grounded and the ground detector is isolated by elevation.

Section 250.178, Instrument Equipment Grounding Conductor, provides that this conductor for secondary circuits of instrument transformers and for instrument cases is not to be smaller than 12 AWG copper or 10 AWG aluminum. Cases of instrument transformers, instruments, meters, and relays that are mounted directly on grounded metal surfaces of enclosures or grounded metal switchboard panels are to be considered grounded, and no equipment-grounding conductor is required.

Part X, Grounding of Systems and Circuits over 1 kV, provides the following:

Section 250.180, General, clarifies that where systems over 1 kV are grounded, they must comply with all applicable provisions of the preceding sections of Article 250 and with this section, which supplements and modifies the preceding sections.

Section 250.184, Solidly Grounded Neutral Systems, states that such systems may be either single-point grounded or multigrounded neutral.

Recall that the Code prohibits regrounding, by which is meant connecting the grounded current-carrying conductor and the equipment-grounding conductor at additional points downstream of the main bonding jumper in the service panel or main disconnect enclosure. (It is permitted to place additional supplementary grounding electrodes anywhere along the equipment-grounding conductor or attached to any otherwise properly grounded enclosure.) It is not permitted, however, to connect a grounding electrode through a grounding-electrode conductor at any point in the wiring in lieu of an equipment-grounding conductor that goes back to the source of the electrical supply. This is called a *floating ground*. It is contrary to the Code and a dangerous practice.

The comments that follow regarding multigrounded neutral systems may appear at variance with the Code prohibition of regrounding. They are possible because the provisions of Part X "supplement and modify" the preceding sections.

(A) Insulation Level provides in (1) that the minimum insulation level for neutral conductors of solidly grounded systems is 600 volts. There are three exceptions:

Number 1: Bare copper conductors are permitted for the neutral conductor of service-entrance conductors, service laterals, and direct-buried portions of feeders.

Number 2: Bare conductors are permitted for the neutral conductor of overhead portions installed outdoors.

Number 3: The grounded neutral conductor is permitted to be a bare conductor if it is isolated from phase conductors and protected from physical damage.

(2) Ampacity states that the neutral conductor is to be of sufficient ampacity for the load imposed on the conductor but not less than one-third the ampacity of the phase conductors. An exception permits this figure to be reduced under engineering supervision to one-fifth in industrial and commercial premises.

(B) Single-Point Grounded Neutral System states that the following apply:

(1) A single-point grounded neutral system may be supplied from a separately derived system or a multigrounded neutral system with an equipment-grounding conductor connected to the multigrounded neutral conductor at the source of the single-point grounded neutral system.

(2) A grounding electrode is to be provided for the system.

(3) A grounding-electrode conductor must connect the grounding electrode to the system neutral conductor.

(4) A bonding jumper is to connect the equipment-grounding conductor to the grounding-electrode conductor.

(5) An equipment-grounding conductor is to be provided to each building, structure, and equipment enclosure.

(6) A neutral conductor is required only where phase-to-neutral loads are supplied.

(7) The neutral conductor, where provided, is to be insulated and isolated from earth except at one location.

(8) An equipment grounding conductor is to be run with the phase conductors and must comply with the following:

    a. May not carry continuous load

    b. May be bare or insulated

    c. Must have sufficient ampacity for fault current duty

(C) Multigrounded Neutral Systems states that the following provisions apply:

(1) The neutral conductor of a solidly grounded neutral system may be grounded at more than one point. Grounding may be at one or more of the following locations:

    a. Transformers supplying conductors to a building or other structure

    b. Underground circuits where the neutral conductor is exposed

    c. Overhead circuits installed outdoors

(2) The multigrounded neutral conductor is to be grounded at each transformer and at other additional locations by connection to a grounding electrode.

(3) At least one grounding electrode is to be installed and connected to the multigrounded neutral conductor every 1,300 feet.

(4) The maximum distance between any two adjacent electrodes may not be more than 1,300 feet.

(5) In a multigrounded shielded-cable system, the shielding is to be grounded at each cable joint that is exposed to personnel contact.

These grounding requirements for systems operating at over 1,000 volts are quite different from the under-1,000-volt-system requirements. A single-point grounded system is characterized by the use of an equipment-grounding conductor run with the circuit conductors. This conductor may not carry a continuous line-to-neutral load. Multigrounded neutral systems, on the other hand, are characterized by grounding-electrode connections at multiple locations. These must be not more than 1,300 feet apart.

Section 250.186, Impedance Grounded Systems, addresses requirements for these systems operating at over 1 kV. We have examined these systems for up to 1,000 volts. These systems, in which a grounding impedance, usually a resistor, limits the ground-fault current, are permitted where all the following conditions are met:

(1) Conditions of maintenance and supervision ensure that only qualified persons service the installation.

(2) Ground detectors are installed on the system.

(3) Line-to-neutral loads are not served.

These conditions are identical to those in Section 250.36 for systems that are attached to high-impedance grounded neutral systems that operate at not greater than 1 kV. For over 1 kV, the following provisions apply:

(A) Location states that the grounding impedance is to be inserted in the grounding-electrode conductor between the grounding electrode of the supply system and the neutral point of the supply transformer or generator.

(B) Identified and Insulated states that the neutral conductor of an impedance-grounded neutral system is to be identified and fully insulated with the same insulation as the phase conductors.

(C) System Neutral Conductor Connection provides that the system neutral conductor is not to be connected to ground, except through the neutral grounding impedance.

(D) Equipment Grounding Conductors states that such conductors are permitted to be bare and are to be connected to the ground bus and grounding-electrode conductor.

Section 250.188, Grounding of Systems Supplying Portable or Mobile Equipment, provides that systems supplying portable or mobile equipment of 1 kV, except for temporary substations, must comply with the following:

(A) Portable and Mobile Equipment provides that this equipment, if over 1 kV, is to be supplied by a system having its neutral conductor grounded through an impedance. Where a delta-connected system over 1 kV supplies portable or mobile equipment, a system neutral point and associated neutral conductor are to be derived.

(B) Exposed Non-Current-Carrying Metal Parts provides that for portable or mobile equipment, the parts are to be connected by an equipment-grounding conductor to the point at which the system neutral impedance is grounded.

(C) Ground-Fault Current states that the voltage developed between the portable or mobile equipment frame and ground by the flow of maximum ground-fault current is not to exceed 100 volts.

Even though the neutral is grounded through an impedance, the equipment bonding jumper bypasses that impedance so that the equipment voltage to ground in the event of a fault is limited.

(D) Ground-Fault Detection and Relaying states that this is to be provided to automatically deenergize any component of a system over 1 kV that has developed a ground fault. The continuity of the equipment-grounding conductor is to be monitored continuously so as to deenergize the circuit of the system over 1 kV automatically to the portable or mobile equipment on loss of continuity of the equipment-grounding conductor.

(E) Isolation provides that the grounding electrode to which the portable or mobile equipment is attached is to be at least 20 feet from any other system- or equipment-grounding electrode and that there is to be no direct connection between the electrodes, such as buried pipe or fence.

(F) Trailing Cable and Couplers states that in systems over 1 kV, they are to meet the requirements of Article 400 for cables and Section 490.55 for couplers. These are treated separately in Chapter 4, Equipment for General Use.

Section 250.190, Grounding of Equipment, contains provisions for installations operating at over 1 kV.

(A) Equipment Grounding states that all non-current-carrying metal parts of fixed, portable, and mobile equipment and associated fences, housings, enclosures, and supporting structures are to be grounded. An exception permits such metal parts to not be grounded where isolated from ground and located such that any person in contact with ground cannot contact these parts.

(B) Grounding Electrode Conductor states that if it connects non-current-carrying metal parts to ground, it is to be sized in accordance with Table 250.66. It is to be no smaller than 6 AWG copper or 4 AWG aluminum.

(C) Equipment Grounding Conductor states that such conductors must comply with the following:

(1) General states that equipment-grounding conductors that are not an integral part of a cable assembly may not be smaller than 6 AWG copper or 4 AWG aluminum.

(2) Shielded Cables states that the metallic shield encircling the current-carrying conductors may be used as an equipment-grounding conductor if it is rated for clearing time of ground-fault current protective device operation without damaging the metallic shield. The metallic tape insulation shield and drain wire insulation shield may not be used as an equipment-grounding conductor for solidly grounded systems.

(3) Sizing provides that equipment-grounding conductors are to be sized in accordance with Table 250.122.

Section 250.191, Grounding System at Alternating-Current Substations, states that the grounding system is to be in accordance with Part III of Article 250, Grounding Electrode System and Grounding Electrode Conductor.

This completes our survey of NEC Article 250, Grounding. This most basic body of knowledge is always a primary ingredient in licensing exams and is central to design and field work. While it may not be possible to memorize everything in this article, the fundamentals should be second nature for the electrician. If you know this article, it means easy exam points. The information that is not committed to memory should be readily available in an open-book setting if you know the structure of the article. Some of the distinctions regarding separately derived systems may seem a little obscure, but in actual practice, it is essential that we get things right consistently. When it comes to grounding, a mistake can be fatal.

Article 280

Part I, General, of Article 280, Surge Arresters, Over 1 kV, states that the article covers general requirements, installation requirements, and connection requirements for surge arrestors installed on premises wiring systems over 1 kV. The purpose of these devices is to negate the destructive effects of voltage surges. These surges can peak out for a very brief period of time at several thousand volts. Nearby combustible material can ignite, persons can be exposed to shock, and electronic equipment, pumps, motors, furnaces, or anything else electrical can be ruined. A surge can be caused by a nearby lightning strike, even if it is not a direct hit. The path through the air is in effect the primary of a transformer, and a segment of wire outside or within becomes the secondary. The energy is transferred to the premises wiring by electromagnetic induction. Another cause of surges is switching inductive circuits on the premises or nearby. An unintended resonant circuit can result in increased susceptibility.

Surge arresters that are located and connected properly can minimize the harmful effects of surges.

Section 280.2, Uses Not Permitted, states that a surge arrester is not to be installed where the rating of the surge arrester is less than the maximum continuous phase-to-ground power frequency voltage available at the point of application.

Section 280.3, Number Required, makes the point that where used at a specific location on a circuit, a surge arrestor is to be connected to each ungrounded conductor.

Section 280.4, Surge Arrestor Selection, specifies that surge arrestors must comply with (A) and (B):

(A) Rating states that the rating of a surge arrester is to be equal to or greater than the maximum continuous operating voltage available at the point of application.

(B) Silicon Carbide Types states that the rating of these units is to be 125 percent or greater than the maximum continuous operating voltage available at the point of application.

Part II, Installation, provides the following:

Section 280.11, Location, states that surge arrestors may be indoors or outdoors. They must be inaccessible to unqualified persons unless listed for installation in accessible locations.

Section 280.12, Routing of Surge Arrestor Grounding Conductors, states that these leads are to be as short as possible with unnecessary bends avoided.

Part III, Connecting Surge Arresters, provides connection details:

Section 280.21, Connection, specifies that the arrester is to be connected to one of the following:

- Grounded service conductor

- Grounding electrode conductor

- Grounding electrode for the service

- Equipment-grounding terminal in the service equipment

Section 280.23, Surge-Arrester Conductors, provides that both ground and line connections to the surge arrestor are to be no smaller than 6 AWG copper or aluminum. It is unusual to see the same size specified for both metals.

Section 280.24, Interconnections, specifies that the surge arrester protecting a transformer that supplies a secondary distribution system is to be interconnected as in (A), (B), or (C):

(A) Metallic Interconnections states that a metallic interconnection is to be made to the secondary grounded circuit conductor or the secondary-circuit grounding-electrode conductor provided that, in addition to the direct grounding connection at the surge arrester, the following occurs:

(1) Additional Grounding Connection states that the grounded conductor of the secondary has elsewhere a grounding connection to

a continuous metal underground water piping system. In urban water pipe areas where there are at least four water pipe connections on the neutral conductor and not fewer than four such connections in each mile of neutral conductor, the metallic interconnection is permitted to be made to the secondary neutral conductor with omission of the direct grounding connection at the surge arrester.

(2) Multigrounded Neutral System Connection states that the grounded conductor of the secondary system is a part of a multigrounded neutral system or static wire of which the primary neutral conductor or static wire has at least four grounding connections in each mile of line in addition to a grounding connection at each service.

(B) Through Spark Gap or Device states that where the surge-arrester grounding-electrode conductor is not connected as in Section 280.24(A), or where the secondary is not grounded as in that section but is otherwise grounded as in Section 250.52, Grounding Electrodes, an interconnection is to be made through a spark gap or listed device.

(1) Ungrounded or Unigrounded Primary System states that the spark gap or listed device is to have a 60-hertz breakdown voltage of at least twice the primary circuit voltage but not necessarily more than 10 kV, and there must be at least one other ground on the grounded conductor of the secondary that is not less than 20 feet from the surge-arrester grounding electrode.

(2) Multigrounded Neutral Primary System states that the spark gap or listed device is to have a 60-hertz breakdown of not more than 3 kV, and there is to be at least one other ground on the grounded conductor of the secondary that is not less than 20 feet from the surge-arrester grounding electrode.

Section 280.25, Grounding Electrode Conductor Connections and Enclosures, provides that other grounding-electrode connections are to comply with Article 250, Parts III and X. Grounding-electrode conductors installed in metal enclosures must comply with Section 250.64(E), Enclosures for Grounding Electrode Conductors.

These are the rules for surge arresters, typically seen on utility lines. In that setting, they are outside of NEC jurisdiction. However, in the event that customer-owned wiring consists of over 1-kV equipment and conductors, this article is applicable.

Article 285

Article 285, Surge-Protective Devices (SPDs) 1 kV or Less, contains requirements for the use and installation of these protective devices. At one time, the devices were divided into two categories: Surge arrestors were connected on the line side of the service or supply-disconnecting means, whereas transient voltage surge suppressors (TVSSs) were connected on the load side. This terminology is obsolete, at least in Code usage. Surge arrestors may be over 1 kV, and surge protective devices (SPDs) are for 1 kV or less. This reorganization of nomenclature means that the two relevant articles, 280 and 285, appear quite different. Currently, TVSSs are known as Type 2 and Type 3 SPDs. Surge arrestors for systems that are less than 1 kV are known as Type 1 SPDs.

Section 285.1, Scope, notes that the article covers general requirements, installation requirements, and connection requirements for these devices permanently installed on premises wiring systems 1 kV or less.

Section 285.3, Uses Not Permitted, states that an SPD is not to be installed on the following:

- Circuits exceeding 1 kV

- On ungrounded systems, impedance-grounded systems, or corner-grounded delta systems unless listed specifically for use on these systems

- Where the rating of the SPD (surge arrester or TVSS) is less than the maximum continuous phase-to-ground power-frequency voltage available at the point of application

Section 285.4, Number Required, provides that where used at a point on a circuit, the SPD (surge arrester or TVSS) is to be connected to each ungrounded conductor.

Section 285.5, Listing, states that an SPD is to be listed.

Section 285.6, Short-Circuit Current Rating, provides that the device is to be marked with a short-circuit current rating and is not to be installed at a point on the system where the available fault current is in excess of that rating.

Part II, Installation, provides the following:

Section 285.11, Location, provides that the devices may be installed indoors or outdoors and must be inaccessible to unqualified persons unless listed otherwise.

Section 285.12, Routing of Connections, states that conductors to line and ground are to be no longer than necessary and are to avoid unnecessary bends.

Part III, Connecting SPDs, provides (in Section 285.21, Connection) the following:

Section 285.23, Type 1 SPDs (Surge Arresters), states that these items are permitted to be connected to the supply side of a service or as specified in Section 285.24 (below). When installed at services, they must be connected to one of the following:

- Grounded service conductor
- Grounding-electrode conductor
- Grounding electrode for the service
- Equipment grounding terminal in the service equipment

Section 285.24, Type 2 SPDs (TVSSs), states that these items are to be installed in accordance with (A) through (C):

(A) Service-Supplied Building or Structure states that they are to be connected anywhere on the load side of a service-disconnect overcurrent device unless installed in accordance with Section 230.82(8).

(B) Feeder-Supplied Building or Structure states that they are to be connected at the building or structure anywhere on the load side of the first overcurrent device.

(C) Separately Derived System states that they are to be connected on the load side of the first overcurrent device.

Section 285.25, Type 3 SPDS, provides that SPDSs are permitted on the load side of branch-circuit overcurrent protection up to the equipment served.

Section 285.26, Conductor Size, states that line and grounding conductors may not be smaller than 14 AWG copper or 12 AWG aluminum.

Section 285.27, Connection Between Conductors, provides that an SPD is permitted to be connected between any two conductors—ungrounded, grounded, equipment-grounding, or grounding-electrode conductor. The grounded conductor and the equipment-grounding conductor are to be interconnected only by the normal operation of the SPD during a surge.

Section 285.28, Grounding Electrode Conductor Connections and Enclosures, states that except as indicated in Article 285, SPD grounding connections are to be made as specified in Article 250, Part III, Grounding Electrode System and Grounding Electrode Conductor. Grounding electrode conductors installed in metal enclosures must comply with Section 250.64(E), Enclosures for Grounding Electrode Conductors.

This completes our study of NEC 2011 Chapter 2, Wiring and Protection. We have looked at this chapter intensively because it is so fundamental to all types of wiring. By its very nature, Chapter 2 will be the subject of more exam questions than any other Code chapter. The provisions are clear and unambivalent, but it is not always easy to figure out how to access them quickly. The Index is helpful, but there is no substitute for learning the structure of the chapter. Subsequent Code chapters will refer back to Chapter 2. Repeated exposure will make the chapter seem like second nature so that whether you are taking an exam, working at the drawing board, or in the field, it should become familiar territory.

NEC® Chapter 3, Wiring Methods and Materials

The National Electrical Code (NEC) has its own unique logic. Totally humorless, rigorously honest and forthright, Aristotelian rather than Platonic, it undergoes revision after revision, always looking to keep up with innovations. It is scrupulously aware of its mandate to promote safety and yet not be afraid to spin out into new regions of electronic knowledge. Each revision seems to carve out a little more cyberterritory—information technology (IT) equipment, sensitive electronic equipment, solar photovoltaic systems, and small wind electric systems have found a home in Chapter 6, Special Equipment. These and other advanced topics lie down the road.

For now, the task is to examine one of the Code's really basic areas. Chapter 3, Wiring Methods and Materials, continues the NEC program of progressing from the general to the more specific. This pattern is evident in the overall structure of the work. We perceive it also within each chapter and within each article. In this spirit, the Code places Chapter 3 after Chapter 2. It progresses from basic safety mandates to the methods and materials to implement these goals.

Chapter 3 begins with a discussion of wiring methods (the general underpinning) and continues with requirements for all wiring installations and extensive ampacity tables. Next are two articles covering various specific devices and enclosures, and finally, there is a comprehensive treatment in 41 separate articles of individual types of cable and raceways. These articles are accessed frequently in licensing exams and for our everyday wiring jobs.

Article 300

Article 300, Wiring Methods, contains general requirements for many common wiring scenarios.

Part I, General Requirements, opens with some introductory statements:

Section 300.1, Scope, states (A) that Article 300 covers wiring methods for all wiring installations unless modified by other articles.

Later chapters, such as Chapter 5, Special Occupancies, treat specific areas that are encountered. Article 517, Health Care Facilities, for example, contains mandates that modify those found in Chapter 3. The best approach is to become very familiar with the earlier chapters, especially Chapters 2 and 3, and then consult the later chapters for specialized knowledge as needed for an open-book exam or work in the field. The later chapters cover more advanced topics, but they are easier, in a sense, because you know right where to go to find the answer. A question based on one of the earlier chapters, in contrast, may be a little baffling if you don't know how to access specific material. To give an example, who would suppose that Article 300 is the place to find the table that provides minimum cover requirements, that is, required depth, for various types and locations of underground wiring? You could fall into the error of flipping around in Article 210, Branch Circuits. Of course, the Index is a big help in locating information of this sort, but such an approach is not always certain. A certain amount of memorization is necessary, but you don't need to memorize exact cover requirements (such as 18 inches under airport runways for all wiring types). Instead, become familiar with the location of the table. All of life is an open-book exam! (Incidentally, don't forget that there is an entirely different table for minimum cover for underground wiring over 600 volts. Where is it? Part II of the same article, titled Requirements for over 600 volts, nominal).

(B) Integral Parts of Equipment states that the provisions of Article 300 do not apply to the conductors that form an integral part of such equipment as motors, controllers, motor control centers, and factory-assembled control equipment or listed utilization equipment.

Internal wiring within these types of equipment must comply with Underwriters Laboratories (UL), or other testing organization standards in order to achieve listing status. It is up to the authority having jurisdiction (AHJ) to decide whether such equipment is to be approved. Since it is impractical for the AHJ to inspect utilization equipment onsite as part of the electrical inspection, listing is a valuable tool that ensures that the equipment will not be a hazard to persons, buildings, or other equipment.

(C) Metric Designators and Trade Sizes states that for conduit, tubing, and associated fittings and accessories, sizes are given in Table 300.1(C). These designators are a means of identifying various sized units and are not intended to be exact measurements.

Section 300.2, Limitations, provides in (A) that wiring methods specified in Chapter 3 are to be used for 600 volts, nominal, or less where not specifically limited in some section of Chapter 3. They are permitted for over 600 volts where specifically permitted elsewhere in the Code.

(B) Temperature provides that temperature limitations must be in accordance with Section 310.15(A)(3), Temperature Limitation of Conductors.

We'll examine this section, and Article 310, Conductors for General Wiring, together with the associated ampacity tables, in detail later. For now, it is worth noting that a number of factors must be considered in choosing the correct conductors for any given application. It is also essential not to exceed termination temperature ratings. (For these temperature ratings, the Celsius scale is used.)

Section 300.3, Conductors, states

(A) Single Conductors specifies in Table 310.104(A), Conductor Applications and Insulations Rated 600 Volts (which lists 24 conductors by trade names and type letters with maximum operating temperatures and other characteristics), that single conductors are to be installed only where part of a recognized wiring method of Chapter 3. With exceptions for overhead conductors and festoon lighting, this means that you don't want to see individual current-carrying conductors outside of cables, raceways, or enclosures.

(B) Conductors of the Same Circuit provides that all conductors of the same circuit and, where used, the grounded conductor and all equipment-grounding conductors and bonding conductors are to be contained within the same raceway, auxiliary gutter, cable tray, cable bus assembly, trench, cable, or cord unless otherwise permitted in (1) through (4):

(1) Paralleled Installations states that conductors are permitted to be run in parallel in accordance with the provisions of Section 310.10(H), Conductors in Parallel (which, as we shall see, discusses the matter in detail). The requirement to run all circuit conductors within the same raceway, auxiliary gutter, cable tray, trench, cable, or cord applies separately to each portion of the paralleled installation, and the equipment-grounding conductors are to comply with the provisions of 250.122, Size of Equipment Grounding Conductors. Parallel runs in cable trays are to comply with the provisions of 392.20(C), Connected in Parallel (for Cable Trays). An exception allows conductors installed in nonmetallic raceways underground to be arranged as isolated phase installations. The raceways must be installed in close proximity.

(2) Grounding and Bonding Conductors states that equipment-grounding conductors are permitted to be installed outside a raceway or cable assembly where in accordance with Section 250.130(C) for certain existing installations or in accordance with 250.134(B), exception number 2, for direct-current (dc) circuits. Equipment bonding conductors are permitted to be installed on the outside of raceways in accordance with 250.102(E).

(3) Nonferrous Wiring Methods states that conductors in wiring methods with a nonmetallic or other nonmagnetic sheath, where run in different raceways, auxiliary gutters, cable trays, trenches,

cables, or cords, must comply with the provisions of 300.20(B), Individual Conductors.

(4) Enclosures states that where an auxiliary gutter runs between a column-width panelboard and a pull box and the pull box includes neutral terminations, the neutral conductors of circuits supplied from the panelboard are permitted to originate in the pull box.

What's the point of all this? If an individual current-carrying conductor of an alternating-current (ac) circuit is run in a steel or other magnetic raceway or enclosure, current will be induced in the raceway or enclosure because it becomes, in effect, the secondary of a transformer. This creates heat, which, added to ambient and conductor heat, may cause the conductor within to exceed its temperature limitation, eventually degrading the insulation. Additionally, there may be a shock hazard. A hot wire and neutral or two or three opposing-phase conductors have opposing polarities at any instant within the ac waveform. Their inductive effects cancel out. Thus it is of great importance that these mandates be observed at all times, and the licensing examiners may want to be sure that you are aware of these issues.

(C) Conductors of Different Systems states in (1) that conductors of ac and dc systems rated 600 volts, nominal, or less are permitted to occupy the same equipment wiring enclosure, cable, or raceway. All conductors must have an insulation rating equal to at least the maximum circuit voltage applied to any conductor within the enclosure, cable, or raceway.

Notwithstanding this permitted mixing of circuits and systems, it should be noted that certain combinations are not permitted. We will see a lot of this sort of prohibition in Chapter 7, Special Conditions, and particularly Article 725, Class 1, Class 2, and Class 3 Remote-Control, Signaling, and Power-Limited Circuits. For example, Section 725.139, Installation of Conductors of Different Circuits in the Same Cable, Enclosure, Cable Tray, or Raceway, provides in (F) Class 2 or Class 3 Conductors or Cables and Audio System Circuits that these are not to be mixed. We'll learn all about these three classes later on. Audio systems can be treacherous because a long, sustained, high-volume sound can translate into conductor heat.

(2) Over 600 Volts, Nominal, provides that conductors of circuits rated over 600 volts are not to occupy the same enclosures, cables, or raceways that conductors 600 volts or less occupy unless permitted by one of the following:

a. Secondary wiring to electric-discharge lamps of 1,000 volts or less, if insulated for the secondary voltage involved, are permitted to occupy the same luminaire, sign, or outline lighting enclosure as the branch-circuit conductors.

b. Primary leads of electric-discharge lamp ballasts insulated for the primary voltage of the ballast, where contained within the individual wiring enclosure, are permitted to occupy the same luminaire, sign, or outline enclosure as the branch-circuit conductors.

c. Excitation, control, relay, and ammeter conductors used in conjunction with any individual motor or starter are permitted to occupy the same enclosure as the motor-circuit conductors.

d. In motors, switchgear, and control assemblies, as well as similar equipment, conductors of different voltage ratings are permitted.

e. In manholes, if the conductors of each system are permanently and effectively separated from the conductors of the other systems and securely fastened to racks, insulators, or other approved supports, conductors of different voltage ratings are permitted.

Conductors having nonshielded insulation and operating at different voltage levels are not to occupy the same enclosure, cable, or raceway.

Section 300.4, Protection Against Physical Damage, outlines a number of measures that must be taken to ensure that conductors are not subject to damage.

(A) Cables and Raceways Through Wood Members provides the following:

(1) Bored Holes states that where a cable or raceway is installed through bored holes in joists, rafters, or wood members, the holes must be bored so that the hole is not less than 1¼ inch from the nearest edge of the wood member. Where this distance cannot be maintained, the cable or raceway is to be protected from penetration by screws or nails by a steel plate or bushing at least 1/16-inch thick and of appropriate length and width installed to cover the area of the wiring. Exceptions state that the steel plate is not required if the raceway is rigid metal conduit, intermediate metal conduit, or rigid nonmetallic or electrical metallic tubing and that a listed and marked steel plate less than 1/16-inch thick that provides equal or better protection against nail or screw penetration is permitted.

Penetration of the conductor by nail or screw installed by other trades after the electrical rough-in has been completed is a big problem. If a conductor is partially severed, a high-impedance hot spot or series arcing fault may occur. If this is not picked up by the arc-fault protection, fire can result. Maintaining the 1¼-inch setback and use of the marked and listed steel plate mitigate this hazard.

Holes in wood for wiring always should be drilled as close as possible to the center of the member, even if far from the 1¼-inch zone. The reason is that a hole near the center takes less strength from a framing member than a hole or notch near the edge. A horizontal joist sagging midway between two points of support is in compression along the top edge and in tension near the bottom edge. Bored holes near these edges tend to weaken the piece. At the center, the joist is neither in compression nor in tension and so is not weakened appreciably. Also, bored holes should be no larger than necessary.

(2) Notches in Wood states that notches are permitted if protected by steel plate.

(B) Non-Metallic-Sheathed Cables and Electrical Nonmetallic Tubing Through Metal Framing Members provides that where non-metallic-sheathed cables pass through either factory- or field-punched, cut, or drilled slots or holes in metal members, the cable is to be protected by listed bushings or listed grommets covering all metal edges that are securely fastened in the opening prior to installation of the cable. Where nails or screws are likely to penetrate non-metallic-sheathed cable or electrical nonmetallic tubing, a steel sleeve, steel plate, or steel clip not less than 1/16 inch in thickness must be used to protect the cable or tubing.

(C) Cables Through Spaces Behind Panels Designed to Allow Access states that these are to be supported according to their applicable articles. The articles referenced are 320 through 399.

Any type of wiring that is installed above suspended ceiling panels is not to be laid directly on the panels. It must be supported according to the requirements of the articles just cited. Otherwise, an accumulation of wiring resting directly on the panels would prevent them from being removed for access. In many commercial and office buildings, a large part of the wiring, including low-voltage, optical fiber, broadband, and communications cables, is routed in this area, and access is necessary for maintenance and additions.

(D) Cables and Raceways Parallel to Framing Members and Furring Strips calls for protection by means of steel plate, sleeve, or equivalent.

(E) Cables, Raceways, or Boxes Installed in or Under Roof Decking states that such items must be 1½ inches from the lowest surface of the roof decking. An exception exempts rigid metal conduit and intermediate metal conduit.

(F) Cables and Raceways Installed in Shallow Grooves provides that when these items are installed in a groove, to be covered by wall-board, siding, paneling, carpeting, or similar finish, they must be protected by steel plate or sleeve or by not less than 1¼ inches of free space for the full length of the groove in which the cable or raceway is installed.

(G) Insulated Fittings provides that where raceways contain 4 American Wire Gauge (AWG) or larger insulated circuit conductors and these conductors enter a cabinet, box, enclosure, or raceway, the conductors must be protected by an identified fitting providing a smoothly rounded insulating surface, unless the conductors are separated from the fitting or raceway by identified insulating material that is fastened securely in place. An exception states that where threaded hubs or bosses that are an integral part of a cabinet, box, enclosure, or raceway provide a smoothly rounded or flared entry for conductors, the identified fitting is not required.

NEC 2011 goes on to say that conduit bushings constructed wholly of insulating material are not to be used to secure a fitting or raceway. For rigid metal conduit, you need two locknuts, one inside and one outside the enclosure. These are to be driven tight to ensure ground continuity. The insulating fitting or insulating material must have a temperature rating not less than the insulation temperature rating of the installed conductors.

(H) Structural Joints provides that a listed expansion/deflection fitting must be used where a raceway crosses a structural joint intended for expansion, contraction, or deflection and used in buildings, bridges, parking garages, or other structures.

A severe hazard occurs when a metal raceway pulls apart or breaks so that ground continuity is lost. If this happens in conjunction with ungrounded conductor chafing, one part of the metal raceway may become energized with respect to ground, and the overcurrent device may fail to interrupt the circuit because there is no return path.

Section 300.5, Underground Installations, contains mandates for underground electrical lines, including the important Table 300.5, Minimum Cover Requirements.

(A) Minimum Cover Requirements provides that direct-buried cable or conduit or other raceways are to be installed to meet the minimum cover requirements of Table 300.5.

Keep in mind that if you do not recall the exact location of this table, the Code uses the phrase *minimum cover requirements* not *burial depth*. The table lists down the left side various locations, such as under a building and under airport runways. Across the top are types of wiring or circuits, such as direct burial and rigid metal conduit. Where solid rock is encountered, note 5 provides relief. The wiring may be installed in metal or nonmetallic raceways permitted for direct burial and covered by a minimum of 2 inches of concrete extending down to rock. Of course, extra fill can be placed to make the finished grade higher. It should never be necessary to blast or jackhammer to comply with Table 300.5. This table is for 0 to 600 volts. Minimum cover requirements for higher voltages are given in Table 300.50 in Part II, Requirements for Over 600 Volts, Nominal.

(B) Wet Locations states that the interiors of enclosures or raceways installed underground are considered wet locations. Insulated conductors and cables must be listed for wet locations. The same applies to connections and splices.

It is assumed that parts of any underground raceway or enclosure at times may be subject to water infiltration. It is not useful to attempt to seal against water infiltration. Instead, use wet-location materials.

(C) Underground Cables Under Buildings provides that these are to be in a raceway. Exceptions state that MI cable and MC cable listed for direct burial may be placed underground beneath a building.

(D) Protection from Damage provides that direct-buried conductors and cables are to be protected from damage in accordance with (1) through (4):

(1) Emerging from Grade states that they are to be protected by enclosures or raceways extending from the minimum cover distance below grade to a point at least 8 feet above finished grade. In no case does the protection below grade have to exceed 18 inches.

(2) Conductors Entering Buildings states that such conductors must be protected to the point of entrance.

(3) Service Conductors states that if not encased in concrete and 18 inches or more below grade, these conductors must be identified by a warning ribbon placed in the trench at least 12 inches above the underground installation. The warning ribbon is not required for feeders and branch circuits because they are protected by overcurrent devices and typically have less available fault current.

(4) Enclosure or Raceway Damage states that where subject to physical damage, the conductors are to be installed in rigid metal conduit, intermediate metal conduit, Schedule 80 polyvinyl chloride (PVC) conduit, or equivalent.

When doing an underground service, check utility specifications. They may consider the run up the pole subject to physical damage and require Schedule 80 PVC conduit for the first 8 feet above grade.

(E) Splices and Taps permits direct-buried conductors and cables to be spliced or tapped without the use of splice boxes. The splicing means must be listed for underground use. You can't just use wire nuts and tape them.

(F) Backfill that contains large rocks, paving materials, cinders, large or sharply angular substances, or corrosive material may not be placed in an excavation where they may damage raceways, cables, or other substructures or prevent adequate compaction of fill or contribute to corrosion. Where necessary, granular or selected material, running boards, sleeves, or other approved means must be used. As always, *approved* means approved by the AHJ.

(G) Raceway Seals states that conduits or raceways through which moisture may contact live parts are to be sealed or plugged at either or both ends.

(H) Bushing or Terminal Fitting states that an integral bushed opening is to be used at the end of a conduit or other raceway that terminates underground where the conductors or cables emerge as a direct-burial wiring method. A seal incorporating the physical protection characteristics of a bushing may be used.

(I) Conductors of the Same Circuit states that such conductors and, where used, the grounded conductor and all equipment-grounding conductors are to be installed in the same raceway or cable or installed in close proximity in the same trench. An exception permits conductors to be installed in parallel in raceways, multiconductor cables, or direct-buried single-conductor cables. Each raceway or multiconductor cable is to contain all conductors of the same circuit, including equipment-grounding conductors. Each direct-buried single-conductor cable is to be located in close proximity in the trench to the other single-conductor cables in the same parallel set of conductors in the same circuit, including equipment-grounding conductors. The purpose of the requirements in (I) is to reduce inductive heating, as discussed previously.

A second exception reminds us that isolated phase, polarity, grounded conductor, and equipment-grounding and bonding conductor installations are permitted in nonmetallic raceways or cables with a nonmetallic covering or nonmagnetic sheaths in close proximity where conductors are paralleled as permitted in Section 310.10(H) and where the conditions in Section 300.20(B) are met.

An isolated-phase installation is one in which there is only one phase per raceway or cable. If the raceway is ferrous, there will be inductive heating of the raceway in proportion to the amount of current flowing through the conductor. Since the energy to make this heat has to come from the electricity in the circuit, it is inevitable that the inductive heating will be accompanied by energy loss, i.e., increased impedance in the conductors involved. This is an instance where nonmetallic conduit is preferred to metal raceway, in which isolated-phase installations are not used.

(J) Earth Movement states that where direct-buried conductors, raceways, or cables are subject to movement by settlement or frost, they are to be arranged so as to prevent damage to the enclosed conductors or to equipment connected to the raceways. An Informational Note recommends S loops in underground direct burial to raceway transitions, expansion fittings in raceway risers to fixed equipment, and flexible connections to equipment subject to settlement or frost heaves.

(K) Directional Boring states that cables or raceways installed using directional boring equipment are to be approved for the purpose. An example is high-density polyethylene conduit (Type HDPE), the subject of Article 353.

Section 300.6, Protection Against Corrosion and Deterioration, provides that raceways, devices, and electrical hardware are to be of materials suitable for their environment. This very general mandate must be interpreted on a case-by-case basis. Specific Code articles, manufacturers' documentation, and past experience provide guidance on steps that must be taken to ensure that there won't be a problem that translates into a hazardous situation down the road.

(A) Ferrous Metal Equipment states that ferrous metal raceways, devices, and hardware are to be suitably protected against corrosion inside and outside, except threads at joints. Where corrosion protection is necessary and conduit is threaded in the field, the threads are to be coated with an approved electrically conductive, corrosion-resistant compound.

Many types of oils or greases will prevent corrosion at threaded joints, but the problem is that they may not be electrically conductive, especially as they age. Teflon tape may be good for PVC threaded joints, but it is not electrically conductive and so would interrupt ground continuity in a run of rigid metal conduit.

(1) Protected from Corrosion Solely by Enamel states that ferrous metal raceways, devices, and hardware so protected are not to be used outdoors or in wet locations.

(2) Organic Coatings on Boxes or Cabinets states that where boxes or cabinets have an approved system of organic coatings and are marked "Raintight," "Rainproof," or "Outdoor Type," they are permitted outdoors.

(3) In Concrete or in Direct Contact with the Earth states that ferrous metal raceways, devices, and hardware must be made of material approved for the condition where installed in direct contact with the earth. In areas subject to severe corrosive influences, the material must be approved for the condition.

(B) Aluminum Metal Equipment states that such equipment in concrete or in direct contact with the earth is to have supplementary corrosion protection.

(C) Nonmetallic Equipment states that such equipment is to be made of material approved for the condition and is to comply with (1) and (2) below as applicable to the specific installation:

(1) Exposed to Sunlight states that materials must be listed or identified as sunlight resistant.

(2) Chemical Exposure states that materials must be inherently resistant to chemicals based on their listing or identified for the specific chemical reagent.

(D) Indoor Wet Locations states that in portions of dairy processing facilities, laundries, canneries, and other indoor wet locations and in locations where walls are washed frequently or where there are surfaces of absorbent materials, such as damp paper or wood, the entire wiring system, where installed exposed, including all boxes, fittings, raceways, and cable used therewith, is to be mounted so that there is at least a ¼ inch of airspace between it and the wall or supporting surface. An exception exempts nonmetallic raceways, boxes, and fittings on concrete, masonry tile, or similar surface.

While rigid PVC conduit is considered generally less robust than its metallic relatives, it is definitely preferable in some settings, notably underground, in concrete, and in indoor wet locations, particularly within the milkhouse portion of a dairy barn, where frequent wash downs, high humidity, and a corrosive atmosphere quickly degrade even galvanized steel.

Section 300.7, Raceways Exposed to Different Temperatures, addresses two problems that exist when raceways exhibit a temperature gradient, either spatially or over time. One problem is condensation, and the other is differential expansion.

(A) Sealing states that where portions of a raceway or sleeve are known to be subjected to different temperatures, and where condensation is known to be a problem, as in cold-storage areas of buildings or where passing from the interior to the exterior of a building, the raceway or sleeve is to be filled with an approved material to prevent the circulation of warm air to a colder section of the raceway or sleeve. An explosion-proof seal is not required for this purpose.

(B) Expansion Fittings provides that these are to be provided for raceways where necessary to compensate for thermal expansion and contraction.

Section 300.8, Installation of Conductors with Other Systems, provides that raceways or cable trays containing electrical conductors are not to contain any pipe or tube for steam, water, air, gas, drainage, or any service other than electrical.

Section 300.9, Raceways in Wet Locations Above Grade, provides that their interiors are considered wet locations. Insulated conductors and cables installed therein are to comply with Section 310.10(C), which lists insulated conductors suitable for wet locations.

Section 300.10, Electrical Continuity of Metal Raceways and Enclosures, states that metal raceways, cable armor, and other metal enclosures for conductors are to be metallically joined together into a continuous electrical conductor. They must be connected to all boxes, fittings, and cabinets so as to provide effective electrical continuity. Unless specifically permitted elsewhere in the Code, raceways and cable assemblies are to be mechanically secured to boxes, fittings, cabinets, and other enclosures. There are two exceptions:

(1) Short sections of raceways used to provide support or protection of cable assemblies

(2) Equipment enclosures to be isolated, as permitted by Section 250.96(B), Isolated Grounding Circuits

Section 300.11, Securing and Supporting, contains general information on securing and supporting electrical equipment, devices, and raceways.

(A) Secured in Place states that raceways, cable assemblies, boxes, cabinets, and fittings are to be securely fastened in place. Support wires that do not provide secure support are not permitted as the sole support. Support wires and associated fittings that provide secure support and that are installed in addition to the ceiling-grid support wires are permitted as the sole support. Where independent support wires are used, they are to be secured at both ends. Cables and raceways are not to be supported by ceiling grids.

It is common practice in commercial settings and office buildings to install most of the wiring above a suspended ceiling. This type of ceiling consists of 2- × 4-foot (or sometimes 2- × 2-foot) ceiling panels that are laid in a grid composed of light steel T-shaped members (L-shaped around the wall perimeter). The panels can be lifted and shifted to the side temporarily while maintenance or additions are in progress. Such wiring is accessible but not *readily accessible* according to Code definitions. There are two overriding principles that need to be followed regarding wiring that is placed in this location. It must never be laid directly on the grid. An accumulation of wires would quickly make it impossible to lift out the panels, and the wiring and equipment above would no longer be accessible. The second principle is that wiring, whether communication or power and light conductors including raceways, is never to be attached to the ceiling-grid support wires. It may be attached to additional support wires that are installed for the purpose, but these are to be secured at both ends. (If it won't fall, it is *supported*. If it won't wiggle, it is *secured*.) The best solution is to attach struts to the framing or old ceiling surface above so that it does not block access for future maintenance on other wiring or systems that share this space.

(1) Fire-Rated Assemblies states that wiring located within the cavity of a fire-rated floor-ceiling or roof-ceiling assembly is not to be secured to or supported by the ceiling assembly, including the ceiling support wires. An independent means of secure support must be provided, and it is permitted to be attached to the assembly. Where independent support wires are used, they are to be distinguishable by color, tagging, or other effective means from those of the fire-rated design. An exception permits support wiring and equipment that have been tested as part of the fire-rated assembly. The purpose of this requirement is so that if work is done in the future on the electrical support system or on the ceiling support system, they will remain separate and independent from one another.

(2) Non-Fire-Rated Assemblies contains similar provisions and exceptions.

(B) Raceways Used as Means of Support (Figure 3-1) provides that this is permitted only as follows:

(1) Where the raceway or means of support is identified for the purpose

(2) Where the raceway contains power-supply conductors for electrically controlled equipment and is used to support Class 2 circuit conductors or cables that are solely for the purpose of connection to the equipment control circuits

(3) Where the raceway is used to support boxes or conduit bodies in accordance with Section 314.23 or to support luminaires in accordance with Section 410.36(E)

This NEC provision is frequently violated by trunk slammers and telecom technicians alike. It is fast and easy to cable-tie wires to raceway but this procedure may give rise to hazards. In the first place, the

FIGURE 3-1 Raceways are permitted to support wiring only under specified conditions, and this grounding jumper is not one of them.

added weight invalidates the securing intervals given in the individual articles that appear later in Chapter 3 on the various raceways. Second, heat dissipation is limited even as more heat may be added to the mix. This requirement often finds its way into licensing exams.

(C) Cables Not Used as Means of Support states that this is similar to (B) but without the exception.

Section 300.12, Mechanical Continuity—Raceways and Cables, provides that raceways, cable armors, and cable sheaths are to be continuous between cabinets, boxes, fittings, or other enclosures or outlets. There are two exceptions:

Number 1: Short sections of raceway used to provide support or protection for cable assemblies (Notice that noncable conductors are not mentioned. This is so because if they were, the implication would be that such conductors were not enclosed as provided elsewhere.)

Number 2: Raceways and cables installed into the bottom of open-bottom equipment, such as switchboards, motor-control centers, and floor- or pad-mounted transformers

Section 300.13, Mechanical and Electrical Continuity—Conductors, provides (in A) that conductors in raceways are to be continuous between outlets, boxes, devices, and so forth. There can be no splice or tap within a raceway unless the raceway has hinged or removable covers or unless the conductors are busway conductors. Splices and taps must be accessible, and wiring within a raceway is inaccessible unless it is pulled from the raceway.

(B) Device Removal provides that in multiwire branch circuits, the continuity of the grounded conductor is not to depend on device connections at lampholders, receptacles, and so forth, where removal of the devices would interrupt the continuity.

It is necessary, when daisy-chaining outlets on a multiwire branch circuit, to make short white jumpers from the splice in the neutral to the device or utilization equipment to be removed. In this way, the device can be removed without interrupting neutral continuity. If the neutral continuity is interrupted in a multiwire branch circuit, unbalanced voltage may be imposed on downstream loads.

Section 300.14, Length of Free Conductors at Outlets, Junctions, and Switch Points, provides that at least 6 inches of free conductor, measured from the point in the box where it emerges from its raceway or cable sheath, is to be left at each outlet, junction, and switch point for splices or the connection of luminaires or devices. Where the opening is less than 8 inches in any dimension, each conductor is to be long enough to extend at least 3 inches outside the opening (Figure 3-2).

This means that for a box that is less than 8 × 8 inches, both the 6- and 3-inch rules apply. For larger boxes, only the 6-inch rule applies. You might think that cutting the free conductors as short as possible would help to minimize box fill, but this is not the case.

FIGURE 3-2 For a box with an opening not 8 inches or greater in any dimensions, two lengths for free conductor are applicable. There must be 6 inches minimum free conductor from the box entry and 3 inches minimum of free conductor extending outside the opening. If the box is 8 inches or greater in any dimension, the 3-inch length is not required.

Section 300.15, Boxes, Conduit Bodies, or Fittings, states that where required and where the wiring method is conduit, tubing, Type AC cable, Type MC cable, Type MI cable, non-metallic-sheathed cable, or other cables, a box or conduit body is to be installed at each conductor splice point, outlet point, switch point, junction point, termination point, and pull point unless otherwise permitted in (A) through (L) below:

(A) Wiring Methods with Interior Access states that the preceding requirement does not apply to wiring methods with removable covers, such as wireways, multioutlet assemblies, auxiliary gutters, and surface raceways. These covers must be accessible after installation.

(B) Equipment states that the preceding requirement does not apply to equipment with an integral junction box or wiring compartment.

(C) Protection states that the preceding requirement does not apply where cables enter or exit from conduit or tubing that is used to provide cable support or protection against physical damage. A fitting must be provided on the end(s) of the conduit or tubing to protect the cable from abrasion.

(D) Type MI Cable states that the preceding requirement does not apply for mineral-insulated, metal-sheathed cable where accessible fittings are used for straight-through splices.

(E) Integral Enclosure states that a wiring device with integral enclosure identified for the use, having brackets that fasten the device securely to walls or ceilings of conventional on-site frame construction, with non-metallic-sheathed cable is permitted in lieu of a box or conduit body.

(F) Fitting states that a fitting identified for the use is permitted in lieu of a box or conduit body where conductors are not spliced or terminated within the fitting. The fitting must be accessible after installation.

(G) Direct-Buried Conductors states that a box or conduit body is not required for splices and taps in direct-buried conductors and cables.

(H) Insulated Devices states that a box or conduit body is not required for insulated devices supplied by non-metallic-sheathed cable.

(I) Enclosures states that a box or conduit body is not required where a splice, switch, terminal, or pull point is in a cabinet or cutout box in an enclosure for a switch or overcurrent device, in a motor controller, or in a motor control center.

(J) Luminaires states that a box or conduit body is not required where a luminaire is used as a raceway.

(K) Embedded states that a box or conduit body is not required for splices where conductors are embedded.

(L) Manholes and Handhole Enclosures states that a box or conduit body is not required for conductors in manholes or handhole enclosures, except where connecting to electrical equipment.

Section 300.16, Raceway or Cable to Open or Concealed Wiring, provides the following:

(A) Box, Conduit Body, or Fitting states that a box, conduit body, or terminal fitting having a separately bushed hole for each conductor is to be used whenever a change is made from conduit, metallic electrical tubing, nonmetallic electrical tubing, non-metal-sheathed cable, Type AC cable, Type MC cable, or mineral-insulated, metal-sheathed cable, and surface raceway wiring to open wiring or to concealed knob-and-tube wiring.

Article 394, Concealed Knob-and-Tube Wiring, permits this archaic wiring type only for extensions of existing installations and elsewhere by special permission. Its continued use in hollow spaces that have been weatherized by inserting insulation is also prohibited because free air circulation is necessary for heat dissipation. Knob-and-tube wiring is concealed. Open wiring (individual conductors on insulators) is required by Article 398 to be exposed.

(B) Bushing states that a bushing is permitted in lieu of a box or terminal where the conductors emerge from a raceway and enter or terminate at equipment.

Section 300.17, Number and Size of Conductors in Raceway, states that such conductors are not to be more than will permit dissipation of the heat and ready installation or withdrawal of the conductors without damage to the conductors or to their insulation. This brief statement has enormous implications for all raceway work. The maximum permitted raceway or conduit fill has to be calculated to make sure that it is not exceeded. Since raceway or conduit installation is often completed before wire is pulled, it is necessary to design the whole job in advance before beginning to lay out the piping (as conduit and tubing are informally called). If all conductors are the same size, this task is fairly simple: consult Informative Annex C. It contains tables for various conduit or raceway types, with maximum fill using various wire types. Frequently, in actual field installations, all conductors are the same size, and Annex C may be used. In other cases, where they are not the same size, a different, somewhat more complex method must be used, and we will show how that is done in a later chapter. Licensing exams surely will have questions on conduit fill using both Annex C and manual calculation, so the latter has to be mastered, which is not too difficult once it has been done a few times. In actual field work, it is common practice to mentally increase the size of one or two smaller conductors to make them all the same size and thus eligible for Annex C application. Also, experienced electricians often will know by recalling previous installations what size raceway to choose, but these approaches will not work in an exam setting.

Section 300.18, Raceway Installations, contains two provisions:

(A) Complete Runs states that raceways, other than busways or exposed raceways having hinged or removable covers, are to be installed complete between outlet, junction, or splicing points prior to installation of the conductors. Where required to facilitate the installation of utilization equipment, the raceway is permitted to be installed initially without a terminating connection at the equipment. Prewired raceway assemblies are permitted only where specifically allowed for the applicable wiring method. An exception allows for short sections of raceways used to protect conductors or cable assemblies.

With very limited exceptions, the basic Code rule is that a raceway installation is to be completed including terminations at both ends before any wires are pulled. However, covers may be left off junction and pull boxes so that wire may be pulled in stages. The best practice for an underground conduit installation is to install it in the trench (with a pull rope in place) and complete backfilling it before pulling wires. You might think it would be better to pull the wires before backfilling in case damage to the raceway occurs, but the fact is that if that damage occurred after the wires were pulled, the loss would be much greater.

(B) Welding states that metal raceways are not to be supported, terminated, or connected by welding to the raceway unless specifically designed to be or otherwise specifically permitted to be in the Code. Heat from welding destroys the galvanized finish on conduit, causing it to rust prematurely in the area of the weld. Also, welding galvanized material generates very toxic fumes and should not be done. For similar reasons, an oxyacetylene torch is not used to cut conduit or tubing.

Section 300.19, Supporting Conductors in Vertical Raceways, provides that in long vertical raceway runs, it is necessary to support conductors at specified intervals. Otherwise, their cumulative weight would pull them out of terminations and cause them to drop. To counter the problem, this section provides in (A) Table 300.19(A), Spacings for Conductor Supports, for various conductor sizes. These intervals are greater for aluminum because it is lighter than copper. To put it in perspective, the maximum spacing of conductor supports for 12 AWG copper is 100 feet, so we can see that these requirements are not applicable in many installations.

(B) Fire-Rated Cables and Conductors states that support methods and spacings for this type of wiring must comply with specific electrical circuit protective system listings but in no case exceed the values in the table.

(C) Support Methods provides four options:

(1) By clamping devices constructed of or employing insulating wedges inserted in the ends of raceways

(2) By inserting boxes at the required intervals in which insulating supports are installed

(3) In junction boxes, by deflecting the cables not less than 90 degrees and carrying them horizontally to a distance not less than twice the

diameter of the cable, the cables being carried on two or more insulating supports and additionally secured thereto by tie wires if desired (Where this method is used, cables are to be supported at intervals not greater than 20 percent of those mentioned in the preceding tabulation.)

(4) By a method of equal effectiveness

Section 300.20, Induced Currents in Ferrous Metal Enclosures or Ferrous Metal Raceways, provides that induced currents are to be minimized by grouping together all phase conductors and, where used, the grounded conductor and all equipment-grounding conductors. Harmful effects resulting from this phenomenon were discussed earlier. As mentioned earlier, there are exceptions:

Section 300.21, Spread of Fire or Products of Combustion, provides that electrical installations in hollow spaces, vertical shafts, and ventilation or air-handling ducts must be made so that the possible spread of fire or products of combustion will not be increased substantially. Openings around electrical penetrations into or through fire-resistant-rated walls, partitions, floors, or ceilings are to be firestopped using approved methods to maintain the fire-resistance rating.

An Informational Note points us to directories of electrical construction materials published by qualified testing laboratories. The UL White Book covers through-penetration firestop systems and firestop devices. The Informational Note also recommends reference to building codes, which contain restrictions on membrane penetrations on opposite sides of a fire wall. Typically, municipal and other building codes require a 24-inch horizontal separation of boxes on opposite sides of a firewall so that smoke and fire won't blast through. Transformer vaults require special attention, especially in regard to the many conduit penetrations up near the ceiling.

Section 300.22, Wiring in Ducts Not Used for Air Handling, Fabricated Ducts for Environmental Air, and Other Spaces for Environmental Air (Plenums), though fairly brief, contains a lot of information of a quite sensitive nature. Wiring in ducts can introduce fire hazards into any environment, especially in a hotel or industrial setting, where it is vital that any fire be confined to a small area where it can be suppressed before injury to persons and damage to real estate occurs.

As a first step in assimilating the information in this section, the definitions for the three terms in the title of the section should be understood and committed to memory. Then the wiring methods, what is allowed and what is not allowed, can be considered.

The Code states that the provisions of Section 300.22 apply to the installation and uses of electrical wiring and equipment in ducts used for dust, loose stock, or vapor removal; ducts specifically fabricated for environmental air; and other spaces used for environmental air (plenums).

Other spaces used for environmental air refers to enclosed spaces in a building that are not made specifically to handle environmental air but nevertheless perform that function. These spaces are also called *plenums*. A duct made out of sheet metal or plywood would not be a plenum. The space above a suspended ceiling would be a plenum if it is purposely used to contain and channel environmental air.

Here are the mandates:

(A) Ducts for Dust, Loose Stock, or Vapor Removal states that these ducts never may contain wiring of any type. Similarly, a shaft containing only such ducts may not contain wiring of any type.

(B) Ducts Specifically Fabricated for Environmental Air states that these ducts are less hazardous, and therefore, wiring within them is not totally prohibited, but special precautions are required to ensure that they do not become a source of or vehicle for the spread of fire and/or smoke. Only wiring methods consisting of Type MI cable, Type MC cable employing a smooth or corrugated impervious metal sheath without an overall nonmetallic covering, electrical metallic tubing, flexible metallic tubing, intermediate metal conduit, or rigid metal conduit without an overall nonmetallic covering may be installed in ducts specifically fabricated to transport environmental air. Flexible metal conduit is permitted, in lengths not to exceed 4 feet, to connect physically adjustable equipment and devices permitted to be in these fabricated ducts. The connectors used with flexible metal conduit must close any openings in the connection effectively. Equipment and devices are permitted within such ducts only if necessary for the direct action on or sensing of the contained air. Where equipment or devices are installed and illumination is necessary to facilitate maintenance and repair, enclosed gasketed-type luminaires are permitted.

(C) Other Spaces Used for Environmental Air (Plenums) states that these are spaces not specifically fabricated to handle environmental air, as are factory-made or field-fabricated ducts, but which handle environmental air nevertheless. The section does not apply to habitable rooms that also serve to transport air. In these areas, ordinary wiring methods suffice. These spaces may be used to supply air, or they may be a vehicle for air return. An exception states that this section does not apply to joist or stud spaces of dwelling units where the wiring passes through perpendicular to the long dimension of the spaces. Notice that this exception is for dwelling units only.

(1) Wiring Methods states that the wiring methods for such other spaces is limited to totally enclosed, nonventilated, insulated busways having no provisions for plug-in connections, Type MI cable, Type MC cable without an overall nonmetallic covering, Type AC cable, or other factory-assembled multiconductor control or power cable that is specifically listed for use within an air-handling space or listed prefabricated cable assemblies of metallic manufactured wiring

systems without nonmetallic sheaths. Other types of cables, conductors, and raceways may be installed in electrical metallic tubing, flexible metallic tubing, intermediate metal conduit, rigid metal conduit without an overall nonmetallic covering, flexible metal conduit, or, where accessible, surface metal raceways or metal wireways with metal covers.

(2) Cable Tray Systems applies to metallic cable tray systems in other spaces used for environmental air (plenums), where accessible:

    a. Metal Cable Tray Systems states that such systems are permitted to support the wiring methods in Section 300.22(C)(1).

    b. Solid Side and Bottom Metal Cable Tray Systems states that such systems with solid metal covers are permitted to enclose wiring methods and cables not already covered in Section 300.22(C)(1) in accordance with Section 392.10(A) and (B), which will be found later in Chapter 3 in the article on cable trays.

(3) Equipment provides that electrical equipment with a metal enclosure or with a nonmetallic enclosure listed for the purpose and having adequate fire resistant and low-smoke-producing characteristics and associated wiring material suitable for the ambient temperature is permitted in such other spaces unless prohibited elsewhere. An exception permits integral fan systems where specifically identified for use within an air-handling space.

(D) Information Technology Equipment provides that electrical wiring in air-handling areas beneath raised floors for information technology (IT) equipment is permitted in accordance with Article 645, Information Technology Equipment.

Section 300.23, Panels Designed to Allow Access, provides that cables, raceways, and equipment behind such panels, including suspended ceiling panels, are to be arranged and secured so as to allow the removal of panels and access to the equipment.

Part II, Requirements for Over 600 Volts, Nominal, contains requirements for this higher voltage level. Note that we are no longer talking about ducts and air-handling spaces. This part pertains to all wiring over 600 volts. The requirements are generic for this voltage level. Specific requirements are located throughout the Code, typically at the end of the relevant article.

Section 300.31, Covers Required, states that suitable covers are to be installed on all boxes, fittings, and similar enclosures to prevent accidental contact with energized parts or physical damage to parts or insulation.

Section 300.32, Conductors of Different Systems, refers back to Section 300.3(C)(2), Over 600 Volts, Nominal, which prohibits the two different voltage levels from occupying the same equipment wiring enclosure, cable, or raceway unless otherwise permitted. There are five permitted conditions. Additionally, it is specified that conductors having nonshielded insulation and operating at different voltage levels are not to occupy the same enclosure, cable, or raceway. This particular rule applies to different voltage levels above 600 volts.

Section 300.34, Conductor Bending Radius, provides that the conductor for over-600-volt wiring is not to be bent to a radius less than 8 times the overall diameter for nonshielded conductors or 12 times the overall diameter for shielded or lead-covered conductors during or after installation. For multiconductor or multiplexed single-conductor cables having individually shielded conductors, the minimum bending radius is 12 times the diameter of the individually shielded conductors or 7 times the overall diameter, whichever is greater.

Too sharp a bend could compromise the insulation. Conductors carrying higher voltage are much more sensitive to any bad spot in the insulation, and the arc-flash and shock hazards are much greater.

Section 300.35, Protection Against Induction Heating, provides that metallic raceways and associated conductors are to be arranged to avoid heating of the raceway in accordance with the applicable provisions of Section 300.20, as we have seen.

The basic idea behind higher-voltage-level wiring is that everything that applies to "regular" wiring is also applicable, only more so. Additionally, there are other rules that must be observed.

Section 300.37, Above-Ground Wiring Methods, states that above-ground conductors are to be installed in rigid metal conduit, in intermediate metal conduit, in electrical metallic tubing, in reinforced thermosetting resin conduit (RTRC) and PVC conduit, in cable trays, in auxiliary gutters, in busways, in cablebus, in other identified raceways, or in exposed runs of metal-clad cable suitable for the use and purpose. In locations accessible to qualified persons only, exposed runs of Type MV cables, bare conductors, and bare busbars are permitted. Busbars may be either copper or aluminum.

Section 300.39, Braid-Covered Insulated Conductors—Exposed Installation, provides that exposed runs of braid-covered insulated conductors must have a flame-retardant braid. If the conductors used do not have this protection, a flame-retardant saturant is to be applied to the braid covering after installation. This treated braid covering is to be stripped back a safe distance at conductor terminals, according to the operating voltage. Where practicable, this distance must be not less than 1 inch for each kilovolt of the conductor-to-ground voltage.

Section 300.40, Insulation Shielding, provides that metallic and semiconducting insulation shielding components of shielded cables must be removed for a distance depending on the circuit voltage and insulation. Stress-reduction means are to be provided at all terminations of factory-applied shielding.

Metallic shielding components such as tapes, wires, or braids or combinations thereof are to be connected to a grounding conductor, grounding busbar, or grounding electrode.

Section 300.42, Moisture or Mechanical Protection for Metal-Sheathed Cables, states that where cable conductors emerge from a metal sheath and where protection against moisture or physical damage is necessary, the insulation of the conductors is to be protected by a cable-sheath-terminating device.

Section 300.50, Underground Installations, contains provisions for buried conductors operating at over 600 volts.

(A) General states that underground conductors are to be identified for the voltage and conditions under which they are installed. Direct-burial cables must comply with the provisions of Section 310.10(F), Direct-Burial Conductors. Underground cables are to be installed in accordance with Section 300.50(A)(1) or (A)(2), below, and the installation is to meet the depth requirements of Table 300.50.

Table 300.50, Minimum Cover Requirements (for Over 600 Volts), is similar to Table 300.5, Minimum Cover Requirements, 0 to 600 Volts. The main difference is that down the left side are different voltage levels rather than location of wiring method or circuits. Across the top is one more column heading and somewhat different conditions, but the idea is the same. The important thing is to know where these two tables are located because there is sure to be at least one question on any licensing exam on burial depths or, more properly, minimum cover requirements.

(1) Shielded Cables and Nonshielded Cables in Metal-Sheathed Cable Assemblies provides that underground cables, including nonshielded, Type MC, and moisture-impervious metal sheathed cables, must have those sheaths grounded through an effective grounding path. They are to be direct buried or installed in raceways identified for that use.

(2) Other Nonshielded Cables states that such cables are to be installed in rigid metal conduit, intermediate metal conduit, or rigid nonmetallic conduit encased in not less than 3 inches of concrete.

(B) Wet Locations says that the interior of enclosures or raceways installed underground are to be considered a wet location. Insulated conductors and cables installed in these enclosures or raceways are to be listed for use in wet locations. Connections or splices in an underground location are to be approved for wet locations.

(C) Protection from Damage states that conductors emerging from the ground are to be enclosed in listed raceways. Raceways on poles must be rigid metal conduit, intermediate metal conduit, reinforced thermosetting resin conduit–extra high impact (RTRC-XW), Schedule 80 PVC conduit, or equivalent extending from the minimum cover depth to a point 8 feet above finished grade. Conductors entering a building are to be protected by an approved enclosure or raceway from the minimum cover depth to the point of entrance. Where direct-buried conductors, raceways, or cables are subject to movement by settlement or frost, they are to be installed to prevent damage to the enclosed conductors or to the equipment connected to the raceways. Metallic enclosures are to be grounded.

(D) Splices states that direct-burial cables are permitted to be spliced or tapped without the use of splice boxes, provided that they are installed using materials suitable for the application. The taps and splices are to be watertight and protected from mechanical damage. Where cables are shielded, the shielding must be continuous across the splice or tap.

(E) Backfill states that backfill containing large rocks, paving materials, cinders, large or sharply angular substances, or corrosive materials may not be placed in an excavation where materials can damage or contribute to the corrosion of raceways, cables, or other substructures or where it may prevent adequate compaction of fill. Protection in the form of granular or selected material or suitable sleeves is to be provided to prevent physical damage to the raceway or cable.

(F) Raceway Seal provides that where a raceway enters from an underground system, the end within the building is to be sealed with an identified compound so as to prevent the entrance of moisture or gases, or it is to be arranged to prevent moisture from contacting live parts.

This concludes Article 300, Wiring Methods. There is a large quantity of information to assimilate. Most of us are unable to memorize all these requirements, but what we can do is learn the structure of the article so that in an open-book setting we can go to the right place quickly to find the answers that we need. Remember that Article 300 goes from the general to the specific, starting with voltage limitations and ending with requirements for over-600-volt wiring. In between is an array of requirements, such as the much-referenced minimum cover table for underground wiring. In the articles yet to come in Chapter 3, we will see increasing degrees of specificity, culminating in Articles 320 through 398, which provide details on individual raceway and conductor types.

Article 310

Article 310, Conductors for General Wiring, also starts with the general, a statement of scope and definitions, and ends in a highly specific manner with ampacity and other tables. This is one of the major Code articles, and electricians refer to the parts of it that they haven't memorized in the course of every job. The ampacity tables are particularly important in job design, and licensing exam questions make frequent reference to them.

Part I, General, begins with an overview:

Section 310.1, Scope, states that the article covers general requirements for conductors and their type designations, insulations, markings, mechanical strengths, ampacity ratings, and uses. These requirements do not apply to conductors that form an integral part of equipment, such as motors, motor controllers, and similar equipment, or to conductors specifically provided for elsewhere in the Code. An Informational Note refers to Article 400 for flexible cords and cables and to Article 402 for fixture wires.

Section 310.2, Definitions, defines two new terms that are specific to this article: *Electrical ducts* are conduits or other raceways, round in cross section, that are suitable for use underground or embedded in concrete. *Thermal resistivity* means heat-transfer capability through a substance by conduction. It is the reciprocal of *thermal conductivity* and is designated lowercase Greek rho (ρ).

Part II, Installation, begins with information on specific conductor nomenclature, contains an important section on conductors in parallel, and concludes with the much-used ampacity tables.

Section 310.10, Uses Permitted, contains general information regarding use of conductors in various settings as well as specialized configurations.

(A) Dry Locations states that insulated conductors and cables used in dry locations are to be any of the types identified in the Code. This means that insulated conductors and cables that are identified for damp and wet locations also may be used in dry locations. Similarly, conductors and cables identified for use in wet locations may be used in damp locations. The hierarchy is apparent.

(B) Dry and Damp Locations provides that insulated conductors and cables used in these settings are Types FEP, FEPB, MTW, PFA, RHH, RHW, RHW-2, SA, THHN, THW, THW-2, THHW, THWN, THWN-2, TW, XHH, XHHW, XHHW-2, Z, or ZW.

(C) Wet Locations states that insulated conductors and cables used in wet locations are to comply with one of the following:

(1) Be moisture-impervious metal-sheathed

(2) Be type MTW, RHW, RHW-2, TW, THW, THW-2, THHW, THWN, THWN-2, XHHW, XHHW-2, or ZW

(3) Be of a type listed for use in wet locations

Notice that types listed for wet locations contain the letter *W*. One or two *H*'s denote the temperature rating. Later, when we consider Table 310.104(A), Conductor Applications and Insulations Rated 600 Volts, we shall see how these letter designations can be decoded.

(D) Locations Exposed to Direct Sunlight provides that insulated conductors or cables used where exposed to direct rays of the sun must comply with (1) or (2) below:

(1) Conductors and cables are to be listed or listed and marked as being sunlight-resistant.

(2) Conductors and cables are to be covered with insulating material, such as tape or sleeving, that is listed or listed and marked as being sunlight-resistant

Sunlight contains ultraviolet radiation. Exposure over a period of time causes degradation of many materials, including conductor insulation and outer jackets of certain conductors and cables. Look for the words *sunlight-resistant* to know that you are safe to install these materials outdoors.

(E) Shielding states that nonshielded, ozone-resistant insulated conductors with a maximum phase-to-phase voltage of 5,000 volts are permitted in Type MC cables in industrial establishments where the conditions of maintenance and supervision ensure that only qualified persons service the installation. For other establishments, solid dielectric insulated conductors operated above 2,000 volts in permanent installations must have ozone-resistant insulation and must be shielded. All metallic insulation shields are to be connected to a grounding electrode conductor, a grounding busbar, an equipment-grounding conductor, or a grounding electrode. Exceptions allow nonshielded insulated conductors under certain very limited conditions.

These higher voltage levels require specialized training, and such work is beyond the scope of most electricians. Some states require s special license or endorsement appended to an electrician's license in order to do what is called *medium-voltage work*. It is best to enter this work with a view to the hazards involved. Throughout the Code, sections on over-600-volt installations provide mandates that, while essential, are no substitute for experience working with individuals who are adept in these areas. Notwithstanding, it is beneficial for all electricians to become familiar with what is involved, if only for the perspective gained.

(F) Direct-Burial Conductors provides that conductors used for direct-burial applications must be of a type identified for such use.

(G) Corrosive Conditions states that conductors exposed to oils, greases, vapors, gases, fumes, liquids, or other substances having a deleterious effect on the conductor or insulation must be of a type suitable for the application.

(H) Conductors in Parallel allows this type of installation only under specified conditions.

(1) General provides that aluminum, copper-clad aluminum, or copper conductors for each phase, polarity, neutral, or grounded circuit are permitted to be connected in parallel (electrically joined at both ends) only in sizes 1/0 AWG and larger where installed in accordance with (2) through (6) below. (Exceptions permit reduced sizes for control power or for frequencies of 360 hertz and higher and for existing installations under engineering supervision.)

(2) Conductor Characteristic states that the paralleled conductors in each phase, polarity, neutral or grounded circuit conductor, equipment-grounding conductor, or equipment bonding jumper must comply with all the following:

    a. Be the same length

    b. Consist of the same conductor material

    c. Be the same size in circular mil area

d. Have the same insulation type

e. Be terminated in the same manner

The purpose of these rules is to ensure that the paralleled conductors have the same impedance. If there is an imbalance, one of the conductors would carry more than its share of the current, and overheating could result.

(3) Separate Cables or Raceways states that where run in separate cables or raceways, the cables or raceways with conductors must have the same number of conductors and are to have the same electrical characteristics. Conductors of one phase, polarity, neutral or grounded circuit conductor, or equipment-grounding conductor are not required to have the same physical characteristics as those of another phase, polarity, neutral or grounded circuit conductor, or equipment-grounding conductor.

(4) Ampacity Adjustment states that conductors in parallel must comply with the provisions of Table 310.15(B)(3)(a), Adjustment Factors for More Than Three Current-Carrying Conductors in a Raceway or Cable. This table mandates that based on the number of conductors actually carrying current in a raceway or cable, the ampacities of conductors given in Tables 310.15(B)(16) through (19) must be adjusted downward by varying percentages.

This is true of all conductors. This subsection is just reminding us that it is true of paralleled conductors as well. If all of this seems difficult, it is because the Code is referring us ahead, in its usual fashion, to material that has not been covered yet. When we get to these tables, we'll have more of an in-depth look at them, and then it will be clear what is involved.

(5) Equipment Grounding Conductors states that these are to be sized in accordance with Section 250.122 and its associated table.

(6) Equipment Bonding Jumpers provides that where parallel equipment bonding jumpers are installed in raceways, they are to be sized and installed in accordance with Section 250.102, Bonding Conductors and Jumpers. The 1/0 minimum size requirement for paralleled conductors does not apply to the equipment bonding jumper.

Section 310.15, Ampacities for Conductors Rated 0–2,000 Volts, is a long section that contains the ampacity tables used in choosing conductors for an installation based on the load.

(A) General contains some preliminary requirements before proceeding to the tables.

(1) Tables or Engineering Supervision informs us that ampacities may be found by means of calculations performed under engineering supervision or by consulting the tables. For an electrician's licensing exam or design and installation work, the preferred method will be to consult the NEC tables. An Informational Note states that the tables do not take into account voltage drop, so for long runs the conductor

size should be increased. In many cases, raceway size will have to be increased accordingly.

(2) Selection of Ampacity states that where more than one ampacity applies for a given circuit length, the lowest value is to be used. However, an exception provides that where two different ampacities apply to adjacent portions of a circuit, the higher ampacity is permitted to be used beyond the point of transition, a distance of 10 feet or 10 percent of the circuit length figured at the higher ampacity, whichever is less. An Informational Note reminds us that conductor temperature limitations owing to termination provisions are applicable.

(3) Temperature Limitation of Conductors provides that no conductor is to be used in such a manner that its operating temperature exceeds that designated for the type of insulated conductor involved. In no case may conductors be associated together in such a way, with respect to type of circuit, wiring method employed, or number of conductors, that the limiting temperature of any conductor is exceeded.

(B) Tables provides that ampacities for conductors rated 0 to 2,000 volts are as specified in the Tables 310.15(B)(16) through (19), Allowable Ampacities, and Tables 310.15(B)(20) and (21), Ampacities, as modified by Tables 310.15(B)(1) through (7).

The temperature correction and adjustment factors are permitted to be applied to the ampacity for the temperature rating of the conductor if the corrected and adjusted ampacity does not exceed the ampacity for the temperature rating of the termination in accordance with the provisions of Section 110.14(C).

(1) General makes reference to Tables 310.104(A) and (B) for an explanation of type letters and recognized sizes of conductors. For installation requirements, Sections 310.1 through 310.15(A)(3) are cited. For flexible cords, reference is made to tables in Chapter 4, Equipment for General Use, which opens with Article 400, Flexible Cords and Cables.

(2) Ambient Temperature Correction Factors states that for temperatures other than those shown in the ampacity tables, refer to Table 310.15(B)(2)(a) or (b). An alternate method is to use the formula shown.

(3) Adjustment Factors states that this is a critical consideration, and the associated table is applicable to a large proportion of jobs and figures prominently in licensing exams.

  a. More Than Three Current-Carrying Conductors in a Raceway or Cable states that where the number of current-carrying conductors in a raceway or cable exceeds three, or where single conductors or multiconductor cables are installed without maintaining spacing for a continuous length longer than 24 inches and are not installed in raceways, the allowable ampacity of each conductor must be reduced as shown in Table 310.15(B)(3)(a). Each current-carrying conductor of a paralleled set of conductors is to be counted as a current-carrying conductor.

Where conductors of different systems are installed in a common raceway or cable, the adjustment factors apply only to the number of power and lighting conductors. An Informational Note refers to Annex B, Table B.310.15(B)(2)(11) for adjustment factors for more than three current-carrying conductors in a raceway or cable with load diversity.

For ambient temperature, the factors are referred to as *correction factors*. For more than three current-carrying conductors, the factors are referred to as *adjustment factors*. Also, notice that no adjustment factor is required for three current-carrying conductors. It says *more than three*, so adjustment factoring begins with four conductors. The following notes are also applicable:

1. Where conductors are installed in cable trays, the provisions of Section 392.80 apply. This is the specific article on cable trays.

2. Adjustment factors do not apply to conductors in raceways having a length not exceeding 24 inches.

3. Adjustment factors do not apply to underground conductors entering or leaving an outdoor trench if those conductors have physical protection in the form of rigid metal conduit, intermediate metal conduit, rigid PVC, or RTRC having a length not exceeding 10 feet and if the number of conductors does not exceed four.

4. Adjustment factors do not apply to Type AC or Type MC cable under the following conditions:

   a. The cables do not have an overall outer jacket.

   b. Each cable has not more than three current-carrying conductors.

   c. The conductors are 12 AWG copper.

   d. Not more than 20 current-carrying conductors are installed without maintaining spacing, are stacked, or are supported on bridle rings.

5. An adjustment factor of 60 percent is to be applied to Type AC or Type MC cable under the following conditions:

   a. The cables do not have an overall jacket.

   b. The number of current-carrying conductors exceeds 20.

   c. The cables are stacked or bundled longer than 24 inches without spacing being maintained.

b. More Than One Conduit, Tube, or Raceway states that spacing between conduits, tubing, or raceways is to be maintained.

c. Circular Raceways Exposed to Sunlight on Rooftops states that where conductors or cables are installed in circular raceways exposed to direct sunlight on or above rooftops, the adjustments shown in Table 310.15(B)(3)(c) are to be added to the outdoor temperature to determine the applicable ambient temperature for application of the correction factors.

It has been found that conductors in raceways installed on or near the surface of a roof experience a substantial temperature rise in direct sunlight. The temperature is greater the closer the raceway is to the roof, and this heat is added to any heat the conductors dissipate when carrying current.

(4) Bare or Covered Conductors provides that where bare or covered conductors are installed with insulated conductors, the temperature rating of the bare or covered conductor is equal to the lowest temperature rating of the insulated conductors for the purpose of determining ampacity.

Conductors may be bare, covered, or insulated. A *covered* conductor is one that has an outer layer that offers some degree of protection in that it is nonconductive but is not to be considered fully insulated. It is not safe to touch a covered conductor that is energized to a voltage level that would present a shock hazard if the conductor were bare.

(5) Neutral Conductor provides the following:

a. A neutral conductor that carries only the unbalanced current from other conductors of the same circuit is not required to be counted as current carrying when figuring adjustment factors for more than three current-carrying conductors in a raceway or cable.

b. In a three-wire circuit consisting of two phase conductors and the neutral conductor of a four-wire, three-phase wye-connected system, a common conductor carries approximately the same current as the line-to-neutral load currents of the other conductors and is to be counted as current carrying when figuring adjustment factors for more than three current-carrying conductors in a raceway or cable.

c. On a four-wire, three-phase wye-connected circuit where the major portion of the load consists of nonlinear loads, harmonic currents are present in the neutral conductor; the neutral conductor therefore must be considered a current-carrying conductor.

(6) Grounding or Bonding Conductor states that a grounding or bonding conductor is not counted as a current-carrying conductor

when figuring adjustment factors for more than three current-carrying conductors in a raceway or cable.

(7) 120/240-Volt, Three-Wire, Single-Phase Dwelling Services and Feeders contains Table 310.15(B)(7), which provides conductor types (copper or aluminum) and sizes for residential services only. These sizes are smaller than conductor sizes in the ampacity tables and allow a substantial savings where long conductor runs are involved. This table may be difficult to access when time is of the essence. Remember that it is located in Article 310 shortly before the main ampacity tables.

(C) Engineering Supervision provides a formula to be used under engineering supervision to calculate conductor ampacities as an alternative to using the tables.

Tables 310.15(B)(16) through (20) are the main conductor ampacity tables in the NEC. These five tables are all organized in the same way. Across the top are temperature ratings of conductors, with a list of conductors in each temperature category. Down the left side are conductor sizes, expressed in AWG and kcmil (kcmil sizes start at 250, where 4/0 leaves off). The sizes are repeated on the right side of the table for ease in reading. The left side of the main body of each table is for copper; the right side for aluminum or copper-clad aluminum. Licensing exams usually focus on copper unless otherwise stated, so to simplify things, you can draw a light line through the aluminum entries if your state allows written notes in the open-book exam. The horizontal lines are to aid in reading the table.

The most-used table is the first one. It is applicable to not more than three current-carrying conductors in raceway, cable, or earth (direct buried), 60 through 90°C.

The second table is for conductors in free air.

The third table is for conductors in raceways or cables with high temperature ratings.

The fourth table is for conductors in free air with high temperature ratings.

The fifth and final table in this series is for aerial conductors supported by a messenger.

As you can see, the second, third, fourth, and fifth tables are used less commonly. Unless an exam question specifically mentions conductors in free air including messenger supported or rated for the higher temperature, you can assume that the first table is the relevant one. You could use a yellow felt-tip pen to highlight the ampacities for copper conductors only in the first table, and this would greatly simplify things.

Now that we know how to read these tables, we can consider how to use the information they contain.

These tables can be used in either of two ways. If you know the size of the load, you can find the conductor size. Or if you know the size of the conductor, you can find the maximum load it will carry. At all times, you have to keep in mind that correction factors for ambient temperatures other than 40°C and adjustment factors for more than three current-carrying conductors in a raceway or cable have to be applied to the ampacity figures in the tables. You get the same result regardless of the order in which these factors are applied. The adjustment factor is a percentage, and the correction factor is a decimal figure that is over 1.00 when the ambient temperature is below 36°C and under 1.00 when the ambient temperature is above 40°C.

These tables, formidable to those starting out in the trade, are really quite simple. They form the basis for much electrical design work, and electricians use them constantly in the field when the job has not been designed in advance. All licensing exams focus on these tables.

One additional table, Table 310.15(B)(21), Ampacities of Bare or Covered Conductors in Free Air, is set up differently. It is applicable to line work not under utility jurisdiction.

Section 310.60. Conductors Rated 2,001 to 35,000 Volts, contains 21 ampacity tables for higher voltage levels. There is a table of ambient temperature correction factors and ampacity tables for conductors in underground electrical ducts and in free air. Also included is Figure 310.60, Cable Installation Dimensions, which shows duct bank and conductor burial details for underground installations.

Part III, Construction Specifications, is the conclusion to Article 310.

Section 310.104, Conductor Constructions and Applications, refers to a series of tables with detailed information on various types of conductors recognized by the Code.

Section 310.106, Conductors, contains Table 310.104(A). Twenty-four conductors by trade name are listed in the first column. Additional columns give type letter, maximum operating temperature, application provisions, insulation, and thickness of insulation and outer covering. Additional tables are for higher voltage levels.

(B) Conductor Material states that conductors in Article 310 are to be aluminum, copper-clad aluminum, or copper unless otherwise specified. (Nickel or nickel-coated copper is used in some high-temperature applications.)

(C) Stranded Conductors provides that where installed in raceways, conductors 8 AWG and larger, unless specifically permitted or required elsewhere in the Code, are to be stranded.

Stranded conductors are more flexible than solid conductors. Where repeated bending of larger size conductors occurs, the strands are finer, as in welding cables and elevator traveling cables. The bonding conductors of a permanently installed swimming pool must be solid copper, 8 AWG or larger.

(D) Insulated states that conductors, not specifically permitted to be covered or bare, are to be insulated.

Section 310.110, Conductor Identification, repeats requirements presented in greater detail in Section 200.6, Means of Identifying Grounded Conductors, and Section 210.5(C), Identification of Ungrounded Conductors.

Section 310.120, Marking, states that all conductors and cables are to be marked to indicate the following information:

1. The maximum rated voltage.

2. The proper type letter or letters for the type of wire or cable

3. The manufacturer's name or trademark

4. The AWG size or circular mil area

In addition, cable assemblies where the neutral conductor is smaller than the ungrounded conductors must be so marked.

Article 312, Cabinets, Cutout Boxes and Meter Socket Enclosures, covers installation and construction specifications.

Part I, Installation, provides the following:

Section 312.2, Damp and Wet Locations, states that in damp or wet locations, surface-type enclosures are to be placed or equipped so as to prevent moisture from entering and accumulating within the cabinet or cutout box. They must be mounted so that there is at least ¼ inch of airspace between the wall and supporting surface. Enclosures installed in wet locations are to be weatherproof. Raceways or cables entering above the level of uninsulated live parts must use fittings listed for wet locations. An exception permits nonmetallic enclosures to be installed without the airspace on concrete, masonry, tile, or similar surface.

Section 312.3, Position in Wall, provides that in walls of concrete, tile, or other noncombustible material, cabinets are to be installed so that the front edge of the cabinet is not set back from the finished surface by more than ¼ inch. In walls constructed of wood or other combustible material, cabinets are to be flush with the finished surface or project from it.

Section 312.4, Repairing Noncombustible Surfaces, provides that surfaces that are broken or incomplete are to be repaired so that there will be no gaps or open spaces greater than 1/8 inch at the edge of the cabinet or cutout box employing a flush-type cover.

Section 312.5, Cabinets, Cutout Boxes and Meter Socket Enclosures (Figure 3-3), provides that conductors entering these enclosures are to be protected from abrasion and must comply with (A) through (C) below:

(A) Openings to Be Closed states that openings through which conductors enter are to be closed adequately.

(B) Metal Cabinets, Cutout Boxes and Meter Socket Enclosures provides that where these are installed with messenger-supported wiring, open wiring on insulators, or concealed knob-and-tube wiring, conductors must enter through insulating bushings or, in dry

FIGURE 3-3 A 100-ampere entrance panel with double-pole main breaker and main bonding jumper. Most Code requirements in Section 312.5 are of concern to the manufacturer and will not be addressed in most licensing exams.

locations, through flexible tubing extending from the last insulating support and secured firmly to the enclosure.

(C) Cables indicates that where cable is used, each cable is to be secured to the cabinet, cutout box, or meter socket enclosure. This rule prohibits the practice of bunching together two or more cables and running them into a box through a single connector. Each cable has to have its own connector unless the connector is identified for use with more than one cable. An exception allows cables with entirely nonmetallic sheaths to enter the top of a surface-mounted enclosure through one or more nonflexible raceways not less than 18 inches and not more than 10 feet in length provided that all the following conditions are met:

(1) Each cable is fastened within 12 inches of the outer end of the raceway.

(2) The raceway extends directly above the enclosure and does not penetrate a structural ceiling.

(3) A fitting is provided on each end of the raceway to protect the cable(s) from abrasion, and the fittings remain accessible after installation.

(4) The raceway is sealed or plugged at the outer end using approved means so as to prevent access to the enclosure through the raceway.

(5) The cable sheath is continuous through the raceway and extends into the enclosure beyond the fitting not less than ¼ inch.

(6) The raceway is fastened at its outer end and at other points in accordance with the appropriate article.

(7) Where installed as conduit or tubing, the allowable cable fill does not exceed that permitted for complete conduit or tubing systems by Table 1 of Chapter 9 and all applicable notes.

If the raceway is longer than 24 inches, the ampacity correction factors apply.

Section 312.6, Deflection of Conductors, provides that conductors at terminals or entering or leaving cabinets or cutout boxes and the like must comply with (A) through (C) below, except for enclosures for motor controllers with provisions for one or two wires per terminal, which must comply with Section 430.10.

(A) Width of Wiring Gutters provides Table 312.6(A), Minimum Wire-Bending Space at Terminals and Minimum Width of Wiring Gutters. This table is organized in terms of wire size and wires per terminal.

(B) Wire-Bending Space at Terminals states rules provided in (1) and (2) below:

(1) Conductors Not Entering or Leaving Opposite Wall states that Table 312.6(A) applies.

(2) Conductors Entering or Leaving Opposite Wall states that Table 312.6(B) applies.

Both these tables are organized in the same way. Table 312.6(A) applies to terminals and gutters. Table 312.6(B) applies to terminals.

(C) Conductors 4 AWG or Larger provides that the installation must comply with Section 300.4(G), Insulated Fittings.

Section 312.7, Space in Enclosures, states that cabinets and cutout boxes are to have sufficient space to accommodate all conductors installed in them without crowding.

Section 312.8, Switch and Overcurrent Device Enclosures with Splices, Taps, and Feed-Through Conductors, provides that the wiring space of enclosures for switches or overcurrent devices is permitted for conductors feeding through, spliced into, or tapping off to other enclosures, switches, or overcurrent devices where all the following conditions are met:

1. The total of all conductors installed at any cross section of the wiring space does not exceed 40 percent of the cross-sectional area of that space.

2. The total area of all conductors, splices, and taps installed at any cross section of the wiring space does not exceed 75 percent of the cross-sectional area of that space.

3. A warning label is applied to the enclosure that identifies the closest disconnecting means for any feed-through conductors.

The remainder of the article covers construction specifications for these enclosures. Major concerns are that adequate mechanical strength and wiring space are provided.

Article 314

Article 314, Outlet, Device, Pull and Junction Boxes; Conduit Bodies; Fittings; and Handhole Enclosures, contains provisions applicable to these structures. Of primary importance are box fill calculations. They are required knowledge for licensing exams and field work. If you provide too small a box, the wiring will be difficult to install; wire nuts may become loose, causing series arcing faults; and conductors may abrade, causing ground faults. In many instances, intuition and memory of past installations suffice in choosing the correct box, but for examinations and large installations, it is necessary to understand the box fill rules. If you know the Code location, you can look up the exact figures on a case-by-case basis. We begin with some basic mandates.

Section 314.1, Scope, notes that this article covers the installation and use of all boxes and conduit bodies used as outlet, device, junction, or pull boxes, depending on their use, and handhole enclosures. Cast, sheet-metal, nonmetallic, and other boxes such as FS, FD, and larger boxes are not classified as conduit bodies. This article also includes installation requirements for fittings used to join raceways and to connect raceways and cables to boxes and conduit bodies.

Section 314.2, Round Boxes, indicates that such boxes are not to be used where conduits or connectors requiring the use of locknuts or bushings are to be connected to the side of the box. Rectangular or octagonal boxes are needed because they have flat bearing surfaces at each knockout. Plastic boxes may be round because locknuts are not an issue.

Section 314.3, Nonmetallic Boxes, states that these are permitted only with open wiring on insulators, concealed knob-and-tube wiring, cabled wiring methods with entirely nonmetallic sheaths, flexible cords, and nonmetallic raceways. Exceptions permit the use of nonmetallic boxes with metal raceways or metal-armored cable where internal bonding means are provided between all entries or where integral bonding means with a provision for attaching an equipment bonding jumper inside the box are provided between all threaded entries in nonmetallic boxes listed for the purpose.

Section 314.4, Metal Boxes, provides that they are to be grounded in accordance with various parts of Article 250, except as permitted in Section 250.112(I), Remote-Control, Signaling, and Fire Alarm Circuits.

Part II, Installation, contains box fill requirements:

Section 314.15, Damp or Wet Locations, states that boxes, conduit bodies, and fittings are to be placed or equipped so as to prevent moisture from entering or accumulating. Those installed in wet locations are to be so listed.

Section 314.16, Number of Conductors in Outlet, Device and Junction Boxes and Conduit Bodies (Figure 3-4), states that boxes and conduit bodies are to be of sufficient size to provide free space for all

Figure 3-4 Conductor and device fill values are given in Section 314.16 and must not exceed the values in Table 314.16(A). This box is in compliance based on the number of 12 AWG conductors and device fill.

conductors. In no case is the volume of the box, as calculated in Section 314.16(A), to be less than the fill calculation, as calculated in Section 314.16(B). The minimum volume for conduit bodies is to be calculated in Section 314.16(C).

The provisions of this section do not apply to terminal housings supplied with motors and generators. These volume requirements are provided separately in Section 430.12.

Boxes and conduit bodies enclosing conductors 4 AWG or larger also must comply with Section 314.28.

Box fill calculations consist of two parts. You have to determine the volume of the box in question, and you have to determine the box fill based on the sum of required fill for all items within the box. This is simple enough. There is one set of rules for figuring volume and another set of rules for figuring box fill requirements. Then you have to ascertain that the volume of the box is equal to or greater than the box fill.

(A) Box Volume Calculations states that the volume of a wiring enclosure (box) is the total volume of the assembled sections and, where used, the space provided by plaster rings, domed covers, extension rings, and so forth that are marked with their volume or are made from boxes the dimensions of which are listed in Table 314.16(A) (Figure 3-5).

(1) The volumes of standard boxes that are not marked with their volume are given in Table 314.16(A). This table also shows the maximum number of conductors, but that is applicable only if the conductors

FIGURE 3-5 A 4 × 4 box with no volume marked. To calculate box fill, find volume in Table 314.16(A).

go straight through with no splices, devices, or other elements that would contribute to a greater box fill.

(2) Other Boxes states that boxes 100 cubic inches or less, other than those described in Table 314.16(A) and nonmetallic boxes, are to be durably and legibly marked by the manufacturer with their volume. Boxes described in Table 314.16(A) that have a volume larger than is designated in the table are permitted to have their volume marked.

(B) Box Fill Calculations states that the volumes in (1) through (5) below are to be added together. No allowance is required for small fittings such as locknuts and bushings.

(1) Conductor Fill states that each conductor that originates outside a box and terminates or is spliced within the box is to be counted once. Each loop or coil of unbroken conductor not less than twice the minimum length required for free conductors is to be counted twice. The conductor fill is calculated using Table 314.16(B). A conductor, no part of which leaves the box, is not required to be counted. An exception states that an equipment conductor or conductors or not over four fixture wires smaller than 14 AWG or both are permitted to be omitted from the calculations where they enter a box from a domed luminaire or similar canopy and terminate within that box.

(2) Clamp Fill states that where one or more internal cable clamps, whether factory or field supplied, are present in the box, a single volume allowance in accordance with Table 314.16(B) is to be made based on the largest conductor present in the box. No allowance is required for a cable connector with its clamping mechanism outside the box.

(3) Support Fittings Fill states that where one or more luminaire studs or hickeys are present in a box, a single volume allowance in accordance with Table 314.16(B) is to be made for each type of fitting based on the largest conductor present in the box.

(4) Device or Equipment Fill states that for each strap or yoke containing one or more devices or equipment, a double volume allowance in accordance with Table 314.16(B) is to be made for each yoke or strap based on the largest conductor connected to a device or equipment supported by that yoke or strap. A device or utilization equipment wider than a single 2-inch device box, as described in Table 314.16(A), is to have double volume allowances provided for each gang required for mounting.

(5) Equipment Grounding Conductor Fill states that where one or more equipment-grounding conductors or equipment bonding jumpers enter a box, a single volume allowance in accordance with Table 314.16(B) is to be made based on the largest equipment-grounding conductor or equipment bonding jumper present in the box. Where an additional set of equipment-grounding conductors (for an isolated receptacle used for reduction of electrical noise) is present in the box, an additional volume allowance is to be made based on the largest equipment-grounding conductor in the additional set.

(C) Conduit Bodies states that such bodies are defined in Article 100, Definitions: A *conduit body* is a separate portion of a conduit or tubing system that provides access through a removable cover(s) to the interior of the system at a junction of two or more sections of the system or at a terminal point of the system. These include LB, LL, LR, C, T, and X versions.

This section provides in (1) that conduit bodies enclosing 6 AWG conductors or smaller, other than short-radius conduit bodies, are to have a cross-sectional area not less than twice the cross-sectional area of the largest conduit or tubing to which they can be attached. The maximum number of conductors permitted is the maximum number permitted by Table 1 of Chapter 9 for the conduit or tubing to which it is attached.

(2) With Splices, Taps, or Devices states that only conduit bodies that are durably and legibly marked by the manufacturer with their volume are permitted to contain splices, taps, or devices (Figure 3-6). The maximum number of conductors is to be calculated in accordance with Section 314.16(B). Conduit bodies are to be supported in a rigid and secure manner.

(3) Short Radius Conduit Bodies provides that conduit bodies such as capped elbows and service-entrance elbows that enclose conductors 6 AWG or smaller and are intended only to enable installation of the raceway and the contained conductors are not to contain splices, taps, or devices and are to be of sufficient size to provide free space for all conductors enclosed in the conduit body.

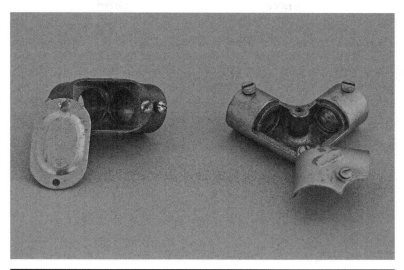

FIGURE 3-6 Two conduit bodies with no volume marked and therefore not permitted to contain splices. These are useful, however, in complying with maximum number of bends between pull points in a raceway installation.

Section 314.17, Conductors Entering Boxes, Conduit Bodies, or Fittings, provides that these are to be protected from abrasion and comply with (A) through (D) below.

(A) Openings to Be Closed states that openings through which conductors enter are to be closed adequately.

(B) Metal Boxes and Conduit Bodies states that where metal boxes or conduit bodies are installed with messenger-supported wiring, open wiring on insulators, or concealed knob-and-tube wiring, conductors are to enter through insulating bushings or, in dry locations, through flexible tubing extending from the last insulating support to not less than ¼ inch inside the box and beyond any cable clamps. Except as provided in Section 300.15(C), the wiring is to be firmly secured to the box or conduit body. Where raceway or cable is installed with metal boxes or conduit bodies, the raceway or cable is to be secured to such boxes and conduit bodies (Figure 3-7).

(C) Nonmetallic Boxes and Conduit Bodies states that these are to be suitable for the lowest temperature-rated conductor entering the box. Where nonmetallic boxes and conduit bodies are used with messenger-supported wiring, open wiring on insulators, or concealed knob-and-tube wiring, the conductors must enter the box through individual holes. Where flexible tubing is used to enclose the conductors, the tubing must extend from the last insulating support to not less than ¼ inch inside the box and beyond any cable clamp. Where non-metallic-sheathed cable or multiconductor Type UF cable is used,

FIGURE 3-7   A 4 × 4 box with two cable entries at the same connector. This is a violation, permitted, however, for nonmetallic boxes.

the sheath must extend not less than ¼ inch inside the box and beyond any cable clamp. In all instances, all permitted wiring methods are to be secured to the boxes. An exception allows non-metallic-sheathed cable or multiconductor Type UF cable under stated circumstances to enter these nonmetallic boxes without being secured. Multiple cable entries through a single knockout are permitted.

Section 314.19, Boxes Enclosing Flush Devices, states that boxes must completely enclose flush devices and that screws for supporting the box are not to be used to attach the device.

Section 314.20, In Wall or Ceiling, provides that in walls or ceilings with a surface of concrete, tile, gypsum, plaster, or other noncombustible material, boxes employing a flush-type cover or faceplate are to be installed so that the front edge of the box, plaster ring, extension ring, or listed extender will not be set back from the finished surface by more than ¼ inch. In walls and ceilings constructed of wood or other combustible surface material, boxes, plaster rings, extension rings, or listed extenders are to be flush with the finished surface or project from it.

Section 314.21, Repairing Noncombustible Surfaces, states that where these are broken or incomplete around boxes employing a flush-type cover or faceplate, they are to be repaired so that there will be no gaps or open spaces greater than 1/8 inch at the edge of the box.

Section 314.22, Surface Extensions, states that such extensions are to be made by mounting and mechanically securing an extension ring over the box. Equipment grounding is to be in accordance with Part VI of Article 250. An exception provides that a surface extension may be made from the cover of a box where the cover is designed so it is unlikely to fall off or be removed if its securing means becomes loose. The wiring method must be flexible for a length sufficient to permit removal of the cover and provide access to the box interior and arranged so that grounding continuity is independent of the connection between the box and cover.

Section 314.23, Supports, provides that enclosures are to be supported in accordance with one or more of the provisions in (A) through (H) below:

(A) Surface Mounting states that an enclosure mounted on a building or other surface is to be rigidly and securely fastened in place. If the surface does not provide rigid and secure support, additional support in accordance with other provisions of this section must be provided.

(B) Structural Mounting states that an enclosure supported from a structural member of a building or from grade is to be rigidly supported either directly or by using a metal, polymeric, or wood brace.

(1) Nails and Screws, states that, where used as a fastening means, nails and screws are to be attached by using brackets on the outside of the enclosure, or they are to pass through the interior within ¼ inch of the back or ends of the enclosure. Screws are not permitted to pass through the box unless exposed threads in the box are protected using approved means to avoid abrasion of conductor insulation.

(2) Braces states that metal braces are to be protected against corrosion and formed from metal that is not less than 0.02 inch thick uncoated. Wood braces must have a cross section not less than nominal 1 × 2 inches. Wood braces in wet locations are to be treated for the conditions. Polymeric braces are to be identified for the use.

(C) Mounting in Finished Surfaces states that an enclosure mounted on a finished surface must be rigidly secured thereto by clamps, anchors, or fittings identified for the application.

(D) Suspended Ceilings states that an enclosure mounted to structural or supporting elements of a suspended ceiling is to be not more than 100 cubic inches in size and must be fastened securely in place in accordance with either (1) or (2) below:

(1) Framing Members states that an enclosure is to be fastened to the framing members by mechanical means such as bolts, screws, or rivets or by the use of clips or other securing means identified for use with the type of ceiling framing member(s) and enclosure(s) employed. The framing members are to be supported adequately and fastened securely to each other and to the building structure.

(2) Support Wires says that the installation must comply with the provisions of Section 300.11(A), Securing and Supporting. The enclosure is to be secured, using methods identified for the purpose, to ceiling support wire(s), including any additional support wire(s)

installed for that purpose. Support wire(s) used for enclosure support are to be fastened at each end so as to be taut within the ceiling cavity.

(E) Raceway Supported Enclosure, Without Devices, Luminaires or Lampholders states that an enclosure that does not contain a device(s) other than splicing devices or support a luminaire(s), lampholder(s), or other equipment and is supported by entering raceways must not exceed 100 cubic inches in size. It is to have threaded entries or have hubs identified for the purpose. It is to be supported by two or more conduits threaded wrench-tight into the enclosure or hubs. Each conduit must be secured within 3 feet of the enclosure or within 18 inches of the enclosure if all conduit entries are on the same side.

Boxes may not be supported by rigid raceways using lock nuts, by electrical metallic tubing (EMT), which is not threaded, or by a single raceway. An exception, however, is made for a conduit body where the trade size is not larger than the trade size of the raceway. This exception is applicable to rigid metallic conduit (RMC), intermediate metallic conduit (IMC), PVC, RTRC, and EMT.

(F) Raceway-Supported Enclosures, with Devices, Luminaires or Lampholders states that an enclosure that contains a device(s), other than splicing devices, or supports a luminaire(s), lampholder, or other equipment and is supported by entering raceways must not exceed 100 cubic inches in size. It is to have threaded entries or have hubs identified for the purpose. It is to be supported by two or more conduits threaded wrench-tight into the enclosure or hubs. Each conduit is to be secured within 18 inches of the enclosure.

Numerous exceptions are attached to this section. They should be consulted prior to an installation. Notice that the securing interval is closer than for the installation without devices, luminaries, or lampholders.

(G) Enclosures in Concrete or Masonry states that an enclosure so supported must be identified as suitably protected from corrosion and securely embedded in concrete or masonry.

(H) Pendant Boxes provides that an enclosure supported by a pendant is to comply with (1) and (2) below:

(1) Flexible Cord states that a box is to be supported from a multiconductor cord or cable in an approved manner that protects the conductors against strain, such as a strain-relief connector threaded into a box with a hub.

(2) Conduit states that a box supporting lampholders or luminaires or wiring enclosures within luminaires used in lieu of boxes are to be supported by rigid or intermediate conduit stems. For stems longer than 18 inches, the stems are to be connected to the wiring system with flexible fittings suitable for the location. At the luminaire end, the conduit(s) is to be threaded wrench-tight into the box or wiring enclosure or into hubs identified for the purpose.

Where supported by only a single conduit, the threaded joints must be prevented from loosening by the use of setscrews or other effective means, or the luminaire, at any point, is to be at least 8 feet above grade or standing area and at least 3 feet measured horizontally to the 8-foot elevation from windows, doors, porches, fire escapes, or similar locations. A luminaire supported by a single conduit must not exceed 12 inches in any horizontal direction from the point of conduit entry.

Section 314.24, Depth of Boxes, provides that outlet and device boxes are to have sufficient depth to allow equipment installed within them to be mounted properly and without likelihood of damage to conductors within the box.

(A) Outlet Boxes Without Enclosed Devices or Utilization Equipment provides that they are to have a minimum internal depth of ½ inch.

(B) Outlet and Device Boxes with Enclosed Devices or Utilization Equipment provides that they are to have a minimum internal depth that accommodates the rearward projection of the equipment and the size of the conductors that supply the equipment. The internal depth includes, where used, that of any extension boxes, plaster rings, or raised covers. The internal depth must comply with all applicable provisions of (1) through (5) below:

(1) Large Equipment states that boxes that enclose devices or utilization equipment that projects more than 1-7/8 inches rearward from the mounting plane of the box must have a depth that is not less than the depth of the equipment plus ¼ inch.

(2) Conductors Larger than 4 AWG states that boxes that enclose devices or utilization equipment supplied by conductors larger than 4 AWG are to be identified for their specific function.

(3) Conductors 8, 6, or 4 AWG states that boxes that enclose devices or utilization equipment supplied by 8, 6, or 4 AWG conductors are to have an internal depth that is not less than 2-1/16 inches.

(4) Conductors 12 or 10 AWG states that boxes that enclose devices or utilization equipment supplied by 12 or 10 AWG conductors are to have an internal depth that is not less than 1-3/16 inches. Where the equipment projects rearward from the mounting plane of the box by more than 1 inch, the box must have a depth not less than the equipment plus ¼ inch.

(5) Conductors 14 AWG and Smaller states that boxes that enclose devices or utilization equipment supplied by 14 AWG or smaller conductors are to have a depth that is not less than 15/16 inch.

Section 314.25, Covers and Canopies, provides that in completed installations, each box must have a cover, faceplate, and lampholder or luminaire canopy, except where the installation complies with Section 410.24(B), Access to Boxes (for electric-discharge and LED luminaries).

(A) Nonmetallic or Metal Covers and Plates states that such items are permitted. Where metal covers or plates are used, they must comply with the grounding requirements of Section 250.110.

(B) Exposed Combustible Wall or Ceiling Finish states that where a luminaire canopy or pan is used, any combustible wall or ceiling finish exposed between the edge of the canopy or pan and the outlet box must be covered with noncombustible material.

The fiberglass pad furnished within the ceiling pan of an indoor luminaire will serve to meet this requirement.

(C) Flexible Cord Pendants states that covers of outlet boxes and conduit bodies having holes through which flexible cord pendants pass are to be provided with bushings designed for the purpose or are to have smooth, well-rounded surfaces on which the cords may bear. So-called hard rubber or composition bushings may not be used.

Section 314.27, Outlet Boxes, contains these provisions:

(A) Boxes at Luminaire or Lampholder Outlets states that outlet boxes or fittings designed for the support of luminaires and lampholders and installed as required by Section 314.23 are permitted to support a luminaire or lampholder.

(1) Wall Outlets states that boxes used at luminaire or lampholder outlets in a wall are to be marked on the interior of the box to indicate the maximum weight of the luminaire that is permitted to be supported by the box in the wall, if other than 50 pounds. An exception states that a wall-mounted luminaire or lampholder weighing not more than 6 pounds is permitted to be supported on other boxes or plaster rings that are secured to other boxes, provided that the luminaire or its supporting yoke or the lampholder is secured to the box with no fewer than two No. 6 or larger screws. This requirement has been seen on a licensing exam.

(2) Ceiling Outlets states that at every outlet used exclusively for lighting, the box is to be designed or installed so that a luminaire or lampholder may be attached. Boxes are required to support a luminaire weighing a minimum of 50 pounds. A luminaire that weighs more than 50 pounds is to be supported independently of the outlet box unless the outlet box is listed and marked for the maximum weight to be supported.

(B) Floor Boxes states that boxes listed specifically for this application must be used for receptacles located in the floor. An exception exempts boxes located in the elevated floors of show windows where the AHJ judges the installation suitable.

(C) Boxes at Ceiling-Suspended (Paddle) Fan Outlets states that outlet boxes or outlet box systems used as the sole support of a ceiling-suspended (paddle) fan must be listed and marked for the purpose and may not support ceiling-suspended fans that weigh more than 70 pounds. For outlet boxes or outlet box systems designed to support ceiling-suspended fans that weigh more than 35 pounds, the required marking is to include the maximum weight to be supported.

Where spare, separately switched ungrounded conductors are provided to a ceiling-mounted outlet box in a location acceptable for a ceiling-supported fan in single or multifamily dwellings, the outlet box or outlet box system is to be listed for sole support of a ceiling-suspended fan.

(D) Utilization Equipment states that boxes used for the support of utilization equipment other than ceiling-suspended (paddle) fans must meet the requirements of Section 314.27(A) for the support of a luminaire that is the same size and weight. An exception allows utilization equipment not over 6 pounds to be supported on boxes, provided that the equipment or its supporting yoke is secured to the box with no fewer than two No. 6 or larger screws.

Section 314.28, Pull and Junction Boxes and Conduit Bodies, states that such items must comply with (A) through (E) below. (An exception exempts terminal housings supplied with motors, which are treated in Section 430.12.)

(A) Minimum Size states that for raceways containing conductors of 4 AWG or larger that are required to be insulated and for cables containing conductors of 4 AWG or larger, the minimum dimensions of pull or junction boxes installed in a raceway or cable run must comply with (1) through (3) below. Where an enclosure dimension is to be calculated based on the diameter of entering raceways, the diameter is to be the metric designator (trade size) expressed in the units of measurement employed.

(1) Straight Pulls states that the length of the box or conduit body must not be less than 8 times the metric designator (trade size) of the largest raceway.

(2) Angle or U Pulls or Splices states that the distance between each raceway entry inside a box or conduit body and the opposite wall of the box or conduit body is to be not less than 6 times the metric designator (trade size) of the largest raceway in a row. This distance must be increased for additional entries by the amount of the sum of the diameters of all other raceway entries in the same row on the same wall of the box. Each row is to be calculated individually, and the single row that provides the maximum distance is to be used. An exception provides that where a raceway or cable entry is in the wall of a box or conduit body opposite a removable cover, the distance from that wall to the cover is permitted to comply with the distance required for one wire per terminal in Table 312.6(A), Minimum Wire-Bending Space at Terminals and Minimum Width of Wiring Gutters.

The minimum distance between raceway entries enclosing the same conductor is to be not less than 6 times the metric designator (trade size) of the larger raceway. When transposing cable size into raceway size in (1) and (2) above, the minimum metric designator (trade size) raceway required for the number and size of conductors in the cable is to be used.

(3) Smaller Dimensions states that boxes or conduit bodies of dimensions less than those required in (1) and (2) above are permitted for installations of combinations of conductors that are less than

the maximum conduit or tubing fill (of conduits or tubing being used) permitted by Table 1 of Chapter 9, provided that the box or conduit body has been listed for and is permanently marked with the maximum number and size of conductors permitted.

(B) Conductors in Pull or Junction Boxes states that in pull boxes or junction boxes having any dimension over 6 feet, all conductors are to be cabled or racked up in an approved manner.

(C) Covers states that all pull boxes, junction boxes, and conduit bodies are to be provided with covers compatible with the box or conduit body construction and suitable for the conditions of use. Where used, metal covers must comply with the grounding requirements of Section 250.110.

(D) Permanent Barriers states that where permanent barriers are installed in a box, each section is considered as a separate box.

(E) Power Distribution Blocks states that such blocks are permitted in pull and junction boxes over 100 cubic inches for connections of conductors where installed in boxes and where the installation complies with (1) through (5) discussed next. (An exception exempts equipment-grounding terminal bars in smaller enclosures.)

(1) Installation states that power-distribution blocks installed in boxes are to be listed.

(2) Size states that in addition to the overall size requirement, the power-distribution block is to be installed in a box with dimensions not smaller than specified in the installation instructions of the power-distribution block.

(3) Wire Bending Space states that the terminals of power-distribution blocks must comply with Section 312.6, Deflection of Conductors.

(4) Live Parts states that power-distribution blocks may not have uninsulated live parts exposed within a box, whether or not the box cover is installed.

(5) Through Conductors states that where pull or junction boxes are used for conductors that do not terminate on power-distribution block(s), the through conductors must be arranged so that the power-distribution block terminals are unobstructed following installation.

Section 314.29, Boxes, Conduit Bodies and Handhole Enclosures To Be Accessible provides that these items are to be installed so that the wiring in them can be rendered accessible without removing any part of the building or, in underground circuits, without excavating sidewalks, paving, earth, or other substance that is used to establish the finished grade. An exception allows listed boxes and handhole enclosures to be covered by gravel, light aggregate, or noncohesive granulated soil if their location is effectively identified and accessible for excavation.

It is a definite Code violation to install a junction box containing splices behind a finished-wall surface where it cannot be accessed.

The box has to be accessible, but not readily accessible. It could be located above a suspended ceiling but not above a drywall ceiling.

Section 314.30, Handhole Enclosures, states that such items are to be designed and installed to withstand all loads likely to be imposed on them. They are to be identified for use in underground systems.

Handhole enclosure is defined in Article 100. It is "an enclosure for use in underground systems, provided with an open or closed bottom, and sized for personnel to reach into but not enter for the purpose of installing, operating or maintaining equipment or wiring or both."

(A) Size states that handhole enclosures are to be sized in accordance with Section 314.28(A) for conductors operating at 600 volts and below and in accordance with Section 314.71 for conductors operating at over 600 volts.

(B) Wiring Entries provides that underground raceways and cable assemblies entering a handhole enclosure must extend into the enclosure, but they are not required to be mechanically connected to the enclosure.

(C) Enclosed Wiring states that such wiring is to be listed as suitable for wet locations. It is assumed that handhole enclosures may be filled with water at times. No wire nut splices are permitted regardless of whether the handhole enclosure is bottomless or totally enclosed.

(D) Covers states that covers must have an identifying mark such as "Electric." They are to require the use of tools to open or weigh over 100 pounds. Metal covers and other conductive surfaces are to be bonded in accordance with 250.92 if the conductors in the handhole are service conductors or in accordance with 250.96(A) if the conductors in the handhole are feeder or branch-circuit conductors.

Part III, Construction Specifications, contains these provisions:

Section 314.40, Metal Boxes, Conduit Bodies, and Fittings, states that such items are to be constructed as follows:

(A) Corrosion Resistant states that such items are to be corrosion-resistant or coated inside and out to prevent corrosion.

(B) Thickness of Metal gives thicknesses for various types of boxes based on the material they are made of.

(C) Metal Boxes Over 100 Cubic Inches states that such boxes are to be constructed so as to be of ample strength and rigidity.

(D) Grounding Provisions states that such provisions must be supplied for each metal box. This may be a tapped hole.

Section 314.41, Covers, provides that covers must be of the same material as the box or conduit body, or they must be lined with insulating material, or they must be listed for the purpose. Metal covers are to be the same thickness as the boxes or conduit bodies with which they are used or listed for the purpose.

Section 314.42, Bushings, states that covers of outlet boxes and conduit bodies having holes through which flexible-cord pendants may pass are to have smooth, well-rounded surfaces on which the cord may bear. Where individual conductors pass through a metal cover, a separate hole equipped with a bushing is to be provided for each conductor.

Section 314.43, Nonmetallic Boxes, requires that provisions for supports or other mounting means are to be outside the box, or the box must be constructed so as to prevent contact between the conductors in the box and the supporting screws.

Section 314.44, Marking, provides that all boxes and conduit bodies, covers, extension rings, plaster rings, and the like are to be durably and legibly marked with the manufacturer's name or trademark.

Part IV, Pull and Junction Boxes, Conduit Bodies, and Handhole Enclosures for Use on Systems Over 600 Volts, Nominal, provides installation and construction requirements.

Section 314.71, Size of Pull and Junction Boxes, Conduit Bodies, and Handhole Enclosures contains these provisions:

(A) For Straight Pulls states that the length of the box must be not less than 48 times the outside diameter, over sheath, of the largest shielded or lead-covered conductor or cable entering the box. The length must not be less than 32 times the outside diameter of the largest unshielded conductor or cable.

(B) For Angle or U Pulls states in (1) that the distance between each cable or conductor entry inside the box and the opposite wall of the box must be not less than 36 times the outside diameter, over sheath, of the largest cable or conductor. This distance is to be increased for additional entries by the amount of the sum of the outside diameters, over sheath, of all other cables or conductor entries through the same wall of the box.

(2) Distance Between Entry and Exit states that the distance between a cable or conductor entry and its exit from a box is not to be less than 36 times the outside diameter, over sheath, of that cable or conductor.

(C) Removable Sides provides that one or more sides of any pull box must be removable.

Section 314.72, Construction and Installation Requirements, contains specifications for over-600-volt boxes.

(A) Corrosion Protection states that boxes are to be made of material inherently resistant to corrosion or protected inside and out by enameling, galvanizing, plating, or other means.

(B) Passing Through Partitions provides for bushings, shields, or fittings with smooth, well-rounded edges where conductors pass through partitions and at other locations where necessary.

(C) Complete Enclosure states that boxes must provide a complete enclosure for conductors or cables.

(D) Wiring Is Accessible provides that boxes and conduit bodies are to be installed so that conductors are accessible without removing any fixed part of the building or structure. Working space is to be provided in accordance with Section 110.34.

(E) Suitable Covers states that boxes are to be closed by suitable covers fastened securely in place. Underground box covers that weigh over 100 pounds meet this requirement. Covers are to be permanently marked "DANGER—HIGH VOLTAGE—KEEP OUT."

(F) Suitable for Expected Handling provides that boxes and their covers are to be capable of withstanding the handling to which they are likely to be subjected.

This concludes Article 314, Outlet, Device, Pull and Junction Boxes; Conduit Bodies; Fittings and Handhole Enclosures. In preparing for a licensing exam, emphasis should be placed on box volume and box fill calculations. Intuition and experience usually suffice in choosing the correct box, and in the smaller sizes, there is no harm in oversizing a bit. For larger sizes and in an exam setting, precise calculations are necessary. Remember the key tables, Table 314.16(A) for minimum volume of metal boxes and Table 314.16(B) for volume allowance required per conductor. Section 314.16(B), Box Fill Calculations, lists items contributing to box fill with values for each.

Articles 320 through 399

Chapter 3, Wiring Methods and Materials, concludes with Articles 320 through 399, which are devoted to specific types of cable and raceway. Each of these articles is organized in the same way. Section numbers are coordinated for various aspects addressed in each of the articles. For example, Uses Permitted for Armored Cable is Section 320.10, For Underground Feeder is Section 340.10, For Flexible Metal Tubing is Section 360.10, and so on. This makes these articles very user friendly.

The articles have grown and expanded over the years; hence their order is not entirely rational. Articles 320 through 340 are devoted to various types of cable. These are in alphabetical order. Next are the conduits, tubings, and raceways, not in alphabetical order. Last we see four more types of wiring. The easiest way to find a given cable or raceway article is to refer to the Contents at the beginning of the Code, as opposed to flipping through the many pages devoted to these articles. (The NEC® Handbook has a separate table of contents for each article, which facilitates internal navigation.) You probably will learn the article numbers of items that you access frequently. Type MC (metal-clad cable) appears in Article 330, for example. You may never have occasion to look up in Article 326 that most exotic of all wiring methods, integrated gas spacer cable (IGS).

Since it is familiar to all, we shall discuss in detail Article 334, Non-Metallic-Sheathed Cable (Type NM), commonly known by its trade name, Romex.

Article 334, Non-Metallic-Sheathed Cable: Types NM, NMC and NMS, contains requirements and specifications for this widely used wiring method (Figure 3-8).

Figure 3-8   Type NM cable (trade name Romex) is used widely in residential and some commercial work. Note "Uses Permitted" and "Uses Not Permitted" in Article 334. This wiring type is frequently misused, for example, in commercial garages.

Part I, General, begins with Section 334.1, Scope, which states that the article covers use, installation, and construction specifications of non-metallic-sheathed cable.

Section 334.2 contains four new definitions that do not occur in Article 100 because they are specific to this article:

- *Non-metallic-sheathed cable* is a factory assembly of two or more insulated conductors enclosed within an overall non-metallic jacket.

- *Type NM* consists of insulated conductors enclosed within an overall nonmetallic jacket.

- *Type NMC* consists of insulated conductors enclosed within an overall corrosion-resistant nonmetallic jacket.

- *Type NMS* consists of insulated power or control conductors with signaling, data, and communications conductors within an overall nonmetallic jacket.

Romex replaced concealed knob-and-tube wiring and exposed open wiring in the 1920s. It is exceptionally user friendly, although not suitable for more sensitive locations, where Type MC or better may be required. It can be concealed or exposed, but it is not appropriate for surface mounting in finished residential or commercial areas.

Section 334.6, Listed, requires that these wiring methods be listed.

Part II, Installation, provides the following:

Section 334.10, Uses Permitted, states that Type NM, Type NMC, and Type NMS are permitted to be used in the following:

1. One- and two-family dwellings and their attached or detached garages and their storage buildings

2. Multifamily dwellings permitted to be of Types III, IV, and V construction except as prohibited in Section 334.12

3. Other structures permitted to be of Types III, IV, and V construction except as prohibited in Section 334.12 (Cables are to be concealed within walls, floors, or ceilings that provide a thermal barrier of material that has at least a 15-minute finish rating, as identified in listings of fire-rated assemblies.)

4. Cable trays in structures permitted to be Types III, IV, or V where the cables are identified for the use

5. Types I and II construction where installed within raceways permitted to be installed in Types I and II construction

An Informational Note refers to Informative Annex E for determination of building types. Type I is the most resistive to fire, going down to Type V, the most combustible. These building types are regulated by the local building code, which considers factors including occupancy, height, and building area.

(A) Type NM cable is permitted:

(1) For both exposed and concealed work in normally dry locations, except as prohibited in item 3 above

(2) To be installed or fished in air voids in masonry block or tile walls

(B) Type NMC cable is permitted:

(1) For both exposed and concealed work in dry, moist, damp, or corrosive locations, except as prohibited by item 3 above

(2) In outside and inside walls of masonry block or tile

(3) In shallow chases in masonry, concrete, or adobe protected against nails or screws by a steel plate at least 1/16 inch thick and covered with plaster, adobe, or similar finish

(C) Type NMS is permitted:

(1) For both exposed and concealed work in normally dry locations, except as prohibited by Section 334.10(3)

(2) To be installed or fished in air voids in masonry block or tile walls

Section 334.12, Uses Not Permitted, provides the following:

(A) Types NM, NMC, and NMS are not permitted:

(1) In any dwelling or structure not permitted in Section 334.10 (1), (2), and (3)

(2) Exposed in dropped or suspended ceilings in other than one- and two-family and multifamily dwellings

(3) As service-entrance cable

(4) In commercial garages having hazardous (classified) locations

(5) In theaters and similar locations

(6) In motion picture studios

(7) In storage-battery rooms

(8) In hoistways or on elevators or escalators

(9) Embedded in poured cement, concrete, or aggregate

(10) In hazardous (classified) locations, except where specifically permitted by other articles of the Code

(B) Types NM and NMS are not permitted:

(1) Where exposed to corrosive fumes or vapors

(2) Where embedded in masonry, concrete, adobe, fill, or plaster

(3) In a shallow chase in masonry, concrete, or adobe and covered with plaster, adobe, or similar finish

(4) In wet or damp locations

Section 334.15, Exposed Work, states that in exposed work, cable is to be installed as specified in (A) through (C) below:

(A) To Follow Surface states that cable is to closely follow the surface of the building finish or running boards.

(B) Protection from Physical Damage states that cable is to be protected from physical damage where necessary by rigid metal conduit, intermediate metal conduit, electrical metallic tubing, Schedule 80 PVC conduit, Type RTRC marked with the suffix –XW, or other approved means. Where passing through a floor, the cable is to be protected as above, extending at least 6 inches above the floor.

Type NMC cable installed in shallow chases or grooves in masonry, concrete, or adobe is to be protected in accordance with the requirements in Section 300.4(F), Cables and Raceways Installed in Shallow Grooves, and covered with plaster, adobe, or similar finish.

(C) In Unfinished Basements and Crawl Spaces states that where cable is run at angles with joists in unfinished basements and crawl spaces, it is permissible to secure cables not smaller than two 6 AWG or three 8 AWG conductors directly to the lower edges of the joists. Smaller cable is to be run either through bored holes in joists or on running boards. Non-metallic-sheathed cable installed on the wall of an unfinished basement is permitted to be installed in a listed conduit or tubing or is to be protected in accordance with Section 300.4. Conduit or tubing is to be provided with a suitable insulating bushing or adapter at the point the cable enters the raceway. The sheath of the non-metallic-sheathed cable must extend through the conduit or tubing and into the outlet or device box not less than ¼ inch. The cable is to be secured within 12 inches of the point where the cable enters the conduit or tubing. Metal conduit, tubing, and metal outlet boxes are to be connected to an equipment-grounding conductor.

Section 334.17, Through or Parallel to Framing Members, provides that Types NM, NMC, or NMS cable are to be protected in accordance with Section 300.4 where installed through or parallel to framing members. Grommets must remain in place and be listed for the purpose of cable protection.

Section 334.23, In Accessible Attics, states that the installation must comply with Section 320.23 in the article on armored cable.

Section 334.24, Bending Radius, states that bends in Types NM, NMC, and NMS cable are to be made so that the cable will not be damaged. The radius of the curve of the inner edge of any bend during or after the installation is not to be less than 5 times the diameter of the cable.

Various cable types have different minimum bending radii. For Type MC with interlocked-type armor or corrugated sheath, in contrast to NM, it is 7 times the external diameter of the metallic sheath.

Section 334.30, Securing and Supporting, provides that non-metallic-sheathed cable must be supported and secured by staples, cable ties, straps, hangers, or similar fittings designed and installed so as not to damage the cable at intervals not exceeding 4½ feet and within 12 inches of every outlet box, junction box, cabinet, or fitting. Flat cables are not to be stapled on edge.

Sections of cable protected from damage by raceways are not required to be secured within the raceway.

(A) Horizontal Runs Through Holes and Notches states that in other than vertical runs, cable installed in accordance with Section 300.4 is considered to be supported and secured where such support does not exceed 4½-foot intervals and the non-metallic-sheathed cable is fastened securely in place by an approved means within 12 inches of each box, cabinet, conduit body, or other non-metallic-sheathed cable termination.

(B) Unsupported Cables permits non-metallic-sheathed cable to be unsupported where the cable

(1) Is fished between access points through concealed spaces in finished buildings or structures and supporting is impracticable

(2) Is not more than 4½ feet from the last point of cable support to the point of connection to a luminaire or other piece of electrical equipment and the cable and point of connection are within an accessible ceiling

(C) Wiring Device Without a Separate Outlet Box states that a wiring device identified for the use, without a separate outlet box and incorporating an integral cable clamp, is permitted where the cable is secured in place at intervals not exceeding 4½ feet and within 12 inches of the wiring device wall opening, and there is to be at least a 12-inch loop of unbroken cable or 6 feet of a cable end available on the interior side of the finished wall to permit replacement.

Section 334.40, Boxes and Fittings, provides the following:

(A) Boxes of Insulating Material states that such boxes are permitted.

(B) Devices of Insulating Material states that switch, outlet, and tap devices of insulating material are permitted to be used without boxes in exposed cable wiring and for rewiring in existing buildings where the cable is concealed and fished. Openings in such devices must form a close fit around the outer covering of the cable, and the device must fully enclose the part of the cable from which any part of the covering has been removed. Where connections to conductors are by binding-screw terminals, there is to be available as many terminals as conductors.

(C) Devices with Integral Enclosures states that devices identified for such use are permitted.

Section 334.80, Ampacity, states that the ampacity of Type NM, NMC, and NMS cable is to be determined in accordance with Section 310.15. The allowable ampacity is not to exceed that of a 60°C rated conductor. The 90°C rating is permitted to be used for ampacity adjustment and correction calculations, provided that the final derated ampacity does not exceed that of a 60°C rated conductor. The ampacity of Type NM, NMC, and NMS cable installed in a cable tray is to be determined in accordance with Section 392.80(A), Ampacity of Cables, Rated 2,000 Volts or Less, in Cable Trays.

Where more than two NM cables containing two or more current-carrying conductors are installed, without maintaining spacing between the cables, through the same opening in wood framing that is to be sealed with thermal insulation, caulk, or sealing foam, the allowable ampacity of each conductor is to be adjusted in accordance with Table 310.15(B)(3)(a) and the provisions of Section 310.15(A)(2). Exceptions do not apply.

Where more than two Type NM cables containing two or more current-carrying conductors are installed in contact with thermal insulation without maintaining spacing between cables, the allowable ampacity of each conductor is to be adjusted in accordance with Table 310.15(B)(3)(a).

Part III, Construction Specifications, provides the following:

Section 334.100, Construction, states that the outer cable sheath of non-metallic-sheathed cable is to be a nonmetallic material.

Section 334.104, Conductors, states that 600-volt insulated conductors are to be sizes 14 AWG through 2 AWG copper or 12 AWG through 2 AWG aluminum.

Section 334.108, Equipment Grounding Conductor, states that in addition to insulated conductors, the cable is to have an insulated, covered, or bare equipment-grounding conductor.

Section 334.112, Insulation, states that insulated power conductors are to be one of the types listed in Table 310.104(A) that are suitable for branch-circuit wiring or one that is identified for use in these cables. Conductor insulation is to be rated 90°C.

Section 334.116, Sheath, states that the outer sheath of non-metallic-sheathed cable must comply with (A), (B), and (C) given next:

(A) Type NM states that the overall covering is to be flame-retardant and moisture-resistant.

(B) Type NMC states that the overall covering is to be flame-retardant, moisture-resistant, fungus-resistant, and corrosion-resistant.

(C) Type NMS states that the overall covering is to be flame-retardant and moisture-resistant. The sheath is to be applied so as to separate the power conductors from the communications conductors.

The other cable and raceway articles of Chapter 3 are similar in structure and scope. These are the articles you want to consult when working with a new type of wiring, first to ensure that it is permitted for the contemplated usage and second to ascertain the correct securing intervals, bending radius, and other installation details. Licensing exams focus intensively on these articles. If you are asked a question on cable and raceway requirements, you may consider it a stroke of good luck because the material is very easy to access and understand.

NEC® Chapter 4, Equipment for General Use

A t this point, the National Electrical Code (NEC) shifts abruptly from the general to the very specific. Previous chapters have laid down generic principles regarding what is to be accomplished in premises wiring and how those precepts are to be accomplished. Now we come to a chapter that contains descriptions and wiring mandates for specific types of equipment, with 21 categories ranging from flexible cords and fixture wires to transformer vaults and storage batteries. The chapter closes, as usual, with an article on equipment operating at over 600 volts.

The good news is that information in this chapter is very easy to access. If you have an exam question or need information for a field installation on refrigeration equipment, for example, consult the Contents and you will quickly find Article 440, Air-Conditioning and Refrigerating Equipment. The article is structured in a thoroughly rational way. As with other Code articles, it begins with a statement of scope, followed by definitions, references to other articles, and so on. The article is fairly short because a lot of the groundwork has been laid in Article 430, Motors, Motor Circuits, and Controllers.

For our study of NEC 2011, we need not go through each article. The information is right there before us and may be consulted on a case-by-case basis. Instead, we'll focus on Article 430. It is unique in that it is by far the longest of all Chapter 4 articles, which is natural because it addresses a complex and fairly technical aspect of the electrician's body of knowledge. Many apprentice electricians find the article somewhat baffling, particularly in regard to overcurrent protection and sizing of conductors. These topics are a little bit counterintuitive until you realize the thinking behind them, which is the reality that motors require a lot more current to start than to run. Most loads are like that, even incandescent light bulbs, but motors are an extreme case and as such require different modes of overcurrent protection. Add to this the fact that motors have specialized tap rules, and it makes for some interesting reading.

For now, the best thing is to start at the beginning and see if we can make this article clear and simple.

Article 430

Part I, General, contains introductory material pertaining to all motor installations.

Section 430.1, Scope, states that the article covers motors, motor branch-circuit and feeder conductors and their protection, motor overload protection, motor control circuits, motor controllers, and motor control centers (Figure 4-1).

Notice that right away a distinction is made between conductor protection and motor overload protection. Unlike most other wiring, these are separate functions, and the protection is at different levels. Conductor protection is at a higher level (less sensitive) than motor overload protection.

Section 430.2, Definitions, contains six new definitions that do not appear in Article 100 because they pertain exclusively to this article:

- *Adjustable-speed drive.* This is a combination of power converter, motor, and motor-mounted auxiliary devices such as encoders, tachometers, thermal switches and detectors, air blowers, heaters, and vibration sensors.

FIGURE 4-1   Three motor-driven compressors are covered by Article 440, Air-Conditioning and Refrigerating Equipment.

- *Adjustable-speed drive system.* This is an interconnected combination of equipment that provides a means of adjusting the speed of a mechanical load coupled to a motor. A drive system typically consists of an adjustable-speed drive and auxiliary electrical apparatus.

- *Controller.* For the purpose of this article, a controller is any switch or device that is normally used to start and stop a motor by making and breaking the motor circuit current.

- *Motor control circuit.* This is the circuit of a control apparatus or system that carries the electric signals directing the performance of the controller but does not carry the main power current.

- *System isolation equipment.* This is a redundantly monitored, remotely operated contactor-isolating system packaged to provide the disconnection/isolation function and capable of verifiable operation from multiple remote locations by means of lockout switches, each having the capability of being padlocked in the "off" (open) position.

- *Valve actuator motor (VAM) assemblies.* This is a manufactured assembly used to operate a valve consisting of an actuator motor and other components such as controllers, torque switches, limit switches, and overload protection.

Section 430.4, Part-Winding Motors, begins with a description of this type of motor. It is one in which the primary (armature) winding is divided into parts that can be energized separately in stages so that the motor starts more easily. In run mode, the entire winding is online.

The section states that where separate overload devices are used, each half of the motor winding is to be individually protected, its trip current being one-half that specified. The article further provides that each motor-winding connection is to have branch-circuit short-circuit and ground-fault protection rated at not more than one-half that specified by 430.52, Rating or Setting for Individual Motor Circuit, which has an associated table providing percentage values of full-load current for various types of motors. This is one of the key tables for Article 430, as we shall see. It allows for much higher ratings or settings for branch-circuit short-circuit and ground-fault protection than you would think possible only because there is separate, much slower-acting overload protection for the motor(s).

Section 430.5, Other Articles, lists the types of equipment and occupancies that have applicable article and section references. These have additional specifications and mandates beyond those in Article 430. An example is fire pumps, covered in Article 695.

Section 430.6, Ampacity and Motor Rating Determination, is a section that is central to this article, and it is the place to begin for many calculations involved in designing a motor installation. It states that the sizes of conductors supplying equipment covered by Article 430 are to be selected from the allowable ampacity tables in accordance with 310.15(B) or are to be calculated in accordance with 310.15(C). Where flexible cord is used, the size of the conductor is to be selected in accordance with 400.5. The required ampacity and motor ratings are to be determined as specified in (A) through (D) below:

(A) General Motor Applications states that the current ratings are to be determined based on (1) and (2) below:

(1) Table Values provides that other than for motors built for low speeds (<1,200 rpm) or high torques and for multispeed motors, the values given in Tables 430.247, 430.248, 430.249, and 430.250 are to be used to determine the ampacity of conductors or ampere ratings of switches and branch-circuit short-circuit and ground-fault protection instead of the actual current rating marked on the motor nameplate. Where a motor is marked in amperes but not in horsepower, the horsepower rating is to be assumed to be that corresponding to the value given in Tables 430.247, 430.248, 430.249, and 430.250, interpolated if necessary. Motors built for low speeds (<1,200 rpm) or high torques may have higher full-load currents, and multispeed motors will have full-load current varying with speed, in which case the nameplate current ratings are to be used.

These tables are in Part XIV, Tables, at the end of the article. Table 430.247 is for direct-current (dc) motors. Table 430.248 is for single-phase alternating-current (ac) motors. Table 430.249 is for two-phase ac motors (rarely used). Table 430.250 is for three-phase ac motors. There are three exceptions:

Number 1 states that multispeed motors are in accordance with Section 430.22(A) for dc motors and Section 430.52 for ac motors.

Number 2 states that a shaded-pole or permanent-split capacitor-type fan or blower motor goes by full-load current.

Number 3 states that a listed motor-operated appliance goes by full-load current.

(2) Nameplate Values states that separate motor overload protection is based on the motor nameplate current rating. To summarize, branch-circuit short-circuit and ground-fault protection is based on the horsepower rating of the motor, subject to exceptions. The overload protective devices are sized based on the full-load current. Both of these values are taken off the nameplate.

(B) Torque Motors states that the rated current is locked-rotor current, and this nameplate current is to be used to determine the ampacity of the branch-circuit conductors covered in Sections 430.22 and 430.24, which addresses the ampere rating of the motor overload protection and the ampere rating of motor branch-circuit short-circuit and ground-fault protection.

(C) Alternating-Current Adjustable Voltage Motors states that for motors used in ac adjustable-voltage, variable-torque drive systems, the ampacity of conductors or ampere rating of switches, branch-circuit short-circuit and ground-fault protection, and so forth is to be based on the maximum operating current marked on the motor or control nameplate or both. If the maximum operating current does not appear on the nameplate, the ampacity determination is to be based on 150 percent of the values given in Tables 430.249 and 430.250.

(D) Valve Actuator Motor Assemblies provides substantially the same as (C) above.

Section 430.7, Marking on Motors and Multimotor Equipment, lists information required to be marked on the nameplate of a motor. The nameplate contains all the information needed to design the installation, including branch circuit and motor controller.

(A) Usual Motor Applications lists 15 items that are to appear on a standard motor nameplate. Certain of these, as we shall see, are crucial in complying with Code mandates. In sizing the branch-circuit conductors, reference must be made to rated horsepower. To size motor overload protection, full-load current is used. Time rating (5, 15, 30, or 60 minutes or continuous) is important in specifying the correct motor for a given application.

A locked-rotor-indicating code letter, not to be confused with the design letter, provides information about an important motor characteristic, which is the number of kilovolt-amperes per horsepower with a locked rotor. Values for code letters A through V are given in Table 430.7(B). To find the locked-rotor current for a specific motor, find the code letter and horsepower on the nameplate. Using the code letter, consult Table 430.7(B), and find the kilovolt-amperes per horsepower with a locked rotor. Multiply that value by the rated horsepower. The result is the locked-rotor current for the motor. Multispeed motors are marked with the code letter designating the locked-rotor kilovolt-amperes per horsepower for the highest speed.

Section 430.8, Marking on Controllers, provides that a controller is to be marked with the manufacturer's name or identification, the voltage, the current or horsepower rating, the short-circuit current rating, and such other necessary data to properly indicate the applications for which the motor is suitable.

Section 430.9, Terminals, provides in (A) that terminals of motors and controllers are to be suitably marked or colored where necessary to indicate the proper connections.

(B) Conductors provides that motor controllers and terminals of control-circuit devices are to be connected with copper conductors unless identified for use with a different conductor.

(C) Torque Requirements states that control-circuit devices with screw-type pressure terminals used with 14 American Wire Gauge (AWG) or smaller copper conductors are to be torqued to a minimum of 7 inch-pounds unless identified for a different torque value.

Section 430.10, Wiring Space in Enclosures, and the associated table specify minimum wire-bending space at the terminals of enclosed motor controllers. This information is of interest primarily to manufacturers.

Section 430.11, Protection Against Liquids, states that suitable guards or enclosures are to be provided to protect exposed current-carrying parts of motors and the insulation of motor leads where the motors are installed directly under equipment or in other locations where dripping or spraying of oil, water, or other liquid is capable of occurring unless the motor is designed for the existing conditions.

Section 430.12, Motor Terminal Housings, contains sections on material, dimensions, wiring connections, and equipment-grounding connections. This information is also of interest primarily to manufacturers.

Section 430.13, Bushings, states that where wires pass through an opening in an enclosure, conduit box, or barrier, a bushing is to be used to protect conductors from sharp edges.

Section 430.14, Location of Motors, states in (A) that motors must be located so that adequate ventilation is provided and so that maintenance, such as lubrication of bearings and replacement of brushes, can be readily accomplished. In a rare display of humor, the Code notes, in an exception, that ventilation is not required for submersible motors (Figure 4-2).

(B) Open Motors states that open motors having commutators or collector rings are to be located or protected so that sparks cannot reach adjacent combustible material. An exception permits installation on wooden floors or supports.

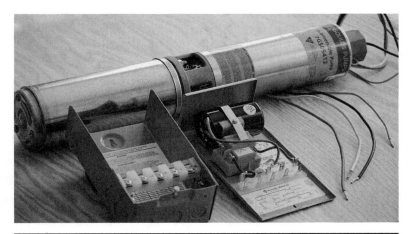

FIGURE 4-2   A hermetically-sealed submersible motor bolts onto a rebuildable pump. The water intake is above the motor so that the motor is always cooled by being underwater.

Section 430.16, Exposure to Dust Accumulations, provides that in locations where flying material collects on or in motors in such quantities as to seriously interfere with ventilation or cooling and thereby causes dangerous temperatures, suitable types of enclosed motors that do not overheat under the prevailing conditions are to be used.

Section 430.17, Highest Rated or Smallest Rated Motor, states that in determining compliance with 430.24, 430.53(B), and 430.53(C), the highest rated or smallest rated motor is to be based on the full-load current, as selected from Tables 430.247, 430.248, 430.249, and 430.250. The articles cited concern group installations of motors.

Section 430.18, Nominal Voltage of Rectifier Systems, provides that the nominal voltage of the ac voltage being rectified is to be used to determine the voltage of a rectifier-derived system. An exception provides that the nominal dc voltage of the rectifier is to be used if it exceeds the peak value of the ac voltage being rectified.

Because the speed and direction of rotation of dc motors can be regulated by changing the input voltage and polarity, these motors are used widely for many applications, including elevators, ski lifts, and so on. The ac supply voltage delivered through the branch circuit is rectified to change it to dc to power the motor. An ac motor runs at a set voltage, which should not be altered to adjust the motor's speed.

The speed of an ac motor is frequency-dependent. Multispeed ac motors have separate windings for each speed, or the speed is regulated by an adjustable-speed drive. Single-phase ac motors cannot have the direction of rotation changed by reversing the two conductors. In a three-phase motor, direction of rotation is changed by switching any two of the three phase connections.

Part I of Article 430, as we have seen, sets the ground rules for the design and field installation of wiring for motors. Subsequent parts build on this information in order to provide details regarding sizing conductors and the many other facets of motor installation.

Part II, Motor Circuit Conductors, begins with a statement of scope.

Section 430.21, General, states that Part II specifies ampacities of conductors that are capable of carrying the motor current without overheating under the conditions specified. It is noted that the provisions of Part II do not apply to motors rated over 600 volts, nominal. These are addressed separately in Part XI.

The provisions of Articles 250, 300, and 310 do not apply to conductors that form an integral part of equipment, such as motors, motor controllers, motor control centers, or other factory-assembled control equipment.

Section 430.22, Single Motor, provides that conductors that supply a single motor used in a continuous-duty application (Figure 4-3) are to have an ampacity of not less than 125 percent of the motor full-load-current rating, as determined by Section 430.6(A)(1), Table Values, or not less than specified in (A) through (G) below:

FIGURE 4-3 A single motor that is considered continuous duty. The disconnecting means is a wall switch mounted within sight.

Most motor applications are continuous duty, unless the nature of the apparatus the motor drives precludes continuous operation with load under any conditions. First, ascertain the horsepower from the nameplate. Second, go to the applicable table at the end of the article and find the full-load current in amperes. Then, for single-motor application in continuous duty, multiply by 1.25. Finally, after applying adjustment and correction factors, consult the applicable ampacity table in Chapter 3 to determine the correct size conductors. This is the case unless one of the following is applicable:

(A) Direct-Current Motor-Rectifier Supplied states that for dc motors operating from a rectified power supply, the conductor ampacity on the input of the rectifier is to be not less than 125 percent of the rated input current to the rectifier. For dc motors operating from a rectified single-phase power supply, the conductors between the field-wiring output terminals of the rectifier and the motor are to have an ampacity of not less than the following percentages of the motor full-load current rating:

(1) Where a rectifier bridge of the single-phase, half-wave type is used, 190 percent.

(2) Where a rectifier bridge of the single-phase, full-wave type is used, 150 percent.

(B) Multispeed Motor states that the selection of branch-circuit conductors on the line side of the controller is to be based on the highest of the full-load current ratings shown on the motor nameplate. The ampacity of the branch-circuit conductors between the controller and the motor must not be less than 125 percent of the current rating of the windings or winding that the conductors energize.

(C) Wye-Start, Delta-Run Motor states that the ampacity of the branch-circuit conductors on the line side of the controller is to be not less than 125 percent of the motor full-load current as determined by Section 430.6(A)(1). The ampacity of the conductors between the controller and the motor must not be less than 72 percent of the motor full-load current rating as determined by Section 430.6(A)(1). This type of motor is used for some air-conditioning compressors and other machinery where reduced-voltage starting is desired. The wiring configurations take place within the controller for the delta-wired motor.

(D) Part-Winding Motor states that the ampacity of the branch-circuit conductors on the line side of the controller is to be not less than 125 percent of the motor full-load current, as determined by Section 430.6(A)(1). The ampacity of the conductors between the controller and the motor is to be not less than 62.5 percent of the motor full-load current, as determined by Section 430.6(A)(1).

(E) Other Than Continuous Duty (Figure 4-4) states that conductors for a motor used in a short-time, intermittent, periodic, or varying-duty application must have an ampacity of not less than the percentage of the motor nameplate current rating shown in Table 430.22(E) unless the authority having jurisdiction (AHJ) grants permission for conductors of lower ampacity.

(F) Separate Terminal Enclosure provides that the conductors between a stationary motor rated 1 horsepower or less and the

FIGURE 4-4 Portable cord-and-plug-connected hand drill is other than continuous duty.

separate terminal enclosure permitted in Section 430.245(B), Separation of Junction Box from Motor, are permitted to be smaller than 14 AWG but not smaller than 18 AWG, provided that they have an ampacity as specified in Section 430.22(A) above.

(G) Conductors for Small Motors states that such conductors may not be smaller than 14 AWG unless otherwise permitted. There follows a number of conditions under which these conductors may be 18 AWG or 16 AWG copper. These should be consulted if such an installation is contemplated.

Section 430.23, Wound-Rotor Secondary, contains provisions relating to the sizing of conductors for this application.

Section 430.24, Several Motors or a Motor(s) and Other Load(s), states that conductors supplying these are to have an ampacity not less than the sum of each of the following:

(1) 125 percent of the full-load current rating of the highest rated motor, as determined by Section 430.6(A)

(2) Sum of the full-load current ratings of all the other motors in the group, as determined by Section 430.6(A)

(3) 100 percent of the noncontinuous nonmotor load

(4) 125 percent of the continuous nonmotor load

This is straightforward. The things to remember are that the motor loads are determined by the tables, not by the full-load current off the nameplate, and that it is 125 percent of just the highest rated motor load, not all the motors. The other motors are taken at 100 percent, even if they are continuous. This is as opposed to nonmotor continuous loads, which are all taken at 125 percent.

There are three exceptions:

Number 1: Where one or more of the motors of the group are used for short-time, intermittent, periodic, or varying duty, the ampere rating of such motors to be used in the summation is to be determined in accordance with 430.22(E), Other Than Continuous Duty. For the highest rated motor, the greater of either the ampere rating from 430.22(E) or the largest continuous-duty-motor full-load current multiplied by 1.25 is to be used in the summation.

Number 2: The ampacity of conductors supplying motor-operated fixed electric space-heating equipment is to comply with 424.3(B), Branch Circuits, which states that fixed electric space-heating equipment and motors are to be considered continuous load.

Number 3: Where the circuitry is interlocked so as to prevent simultaneous operation of selected motors or other loads, the conductor ampacity is permitted to be based on the summation of the currents of the motors and other loads to be operated simultaneously that results in the highest total current.

Section 430.25, Multimotor and Combination-Load Equipment, provides that the ampacity of the conductors supplying multimotor and combination-load equipment is to be not less than the minimum circuit ampacity marked on the equipment. Where the equipment is not factory wired and the individual nameplates are visible, the conductor ampacity is to be determined in accordance with 430.24 above.

Section 430.26, Feeder Demand Factor, states that the AHJ may allow a reduction in feeder ampacity if reduced heating of the conductors results from motors operating on duty cycle, intermittently, or from motors not operating at one time.

Section 430.27, Capacitors with Motors, states that where capacitors are installed in motor circuits, conductors are to comply with 460.8, Conductors (for capacitors), and 460.9, Rating or Setting of Motor Overload Device.

Section 430.28, Feeder Taps, provides that feeder-tap conductors must have an ampacity not less than that required by Part II, must terminate in a branch-circuit protective device, and in addition, must meet one of the following requirements:

(1) Be enclosed by an enclosed controller or by a raceway, be not more than 10 feet in length, and for field installation, be protected by an overcurrent device on the line side of the tap conductor, the rating of which is not to exceed 1,000 percent of the tap-conductor ampacity

(2) Have an ampacity of a least one-third that of the feeder conductors, be suitably protected from physical damage or enclosed in a raceway, and be not more than 25 feet in length

(3) Have an ampacity not less than the feeder conductors

This series of alternate feeder-tap requirements illustrates perfectly the basic idea: As the allowed ampacity decreases, the required protection also decreases. Protection can take the form of enclosure in a raceway or enclosure by an enclosed controller, shorter permitted length, or load-side overcurrent protection. Where there is no length restriction, as in (3), the ampacity of the feeder may not be increased. (Increased ampacity equates to increased vulnerability and must be mitigated.)

An exception states that feeder taps over 25 feet long are permitted in high-bay manufacturing buildings over 35 feet high at walls where conditions of maintenance and supervision ensure that only qualified persons service the systems. Conductors tapped to a feeder are permitted to be not over 25 feet long horizontally and not over 100 feet in total length where all the following conditions are met:

(1) The ampacity of the tap conductors is not less than one-third that of the feeder conductors.

(2) The tap conductors terminate with a single circuit breaker or a single set of fuses complying with (1) Part IV where the load-side conductors are a branch circuit or (2) Part V where the load-side conductors are a feeder.

(3) The tap conductors are suitably protected from physical damage and are installed in raceways.

(4) The tap conductors are continuous from end to end and contain no splices.

(5) The tap conductors are 6 AWG copper or 4 AWG aluminum or larger.

(6) The tap conductors must not penetrate walls, floors, or ceilings.

(7) The tap is not made more than 30 feet from the floor.

Section 430.29, Constant Voltage Direct-Current Motors—Power Resistors, concerns the ampacity of conductors connecting the motor controller to separately mounted power accelerating and dynamic breaking resistors in the armature circuit and armature shunt resistors. The values are given in Table 430.29, Conductor Rating Factors for Power Resistors. These values are based on time off and time on, in seconds, with a category for continuous duty. It is unlikely that this section would be the subject of a licensing exam question, but the information is readily available because it occurs in Part II, Motor Circuit Conductors.

This concludes Part II. Along with Part III, it comprises one of the key sections of the entire Code. It may be somewhat challenging, but the important thing to keep in mind is the unconventional way in which motor supply-circuit conductors are sized. Their ampacity is usually based not on nameplate full load current values but on table values found at the end of the article and based on horsepower taken from the nameplate.

Part III, Motor and Branch-Circuit Overload Protection, is linked to Parts II and IV. Because this protection is in place, the motor branch-circuit short-circuit and ground-fault protection is permitted to have quite hefty maximum ratings or settings. It is because of this unique relationship that motors, with their high-starting-current requirements, are able to operate without risk of burning up the supply conductors.

Section 430.31, General, states that Part III specifies overload devices intended to protect motors, motor control apparatus, and motor branch-circuit conductors against excessive heating owing to motor overloads and failure to start. It is further stated that these provisions do not require overload protection where a power loss would cause a hazard, such as in the case of fire pumps.

It is desirable that fire pumps continue to pump water for fire-fighting purposes even if an overload condition could cause damage to the fire-pump motor. The value of the fire-pump motor is much less than the value of the building that is threatened.

It is also provided that the provisions of Part III do not apply to motor circuits rated over 600 volts, nominal, which are covered in Part XI.

The purpose of overload protection is to prevent damage to a motor from excessive current. It is not designed for nor is it intended to protect the supply conductors from short-circuit or ground-fault current. This protection is furnished separately at the upstream end of those conductors, albeit at a higher current level than is the case for nonmotor supply wiring.

Section 430.32, Continuous-Duty Motors, states that these require a greater level of protection owing to possible heat rise over an extended period of time and therefore are treated separately.

(A) More Than 1 Horsepower states that each motor used in a continuous-duty application and rated more than 1 horsepower is to be protected against overload by one of the means in (1) through (4) below:

(1) A separate overload device that is responsive to motor current. This device is to be selected to trip or be rated at no more than the following percentage of the motor nameplate full-load current rating:

- Motors with a marked service factor of 1.15 or greater— 125 percent
- Motors with a marked temperature rise of 40°C or less— 125 percent
- All other motors—115 percent

Service factor and temperature rise appear on the motor nameplate. Notice that in determining motor and branch-circuit overload protection, we are no longer looking at Tables 430.247 through 430.250. For this determination, we use the motor full-load current taken from the nameplate. Also notice that unlike branch-circuit short-circuit and ground-fault protection, which uses a figure of 125 percent, the percentage applied to nameplate full-load current (in order to determine the overload-device rating) will vary depending on service factor and temperature rise. Modification of this level of protection is permitted in (C) below, as we shall see.

(2) Thermal protector integral with the motor and approved for use with the motor it protects on the basis that it will prevent dangerous overheating of the motor owing to overload and failure to start. The ultimate trip current of a thermally protected motor is not to exceed the following percentage of motor full-load current given in Tables 430.248 through 430.250:

- Motor full-load current of 9 amperes or less—170 percent
- Motor full-load current from 9.1 to and including 20 amperes— 156 percent
- Motor full-load current greater than 20 amperes—140 percent

For determining ultimate trip current of a thermally protected motor, the tables at the end of Article 430 are used. Because the thermal protector is integral with the motor, it is the manufacturer who performs this selection.

(3) Integral with the motor. A protective device integral with the motor that will protect the motor against damage owing to failure to

start is permitted if the motor is part of an approved assembly that does not normally subject the motor to overloads.

(4) Larger than 1,500 horsepower. For motors larger than 1,500 horsepower, this means a protective device having embedded temperature detectors that cause current to the motor to be interrupted when the motor reaches a temperature greater than marked on the nameplate in an ambient temperature of 40°C.

(B) One Horsepower or Less, Automatically Started, states that any motor of 1 horsepower or less that is started automatically is to be protected against overload by substantially the same means as (A) above, with addition of the following: Impedance-Protected. If the impedance of the motor windings is sufficient to prevent overheating owing to failure to start, the motor is permitted to be protected as specified in Section 430.32(D)(2)(a) below for manually started motors if the motor is part of an approved assembly in which the motor will limit itself so that it will not be dangerously overheated.

(C) Selection of Overload Device states that where the sensing element or setting or sizing of the overload device is not sufficient to start the motor or to carry the load, higher-size sensing elements or incremental settings or sizing are permitted to be used, provided that the trip current of the overload device does not exceed the following percentages of motor nameplate full-load current rating:

- Motors with a marked service factor of 1.15 or greater—140 percent
- Motors with a marked temperature rise of 40°C or less—140 percent
- All other motors—130 percent

If not shunted during the starting period of the motor as provided in Section 430.35 below, the overload device is to have sufficient time delay to permit the motor to start and accelerate its load. This section permits the designer to substitute a higher-rated overload device, within limits, if the motor will not start or run when the overload device is sized based on full-load current. This option applies to the overload device, not to branch-circuit short-circuit and ground-fault protection.

(D) One Horsepower or Less, Nonautomatically Started, states that such motors fall into two categories:

(1) Permanently installed, in which case overload protection is to be in accordance with Section 430.32(B) above

(2) Not permanently installed

    a. Within Sight from Controller: Overload protection is to be permitted to be furnished by the branch-circuit short-circuit and ground-fault protective device; such device, however, is not to be larger than specified in Part IV of Article 430. An exception provides that any such motor is permitted on a nominal 120-volt branch circuit protected at not over 20 amperes.

b. Not Within Sight from Controller: Overload protection is to be in accordance with Section 430.32(B) above.

To summarize, a nonautomatically started continuous-duty motor of 1 horsepower or less that is within sight of the controller does not require separate overload protection. The standard branch-circuit short-circuit and ground-fault protection will do, but it may not be oversized, as permitted in Part IV.

Section 430.33, Intermittent and Similar Duty, provides that a motor used for a condition of service that is inherently short-time, intermittent, periodic, or varying duty is permitted to be protected against overloads by the branch-circuit short-circuit and ground-fault protective device, provided that the protective-device rating or setting does not exceed that specified in Table 430.52, Maximum Rating or Setting of Motor Branch-Circuit Short-Circuit and Ground-Fault Protective Devices. It is further specified that any motor application is to be considered continuous duty unless the nature of the apparatus it drives is such that the motor cannot operate continuously with load under any condition of use.

Section 430 provides a further way of dealing with high starting current in a motor circuit.

(A) Nonautomatically Started states that for a nonautomatically started motor, the overload protection is permitted to be shunted or cut out of the circuit during the starting period of the motor if the device by which the overload protection is shunted or cut out cannot be left in the starting position and if fuses or inverse-time circuit breakers rated or set at not over 400 percent of the full-load current of the motor are located in the circuit so as to be operative during the starting period of the motor.

(B) Automatically Started states that the motor overload protection may not be shunted or cut out during the starting period if the motor is automatically started. An exception allows the motor overload protection to be shunted or cut out during the starting period of an automatically started motor where the following apply:

(1) The motor starting period exceeds the time delay of available motor overload protective devices and

(2) Listed means are provided to perform the following:

a. Sense motor rotation and automatically prevent the shunting or cutout in the event that the motor fails to start.

b. Limit the time of overload shunting or cutout to less than the locked-rotor time rating of the protected motor.

c. Provide for shutdown and manual restart if motor running condition is not reached.

Section 430.36, Fuses—In Which Conductor, states that where fuses are used for motor overload protection, a fuse is to be inserted in each ungrounded conductor and also in the grounded conductor if the supply system is three-wire, three-phase ac with one conductor grounded.

Section 430.37, Devices Other Than Fuses—In Which Conductor, states that where devices other than fuses are used for motor overload protection, Table 430.37, Overload Units, governs the minimum allowable number and location of overload units such as trip coils and relays.

Section 430.38, Number of Conductors Opened by Overload Device, provides that motor overload devices, other than fuses or thermal protectors, must simultaneously open a sufficient number of ungrounded conductors to interrupt current flow to the motor.

(A) Motor Controller as Overload Protection states that a motor controller is also permitted to serve as an overload device if the number of overload units complies with Table 430.37 and if these units are operative in both the starting and running position in the case of a dc motor and in the running position in the case of an ac motor.

Section 430.40, Overload Relays, states that overload relays and other devices for motor overload protection that are not capable of opening short circuits or ground faults must be protected by fuses or circuit breakers with ratings or settings in accordance with 430.52 or by a motor short-circuit protector in accordance with 430.52. An exception states that where approved for group installation and marked to indicate the maximum size of fuse or inverse-time circuit breaker by which they must be protected, the overload devices are to be protected in accordance with this marking.

Section 430.42, Motors on General-Purpose Branch Circuits, provides that overload protection for motors on general-purpose branch circuits is to be provided as specified in (A) through (D) below:

(A) Not over One Horsepower states that one or more motors without individual overload protection are permitted to be connected to a general-purpose branch circuit only where the installation complies with the limiting conditions specified in 430.32(B) and (D) above and 430.53(A)(1) and (2) below.

(B) Over One Horsepower states that motors of ratings larger than specified in 430.53(A) below are permitted to be connected to general-purpose branch circuits only where each motor is protected by overload protection selected to protect the motor as specified in 430.32. Both the controller and the motor overload device are to be approved for group installation, with the short-circuit and ground-fault protective device selected in accordance with 430.53.

(C) Cord-and-Plug-Connected states that where a motor is connected to a branch circuit by means of an attachment plug and a receptacle or a cord connector and individual overload protection is omitted as provided in 430.42(A), the rating of the attachment plug and receptacle or cord connector may not exceed 15 amperes at 125 or 250 volts. Where individual overload protection is required as provided in 430.42(B) for a motor or motor-operated appliance that is attached to the branch circuit through an attachment plug and a receptacle or a cord connector, the overload device is to be an integral part of the motor or the appliance. The rating of the attachment plug and receptacle or the cord connector is to determine the rating of the circuit to which the motor may be connected.

(D) Time Delay states that the branch-circuit short-circuit and ground-fault protective device protecting a circuit to which a motor or motor-operated appliance is connected is to have sufficient time delay to permit the motor to start and accelerate its load.

Section 430.43, Automatic Restarting, provides that a motor overload device that can restart a motor automatically after overload tripping may not be installed unless approved for use with the motor it protects. A motor overload device that can restart a motor automatically after overload tripping may not be installed if automatic restarting of the motor can result in injury to persons.

Section 430.44, Orderly Shutdown, states that if immediate automatic shutdown of a motor by a motor overload protective device(s) would introduce additional or increased hazard(s) to a person(s) and continued motor operation is necessary for safe shutdown of equipment or process, a motor overload-sensing device(s) complying with the provisions of Part III of Article 430 is permitted to be connected to a supervised alarm instead of causing immediate interruption of the motor circuit so that corrective action or an orderly shutdown can be initiated.

Part III shows us how to select and size overload protection for a motor. It is important to understand the differences from and similarities to conventional overcurrent protection as applied to motor circuits and branch circuits in general.

Part IV, Motor Branch-Circuit Short-Circuit and Ground-Fault Protection, covers selection and sizing of branch-circuit overcurrent protection for motor circuits. Because motors have comparatively high starting current, it is necessary in many cases to overfuse the branch-circuit conductors. Otherwise, the overcurrent protection would deenergize the circuit before the motor could attain running speed. It is possible to set the branch-circuit short-circuit and ground-fault protection at a high level because motor overload protection is also present, even if it is at the "wrong" end of the supply wiring.

Section 430.51, General, contains an overview. It states that Part IV specifies devices intended to protect the motor branch-circuit conductors, the motor control apparatus, and the motor(s) against overcurrent owing to short circuits or ground faults. These rules add to or amend the provisions of Article 240.

The provisions in Part IV do not apply to motor circuits over 600 volts, nominal, which are covered in Part XI.

Section 430.52, Rating or Setting for Individual Motor Circuit, includes a table for sizing the overcurrent device.

(A) General states that the motor branch-circuit short-circuit and ground-fault protective device must comply with (B) and either (C) or (D) below as applicable.

(B) All Motors states that the motor branch-circuit short-circuit and ground-fault protective device must be capable of carrying the starting current of the motor.

(C) Rating or Setting provides the following:

(1) In accordance with Table 430.52, a protective device that has a rating or setting not exceeding the value calculated according to the values given in Table 430.52 is to be used. The table lists seven types of motors and gives percentage of full-load current for various types of overcurrent devices. These range from 150 percent for a wound-rotor motor fed through an inverse-time breaker to 1,100 percent for a Design B energy-efficient motor fed through an instantaneous-trip breaker. There are two exceptions:

Number 1: Where the values for branch-circuit short-circuit and ground-fault protective devices determined by Table 430.52 do not correspond to the standard sizes or ratings of fuses, nonadjustable circuit breakers, thermal protective devices, or possible settings of adjustable circuit breakers, a higher size, rating, or possible setting that does not exceed the next-higher standard ampere rating is permitted.

Number 2: Where the rating specified in Table 430.52 or the rating modified by exception number 1 is not sufficient for the starting current of the motor:

a. The rating of a non-time-delay fuse not exceeding 600 amperes or a time-delay Class CC fuse is permitted to be increased but must in no case exceed 400 percent of the full-load current.

b. The rating of a time-delay (dual-element) fuse is permitted to be increased but must in no case exceed 225 percent of the full-load current.

c. The rating of an inverse-time circuit breaker is permitted to be increased but must in no case exceed 400 percent for full-load currents of 100 amperes or less or 300 percent for full-load currents greater than 100 amperes.

d. The rating of a fuse 601–6,000 ampere classification is permitted to be increased but must in no case exceed 300 percent of the full-load current.

(2) Overload Relay Table states that where maximum branch-circuit short-circuit and ground-fault protective device ratings are shown in the manufacturer's overload relay table for use with a motor controller or are otherwise marked on the equipment, they are

not to be exceeded even if higher values are allowed as shown above.

(3) Instantaneous Trip Circuit Breaker states that such a device may be used only if adjustable and if part of a listed combination motor controller having coordinated motor overload and short-circuit and ground-fault protection in each conductor, and the setting is adjusted to no more than the value specified in Table 430.52. There are two exceptions:

Number 1: Where the setting specified in Table 430.52 is not sufficient for the starting current of the motor, the setting of an instantaneous-trip circuit breaker is permitted to be increased but must in no case exceed 1,300 percent of the motor full-load current for other than Design B energy-efficient motors and no more than 1,700 percent of full-load motor current for Design B energy-efficient motors. Trip settings above 800 percent for other than Design B energy-efficient motors and above 1,100 percent for Design B energy-efficient motors are permitted where the need has been demonstrated by engineering evaluation. In such cases, it is not necessary to first apply an instantaneous-trip circuit breaker at 800 or 1,100 percent.

Number 2: Where the motor full-load current is 8 amperes or less, the setting of the instantaneous-trip circuit breaker with a continuous current rating of 15 amperes or less in a listed combination motor controller that provides coordinated motor branch-circuit overload and short-circuit and ground-fault protection is permitted to be increased to the value marked on the controller.

(4) Multispeed Motor states that for a multispeed motor, a single short-circuit and ground-fault protective device is permitted to be used and sized according to the full-load current of the highest current winding, where all the following conditions are met:

    a. Each winding is equipped with individual overload protection sized according to its full-load current.

    b. The branch-circuit conductors supplying each winding are sized according to the full-load current of the highest full-load current winding.

    c. The controller for each winding has a horsepower rating not less than that required for the winding having the highest horsepower rating.

(5) Power Electronic Devices states that suitable fuses are permitted in lieu of devices listed in Table 430.52 for power electronic devices in a solid-state motor controller system, provided that the marking for replacement fuses is provided adjacent to the fuses.

(6) Self-Protected Combination Controller states that a listed self-protected combination controller is permitted in lieu of the devices specified in Table 430.52. Adjustable instantaneous-trip settings are not to exceed 1,300 percent of full-load motor current for other than Design B energy-efficient motors and not more than 1,700 percent of full-load motor current for Design B energy-efficient motors.

(7) Motor Short-Circuit Protector states that such a protector is permitted in lieu of devices listed in Table 430.52 if the motor short-circuit protector is part of a listed combination motor controller having coordinated motor overload protection and short-circuit and ground-fault protection in each conductor and that it will open the circuit at currents exceeding 1,300 percent of motor full-load current for other than Design B energy-efficient motors and 1,700 percent of motor full-load current for Design B energy-efficient motors.

(D) Torque Motors states that branch circuits must be protected at the motor nameplate current rating in accordance with 240.4(B), Overcurrent Devices Rated 800 Amperes or Less, which discusses going to the next-higher standard overcurrent device rating.

Section 430.53, Several Motors or Loads on One Branch Circuit, states that two or more motors or one or more motors and other loads are permitted to be connected to the same branch circuit under conditions specified in (A) through (D) below:

(A) Not Over One Horsepower states that several motors, each not exceeding 1 horsepower in rating, are permitted on a nominal 120-volt branch circuit protected at not over 20 amperes or a branch circuit of 600 volts, nominal, or less protected at not over 15 amperes if all of the following conditions are met:

(1) The full load-rating of each motor does not exceed 6 amperes.

(2) The rating of the branch-circuit short-circuit and ground-fault protective device marked on any of the controllers is not exceeded.

(3) Individual overload protection conforms to 430.32, Continuous Duty Motors.

(B) If Smallest Rated Motor Protected states that if the branch-circuit short-circuit and ground-fault protective device is selected not to exceed that allowed by 430.52 for the smallest rated motor, two or more motors or one or more motors and other load(s), with each motor having individual overload protection, are permitted to be connected to a branch circuit where it can be determined that the branch-circuit short-circuit and ground-fault protective device will not open under the most severe normal conditions of service that might be encountered.

(C) Other Group Installations provides specifications for group installations of any rating, that is, over 1 horsepower. A large number of conditions accompany the permitted installation.

(D) Single Motor Taps pertains to group installations. The conductors of any tap supplying a single motor are not required to have an individual branch-circuit short-circuit and ground-fault protective device, provided that the conductors comply with one of the following:

(1) No conductor to the motor is to have an ampacity less than that of the branch-circuit conductors.

(2) No conductor to the motor is to have an ampacity less than one-third that of the branch-circuit conductors, the conductors to the motor overload device being not more than 25 feet long and being protected from physical damage by being enclosed in an approved raceway or by use of other approved means.

(3) Conductors from the branch-circuit short-circuit and ground-fault protective device to a listed manual motor controller additionally marked "Suitable for Tap Conductor Protection in Group Installations" or to a branch-circuit protective device are permitted to have an ampacity not less than one-tenth the rating or setting of the branch-circuit short-circuit and ground-fault protective device. The conductors from the controller to the motor are to have an ampacity in accordance with Section 430.22. The conductors from the branch-circuit short-circuit and ground-fault protective device to the controller must (1) be suitably protected from physical damage and enclosed either by an enclosed controller or by a raceway and be not more than 10 feet long or (2) have an ampacity not less than that of the branch-circuit conductors.

Section 430.54, Multimotor and Combination-Load Equipment, states that the rating of the branch-circuit short-circuit and ground-fault protective device for multimotor and combination-load equipment is not to exceed the rating marked on the equipment. This section pertains to a single piece of equipment that contains more than one motor or one or more motors plus an additional nonmotor load(s).

Section 430.55, Combined Overcurrent Protection, provides that motor branch-circuit short-circuit and ground-fault protection and motor overload protection are permitted to be combined in a single protective device where the rating or setting of the device provides the overload protection specified in 430.32. This may consist of an inverse-time circuit breaker or a dual-element fuse.

Section 430.56, Branch-Circuit Protective Devices—In Which Conductor, states that branch-circuit protective devices must comply with 240.15, Ungrounded Conductors, which requires that an overcurrent device, fuse, or circuit breaker must be connected in series with each ungrounded conductor. Additional provisions address the issue of handle ties for various systems. Section 430.56 is saying that motors are no different in this respect.

Section 430.57, Size of Fuseholder, states that where fuses are used for motor branch-circuit short-circuit and ground-fault protection, the fuseholders are not to be smaller than required to accommodate the fuses specified in Table 430.52. An exception provides that fuseholders sized to fit the fuses that are used are permitted for fuses having time delay appropriate for the starting characteristics of the motor.

Section 430.58, Rating of Circuit Breaker, provides that a circuit breaker for motor branch-circuit short-circuit and ground-fault protection must have a current rating in accordance with 430.52 and 430.110. The first of these sections refers to the associated table, which gives maximum rating or setting of motor branch-circuit short-circuit and ground-fault protective devices, including instantaneous-trip and inverse-time breakers. Notice that the values for these two are not the same. The second of these sections, as we shall see, concerns disconnecting means.

Part V, Motor Feeder Short-Circuit and Ground-Fault Protection, is fairly brief because feeder overcurrent protection for the most part is based on branch-circuit values.

Section 430.61, General, says that Part V specifies protective devices intended to protect feeder conductors supplying motors against overcurrents owing to short circuits and grounds.

Section 430.62, Rating or Setting—Motor Load, gives the method for calculating the size of the overcurrent device required at the upstream end of a feeder supplying one or more motor loads.

(A) Specific Load provides that a feeder supplying a specific fixed motor load is to have a protective device that has a rating or setting not greater than the largest rating or setting of the branch-circuit short-circuit and ground-fault protective device for any motor supplied by the feeder plus the sum of the full-load currents of the other motors of the group. It is further stipulated that where the same rating or setting of the branch-circuit short-circuit and ground-fault protective device is used on two or more of the branch circuits supplied by the feeder, one of the protective devices is to be considered the largest for the preceding calculations.

Remember that the rating or setting of the branch-circuit short-circuit and ground-fault protective device is determined not by the full-load current marked on the nameplate in most cases but by the values given in the tables at the end of Article 430. To find the rating or setting of the overcurrent device at the upstream end of a feeder supplying a motor group installation, you add that rating or setting to the sum of the full-load currents of the remaining motors.

(B) Other Installations states that where feeder conductors have an ampacity greater than required, the rating or setting of the feeder overcurrent protective device is permitted to be based on the ampacity of the feeder conductors. This section recognizes the occasional practice of oversizing the feeder conductors to mitigate voltage drop or for any other reason and permits the use in that case of a correspondingly higher-rated overcurrent device at the upstream end of the feeder.

Section 430.63, Rating or Setting—Motor Load and Other Loads, provides provides mandates regarding short-circuit and ground-fault protection for situations in which the feeder supplies a motor load plus other loads:

(A) For a single motor, the rating permitted by Section 430.52 and associated Table 430.52, Maximum Rating or Setting of Motor Branch-Circuit Short-Circuit and Ground-Fault Protective Devices.

(B) For a single hermetic refrigerant motor compressor, the rating permitted by Section 440.22. Hermetic refrigerant motor compressors are covered in Article 440, Air-Conditioning and Refrigerating Equipment. Provisions in that article are related to those in Article 430, but sizing of branch circuits is based on branch-circuit selection current, which is always equal to or greater than the marked rated-load current.

(C) For two or more motors, the rating permitted by Section 430.62 above. An exception states that where the feeder overcurrent device provides the overcurrent protection for a motor control center, the provisions of 430.94 apply.

Part VI, Motor Control Circuits, states that such circuits are circuits that turn off and on a motor and otherwise modify its operation. They do not carry the full current that powers a motor and are, accordingly, characterized by having much smaller conductors. Incorrect wiring of motor control circuits can introduce hazards into the installation as a whole. In an industrial setting, the safety of the machine operators largely depends on the design and installation of these circuits.

Section 430.71, General, states that Part VI contains modifications of the general requirements and applies to the particular conditions of motor control circuits.

Section 430.72, Overcurrent Protection, provides that overcurrent protection for these small conductors is a central concern of Part VI.

(A) General outlines the two ways in which a motor control circuit may receive its power—either from the load side of the motor short-circuit and ground-fault protective device or from a separate branch circuit originating in the service entrance panel or some other load center. In all cases, it has to have overcurrent protection. This section provides that a motor control circuit tapped from the load side of a motor branch-circuit short-circuit and ground-fault protective device(s) and functioning to control the motor(s) connected to that branch circuit is to be protected against overcurrent in accordance with 430.72. Such a tapped control circuit is not considered a branch circuit and is permitted to be protected by either a supplementary or branch-circuit overcurrent protective device(s). A motor control circuit other than such a tapped control circuit is to be protected against overcurrent in accordance with 725.43, Class 1 Circuit Overcurrent Protection, or the notes to Tables 11(A) and 11(B) in Chapter 9. These tables provide Class 2 and Class 3 power-source limitation for ac and dc, respectively. Note that the preceding is applicable to overcurrent protection for the controller, not for the conductors.

(B) Conductor Protection states that the overcurrent protection for motor control circuit conductors is to be provided as specified in (1) or (2) below:

(1) Separate Overcurrent Protection states that where the motor branch-circuit short-circuit and ground-fault protective device does not provide protection in accordance with (2) below, separate overcurrent protection is to be provided. The overcurrent protection must not exceed the values specified in Column A of Table 430.72(B), Maximum Rating of Overcurrent Protective Device in Amperes.

(2) Branch-Circuit Overcurrent Protective Device states that conductors are permitted to be protected by the motor branch-circuit short-circuit and ground-fault protective device and require only short-circuit and ground-fault protection. Where the conductors do not extend beyond the motor control equipment enclosure, the rating of the protective device(s) is not to exceed the value specified in Column B of Table 430.72(B). Where the conductors extend beyond the motor control equipment enclosure, the rating of the protective device(s) is not to exceed the value specified in Column C of Table 430.72(B).

These permitted values are higher than those we are used to seeing because it is short-circuit and ground-fault protection we are furnishing, not overcurrent protection in general. Note the blank entries for aluminum and copper-clad aluminum, where copper conductors are required.

(C) Control Circuit Transformer states that where provided, such a transformer must be protected in accordance with one of the following:

(1) Compliance with Article 725: Where the transformer supplies a Class 1 power-limited circuit, a Class 2 or Class 3 remote-control circuit complying with Article 725 is necessary.

(2) Compliance with Article 450: Protection is permitted to be provided in accordance with Section 450.3, Overcurrent Protection (for Transformers).

(3) Less Than 50 Volt-Amps: Control-circuit transformers rated less than 50 volt-amperes and that are an integral part of the motor controller and located within the motor controller enclosure are permitted to be protected by primary overcurrent devices, impedance-limiting means, or other inherent protective means.

(4) Primary Less Than 2 Amperes: Where the control-circuit transformer rated primary current is less than 2 amperes, an overcurrent device rated or set at not more than 500 percent of the rated primary current is permitted in the primary circuit.

(5) Other means for protection are permitted.

Section 430.73, Protection of Conductors from Physical Damage, states that where damage to a motor control circuit would constitute a hazard, all conductors of such a remote motor control circuit that are outside the control device itself are to be installed in a raceway or be otherwise protected from physical damage. Usually in a setting where motor controllers are installed, all wiring is in raceways anyway.

Section 430.74, Electrical Arrangement of Control Circuits, provides that where one conductor of the motor control circuit is grounded, the motor control circuit is to be arranged so that a ground fault in the control circuit remote from the motor controller will (1) not start the motor and (2) not bypass manually operated shutdown devices or automatic safety shutdown devices.

In a 120-volt, single-phase control circuit derived from the 208-volt, three-phase wye system supplying the motor, if one side of the control circuit is the grounded neutral and the start button is connected to the grounded neutral, the motor can start inadvertently if there is a ground fault on the coil side of the start button. This is a hazard, and it is prohibited by Section 430.74. The start button should be placed in the ungrounded side of the control circuit.

Section 430.75, Disconnection, contains two requirements regarding source of supply:

(A) General states that motor control circuits are to be arranged so that they will be disconnected from all sources of supply when the disconnecting means is in the open position. The disconnecting means is permitted to consist of two or more separate devices, one of which disconnects the motor and the controller from the source(s) of power supply for the motor, and the other(s) disconnects the motor control circuit(s) from its power supply. Where separate devices are used, they are to be located immediately adjacent to each other. Exceptions apply where more than 12 motor control circuit conductors are required to be disconnected and where the opening of the disconnecting means is capable of resulting in potentially unsafe conditions.

(B) Control Transformer in Controller Enclosure states that where a transformer or other device is used to obtain a reduced voltage for the motor control circuit and is located in the controller enclosure, it is to be connected to the load side of the disconnecting means for the motor control circuit. Trunk slammers often create noncompliant hookups that appear functional but leave some part of the circuit energized and thus a hazard to maintenance workers.

Part VII, Motor Controllers, covers selection, design, and sizing for motor controllers.

Section 430.81, General, states that Part VII is intended to require suitable controllers for all motors. For some motors, this is nothing more elaborate than the existing branch-circuit disconnect or attachment plug and receptacle.

(A) Stationary Motor of 1/8 Horsepower or Less states that for one of these units that is normally left running and is constructed so that it cannot be damaged by overload or failure to start, such as clock motors and the like, the branch-circuit disconnecting means is permitted to serve as the controller.

(B) Portable Motor of 1/3 Horsepower or Less states that for one of these units, the controller is permitted to be an attachment plug and receptacle or cord connector.

Section 430.82, Controller Design, provides the following:

(A) Starting and Stopping provides that each controller must be capable of starting and stopping the motor it controls and must be capable of interrupting the locked-rotor current of the motor.

(B) Autotransformer provides that an autotransformer starter must provide an "off" position, a running position, and at least one starting position. It is to be designed so that it cannot rest in the starting position or in any position that will render the overload device in the circuit inoperative.

(C) Rheostats states that rheostats are to comply with the following:

(1) Motor-starting rheostats are to be designed so that the contact arm cannot be left on intermediate segments. The point or plate on which the arm rests when in the starting position must have no electrical connection with the resistor.

(2) Motor-starting rheostats for dc motors operated from a constant voltage supply are to be equipped with automatic devices that will interrupt the supply before the speed of the motor has fallen to less than one-third its normal rate.

Section 430.83, Ratings, states that the controller is to have a rating as specified in (A) unless otherwise permitted in (B) or (C) or as specified in (D) below under the conditions specified:

(A) General contains provisions for motor controller ratings under specified conditions:

(1) Horsepower Ratings states that controllers, other than inverse-time breakers and molded case switches, are to have horsepower ratings at the application voltage not lower than the horsepower rating of the motor.

(2) Circuit Breaker states that a branch-circuit inverse-time circuit breaker rated in amperes is permitted as a controller for all motors. Where this circuit breaker is also used for overload protection, it must conform to the appropriate provisions of Article 430 governing overload protection.

(3) Molded Case Switch states that such a switch rated in amperes is permitted as a controller for all motors.

(B) Small Motors states that devices specified in 430.81(A) and (B) above are permitted as a controller.

(C) Stationary Motors of Two Horsepower or Less states that the controller is permitted to be either of the following:

(1) A general-use switch having an ampere rating not less than twice the full-load current rating of the motor

(2) On ac circuits, a general-use snap switch suitable for use only on ac (not a general-use ac-dc snap switch) where the motor full-load current rating is not more than 80 percent of the ampere rating of the switch

(D) Torque Motors states that the controller is to have a continuous-duty, full-load current rating not less than the nameplate current rating of the motor. For a motor controller rated in horsepower but not marked with the foregoing current rating, the equivalent current rating is to be determined from the horsepower rating by using Tables 430.247 through 430.250.

(E) Voltage Rating lays out the method for interpreting two modes of voltage nomenclature. A controller with a straight voltage rating, for example, 240 or 480 volts, is permitted to be applied in a circuit in which the nominal voltage between any two conductors does not exceed the controller's voltage rating. A controller with a slash rating, for example, 120/240 volts or 480Y/277 volts, may be applied only in a solidly grounded circuit in which the nominal voltage to ground from any conductor does not exceed the lower of the two values of the controller's voltage rating, and the nominal voltage between any two conductors does not exceed the higher value of the conductor's voltage rating.

Section 430.84, Need Not Open All Conductors, states that the controller is not required to open all conductors to the motor. An exception states that when the controller also serves as a disconnecting means, it must open all ungrounded conductors to the motor. This is somewhat counterintuitive because we are used to switching all ungrounded legs of any load. But it makes sense when you realize that during maintenance, the worker should know to shut off the disconnect rather than rely on the controller to deenergize the equipment. For a three-phase motor, it is necessary for the controller to interrupt two phases; otherwise, the motor will "single phase," possibly causing damage to the motor.

Section 430.88, Adjustable-Speed Motors, provides that such motors that are controlled by means of field regulation are to be equipped and connected so that they cannot be started under a weakened field. An exception states that starting under a weakened field is permitted where the motor is designed for such starting. This requirement is applicable to dc motors.

Section 430.89, Speed Limitation, provides that machines of the following types must be provided with speed-limiting devices:

- Separately excited dc motors
- Series motors
- Motor generators and converters that can be driven at excessive speed from the dc end, as by a reversal of current or decrease in load

An exception states that separate speed-limiting devices or means are not required under either of the following conditions:

- Where the inherent characteristics of the machines, system, or load and the mechanical connection thereto are such as to safely limit the speed

- Where the machine is always under the manual control of a qualified operator

Section 430.90, Combination Fuseholder and Switch as Controller, states that the rating of a combination fuseholder and switch used as a motor controller is to be such that the fuseholder will accommodate the size of the fuse specified in Part III of Article 430 for motor overload protection. An exception provides that where fuses having time delay appropriate for the starting characteristics of the motor are used, fuseholders of smaller size are permitted.

Part VIII, Motor Control Centers, is a fairly short part. Many requirements, such as working space and dedicated space, are found in Chapter 1.

Motor control centers are made up of a large typically free-standing enclosure containing numerous motor starters, controls, and disconnects. A motor control center is permitted to be used as service equipment if so designed and listed, in which case it must have a single main disconnecting means and main bonding jumper.

Section 430.94, Overcurrent Protection, states that overcurrent protection is required for motor control centers in accordance with Article 240. The rating or setting of the overcurrent protective device must not exceed the rating of the common power bus. The protection may be provided either ahead of or within the motor control center.

Section 430.95, Service Equipment, states that where used as service equipment, each motor control center is to be provided with a main disconnecting means to disconnect all ungrounded service conductors. Where a grounded conductor is provided, the motor control center is to be provided with a main bonding jumper sized in accordance with 250.28(D) within one of the sections for connecting the grounded conductor, on its supply side, to the motor control center equipment-ground bus. An exception states that high-impedance grounded neutral systems are permitted to be connected as provided in 250.36.

Section 430.96, Grounding, provides that multisection motor control centers are to be connected together with an equipment-grounding conductor or an equivalent equipment-grounding bus sized in accordance with Table 250.122. Equipment-grounding conductors are to be connected to this equipment-grounding bus or to a grounding termination point provided in a single-section motor control center.

Section 430.97, Busbars and Conductors, contains the following provisions:

(A) Support and Arrangement states that busbars must be protected from physical damage and be held firmly in place. Other than for required interconnections and control wiring, only conductors that are intended for termination in a vertical section are to be located in that section. An exception states that conductors are permitted to travel horizontally through vertical sections where such conductors are isolated from the busbars by a barrier.

(B) Phase Arrangement states that on three-phase horizontal common power and vertical buses, phase arrangement is to be A, B, C from front to back, top to bottom, or left to right, as viewed from the front of the motor control center. The B phase is to be that phase having the higher voltage to ground on three-phase, four-wire, delta-connected systems. Other arrangements are permitted for additions to existing installations and must be marked. An exception states that rear-mounted units connected to a vertical bus that is common to front-mounted units are permitted to have a C, B, A phase arrangement where identified properly.

(C) Minimum Wire-Bending Space states that such a space at the motor control terminals and minimum gutter space is to be as required in Article 312.

(D) Spacings states that spacings between motor control center bus terminals and other bare metal parts is not to be less than specified in Table 430.97.

(E) Barriers states that barriers are to be placed in all service-entrance motor control centers to isolate service busbars and terminals from the remainder of the motor control center.

Section 430.98, Marking, provides the following:

(A) Motor Control Centers states that such centers are to be marked according to Section 110.21, and such marking must be plainly visible after installation. Marking also must include common power bus current rating and motor control center short-circuit rating.

(B) Motor Control Units states that such units in a motor control center are to be marked to comply with 430.8, Marking on Controllers.

Part IX, Disconnecting Means, provides detailed mandates for motor disconnects. Note that we are no longer focusing on motor control centers but rather on the total motor installation.

Section 430.101, General, states that Part IX is intended to require disconnecting means capable of disconnecting motors and controllers from the circuit as required for maintenance on the motor and/or the driven machinery.

Section 430.102, Location, provides the following:

(A) Controller states that an individual disconnecting means is to be provided for each controller and must disconnect the controller. The disconnecting means must be within sight from the controller location. There are three exceptions:

Number 1: For motor circuits over 600 volts, nominal, a controller disconnecting means capable of being locked in the open position is permitted out of sight of the controller, provided that the controller is marked with a warning label giving the location of the disconnecting means.

Number 2: A single disconnecting means is permitted for a group of coordinated controllers that drive several parts of a single machine. The disconnecting means must be located in sight from the controllers, and both the disconnecting means and the controllers must be located in sight from the machine.

Number 3: This involves valve actuator motors and states that a disconnecting means does not have to be in sight, with conditions.

(B) Motor states that a disconnecting means must be provided for a motor in accordance with (1) or (2) below:

(1) Separate Motor Disconnect states that such a disconnect must be in sight from the motor and the driven machinery.

(2) Controller Disconnect states that such a disconnect is permitted to serve as the disconnecting means for the motor if it is in sight from the motor and the driven machinery.

An exception applies to both (1) and (2) above. It states that the disconnecting means for the motor is not required under either condition (a) or condition (b) below, provided that the controller disconnecting means is individually capable of being locked in the open position. The provision for locking or for adding a lock to the controller disconnecting means provides that such a device is to be installed on or at the switch or circuit breaker used as the disconnecting means and must remain in place with or without the lock installed.

a. Where such a location of the disconnecting means for the motor is impracticable or introduces additional or increased hazards to persons or property

b. In industrial installations, with written safety procedures, where conditions of maintenance and supervision ensure that only qualified persons service the equipment

Section 430.103, Operation, provides that the disconnecting means must open all ungrounded supply conductors and is to be designed so that no pole can be operated independently. The disconnecting means is permitted in the same enclosure with the controller. The disconnecting means must be designed so that it cannot be closed automatically. Notice that the disconnect must open all ungrounded conductors simultaneously, whereas the controller must open only enough conductors to stop the motor.

Section 430.104, To Be Indicating, requires that the disconnecting means must plainly indicate whether it is in the open (off) or closed (on) position.

Section 430.105, Grounded Conductors, states that one pole of the disconnecting means is permitted to disconnect a permanently grounded conductor, provided that the disconnecting means is designed so that the pole in the grounded conductor cannot be opened without simultaneously disconnecting all conductors of the circuit.

Section 430.107, Readily Accessible, provides that at least one of the disconnecting means is to be readily accessible.

Section 430.108, Every Disconnecting Means, states that every disconnecting means in the motor circuit between the point of attachment to the feeder or branch circuit and the point of connection to the motor must comply with the requirements of Sections 430.109 and 430.110 below.

Section 430.109, Type, states that the disconnecting means is to be a type specified in (A) through (G) below:

(A) General provides the following:

(1) A motor circuit switch must be listed and rated in horsepower.

(2) A molded-case circuit breaker must be listed.

(3) A molded-case switch must be listed. A molded-case switch has the same form as a molded-case circuit breaker, but it does not trip to provide overcurrent protection. It just turns on or off like any switch.

(4) An instantaneous-trip circuit breaker must be part of a listed combination motor controller.

(5) A self-protected combination controller must be listed.

(6) Listed manual motor controllers additionally marked "Suitable as Motor Disconnect" are permitted where installed between the final motor branch-circuit short-circuit protective device and the motor. Listed manual motor controllers additionally marked "Suitable as Motor Disconnect" are permitted as disconnecting means on the line side of the fuses permitted in Section 430.52(C)(5). In this case, they are considered supplementary fuses, and suitable branch-circuit short-circuit and ground-fault protective devices are to be installed on the line side of the manual motor controller additionally marked "Suitable as Motor Disconnect."

(7) System isolation equipment is to be listed for disconnection purposes. It is to be installed on the load side of the overcurrent protection and its disconnecting means. The disconnecting means is to be one of the types permitted by (1) through (3) above.

(B) Stationary Motors of 1/8 Horsepower or Less states that the branch-circuit overcurrent device is permitted to serve as the disconnecting means.

(C) Stationary Motors of 2 Horsepower or Less states that if also 300 volts or less, the disconnecting means is permitted to be one of the devices specified in (1), (2), or (3) below:

(1) A general-use switch having an ampere rating not less than twice the full-load current rating of the motor

(2) On ac circuits, a general-use snap switch suitable for use only on ac (not a general-use ac-dc snap switch) where the motor full-load current rating is not more than 80 percent of the ampere rating of the switch

(3) A listed manual motor controller having a horsepower rating not less than the rating of the motor and marked "Suitable as Motor Disconnect"

(D) Autotransformer-Type Controlled Motors states that for motors of over 2 horsepower up to and including 100 horsepower, the separate disconnecting means required for a motor with an auto-transformer-type controller is permitted to be a general-use switch where all the following provisions are met:

(1) The motor drives a generator that is provided with overload protection.

(2) The controller is capable of interrupting the locked-rotor current of the motors, is provided with a no voltage release, and is provided with running overload protection not exceeding 125 percent of the motor full-load current rating.

(3) Separate fuses or an inverse-time circuit breaker rated or set at no more than 150 percent of the motor full-load current is provided in the motor branch circuit.

(E) Isolating Switches states that for stationary motors rated at more than 40 horsepower dc or 100 horsepower ac, the disconnecting means is permitted to be a general-use or isolating switch where plainly marked "Do Not Operate Under Load."

(F) Cord-and-Plug-Connected Motors states that a horsepower-rated attachment plug and receptacle, a flanged surface inlet and cord connector, or an attachment plug-and-cord connector having ratings no less than the motor ratings is permitted to serve as the disconnecting means. Horsepower-rated attachment plugs, flanged surface inlets, receptacles, or cord connectors are not required for cord-and-plug-connected appliances, room air conditioners, or portable motors rated 1/3 horsepower or less (Figure 4-5).

(G) Torque Motors states that the disconnecting means for such motors is permitted to be a general-use switch.

Section 430.110, Ampere Rating and Interrupting Capacity, contains information on rating various types of motors not covered by the tables at the end of Article 430.

(A) General states that the disconnecting means for motor circuits rated 600 volts, nominal, or less must have an ampere rating not less than 115 percent of the full-load current rating of the motor. This is a good figure to memorize because it frequently appears in licensing exams and has frequent application on the job. Why is it 115 percent rather than the ubiquitous 125 percent? Because when the contacts are closed, very little heat is generated, and the make/break action is designed to be very brief.

(B) For Torque Motors states that the disconnecting means for such motors must have an ampere rating of at least 115 percent of the nameplate current.

(C) For Combination Loads states that where two or more motors are used together or where one or more motors are used in combination with other loads, such as resistance heaters, and where the combined load may be simultaneous on a single disconnecting means, the

FIGURE 4-5   Cord-and-plug-connected stationary tool. If it is 1/3 horsepower or less, a horsepower-rated attachment plug is not required.

ampere and horsepower ratings of the combined load are to be determined as follows:

(1) Horsepower Rating states that the rating of the disconnecting means is to be determined from the sum of all currents, including

resistance loads, at the full-load condition and also at the locked-rotor condition. The combined full-load current and the combined locked-rotor current so obtained are considered as a single motor for the purpose of this requirement as follows:

- The full-load current equivalent to the horsepower rating of each motor is to be selected from Tables 430.247 through 430.250. These full-load currents are to be added to the rating in amperes of other loads to obtain an equivalent full-load current for the combined load.

- The locked-rotor current equivalent to the horsepower rating of each motor is to be selected from Table 430.251(A) or (B). The locked-rotor currents are to be added to the rating in amperes of other loads to obtain an equivalent locked-rotor current for the combined load. Where two or more motors or other loads cannot be started simultaneously, the largest sum of locked-rotor currents of a motor or group of motors that can be started simultaneously and the full-load currents of other concurrent loads are permitted to be used to determine the equivalent locked-rotor current for the simultaneous combined loads. In cases where different current ratings are obtained when applying these tables, the largest value is to be used.

An exception states that where part of the concurrent load is a resistance load, and where the disconnecting means is a switch rated in horsepower and amperes, the switch used is permitted to have a horsepower rating that is not less than the combined load of the motor(s) if the ampere rating of the switch is not less than the locked-rotor current of the motor(s) plus the resistance load.

(2) Ampere Rating states that the ampere rating of the disconnecting means must not be less than 115 percent of the sum of all currents at the full-load condition determined by (1) above.

(3) Small Motors states that for small motors not covered by Tables 430.247 through 430.250, the locked-rotor current is assumed to be six times the full-load current.

Section 430.111, Switch or Circuit Breaker as Both Controller and Disconnecting Means, provides that this is permitted if it complies with (A) below and is one of the types specified in (B) below (Figure 4-6).

(A) General states that the switch or circuit breaker must comply with the requirements for controllers, must open all ungrounded conductors to the motor, and must be protected by an overcurrent device in each ungrounded conductor (which is permitted to be the branch-circuit fuses). The overcurrent device protecting the controller is permitted to be either part of the controller assembly or separate. An autotransformer-type controller is to be provided with a separate disconnecting means.

FIGURE 4-6 Switches within sight of these pump motors serve as disconnecting means.

(B) Type states that the device should be one of those specified in (1) through (3) below:

(1) Air-Break Switch: This should be operable directly by applying the hand to a lever or handle.

(2) Inverse-Time Circuit Breaker: This should be operable directly by applying the hand to a lever or handle. The circuit breaker is permitted to be both power and manually operable.

Section 430.112, Motors Served by Single Disconnecting Means, states that each motor is to be provided with an individual disconnecting means. An exception provides broad exemptions from this requirement: A single disconnecting means is permitted to serve a group of motors under any of the following conditions:

- Where a number of motors drive several parts of a single machine

- Where a group of motors is under the protection of one set of branch-circuit protective devices

- Where a group of motors is in a single room within sight from the location of the disconnecting means

Section 430.113, Energy from More Than One Source, states that motor and motor-operated equipment receiving electrical energy from more than one source are to be provided with disconnecting means from each source of electrical energy immediately adjacent to the equipment served. Each source is permitted to have a separate disconnecting means. Where multiple disconnecting means are provided, a permanent warning sign is to be provided on or adjacent to each disconnecting means. There are two exceptions:

Number 1: Where a motor receives electrical energy from more than one source, the disconnecting means for the main power supply to the motor is not required to be immediately adjacent to the motor, provided that the controller disconnecting means is capable of being locked in the open position.

Number 2: A separate disconnecting means is not to be required for a Class 2 remote-control circuit conforming with Article 725 that is rated not more than 30 volts and is isolated and ungrounded.

So much for Part IX. It is complex material and difficult to keep in mind all at one time. The reader can take comfort from the fact that in the context of an open-book exam, the mandates are available for easy access, and it should be possible to pull out answers to questions as needed. When doing design work or an actual field installation, the idea is to research the individual job before specifying any equipment. Focus narrowly on the scope of the job. As more jobs are done, you will be able to recall previous installations, and this will make the work much easier.

Part X, Adjustable-Speed Drive Systems, provides that an ac motor's speed should not be varied simply by varying the input voltage by means of a rheostat. Multispeed ac motors have separate windings for each speed, with black, red, or other input wires controlled by a switch or other means, the white being common. For many applications, however, it is desirable to be able to adjust the speed across a broad spectrum. An adjustable-speed drive system, also known as a *variable-frequency drive* (VFD), fills this need and is used widely in industrial settings. Maintenance electricians working in these facilities should be familiar with the Code requirements so that if a new unit must be installed or alterations made, the work can be done in a compliant fashion. While licensing exams may not focus on VFDs, knowledge of Code mandates for them is desirable if only for the perspective it provides on other motor installations.

There are several issues that have to be addressed in this area. If an ac motor is made to run at less than the speed for which it was designed, any integral fan will be turning slower, and cooling will be diminished. Second, as an Informational Note points out, electrical resonance can result from the interaction of the nonsinusoidal currents from this type of load with power-factor-correction capacitors. Watch out for radio-frequency interference with sensitive data-transmission lines and excessive heating of the neutral conductor that supplies the drive.

Section 430.120, General, notes that the installation provisions of Parts I through IX are applicable unless modified or supplemented by Part X.

Section 430.122, Conductors—Minimum Size and Ampacity, contains two provisions:

(A) Branch/Feeder Circuit Conductors provides that circuit conductors supplying power-conversion equipment included as part of an adjustable-speed drive system must have an ampacity not less than 125 percent of the rated input current to the power-conversion equipment.

(B) Bypass Device states that for an adjustable-speed drive system that uses a bypass device, the conductor ampacity is not to be less than required by Section 430.6. The ampacity of circuit conductors supplying power-conversion equipment included as part of an adjustable-drive system that uses a bypass device is to be the larger of the following:

(1) 125 percent of the rated input current to the power-conversion equipment

(2) 125 percent of the motor full-load current rating as determined by Section 430.6

Section 430.124, Overload Protection, states that overload protection is to be provided.

(A) Included in Power Conversion Equipment states that where the power-conversion equipment is marked to indicate that motor overload protection is included, additional overload protection is not required.

(B) Bypass Circuits states that for adjustable-speed drive systems that use a bypass device to allow motor operation at rated full-load speed, motor overload protection as described in Article 430, Part III, is to be provided in the bypass circuit.

(C) Multiple Motor Applications states that individual motor overload protection is to be provided in accordance with Article 430, Part III.

Section 430.126, Motor Overtemperature Protection, provides the following:

(A) General states that adjustable-speed drive systems must protect against motor overtemperature conditions where the motor is not rated to operate at the nameplate rated current over the speed range required by the application. This protection is to be provided in addition to the conductor protection required in 430.32. Protection is to be provided by one of the following means:

(1) A motor thermal protector in accordance with 430.32

(2) An adjustable-speed drive system with load- and speed-sensitive overload protection and thermal memory retention on shutdown or power loss (An exception states that thermal memory retention on shutdown or power loss is not required for continuous-duty loads.)

(3) An overtemperature-protection relay using thermal sensors embedded in the motor

(4) A thermal sensor embedded in the motor whose communications are received and acted on by an adjustable-speed drive system

(B) Multiple Motor Applications states that individual motor overtemperature protection is to be provided as required in (A) above.

(C) Automatic Restarting and Orderly Shutdown states that the provisions of Sections 430.43 and 430.44 apply to the motor overtemperature-protection means.

Section 430.128, Disconnecting Means, provides that the disconnecting means is permitted to be in the incoming line to the conversion equipment and must have a rating not less than 115 percent of the rated input current of the conversion unit.

Part XI, Over 600 Volts, Nominal, provides the following:

Section 430.221, General, says that Part XI recognizes the additional hazard owing to the use of higher voltages. It adds to or amends the other provisions of Article 430.

Section 430.222, Marking on Controllers, states that in addition to the marking required by 430.8, a controller is to be marked with the control voltage.

Section 430.223, Raceway Connection to Motors, provides that flexible metal conduit or liquid-tight flexible metal conduit not exceeding 6 feet in length is permitted to be employed for raceway connection to a motor terminal enclosure.

Section 430.224, Size of Conductors, states that conductors supplying motors are to have an ampacity not less than the current at which the motor overload-protective device is selected to trip.

Section 430.225, Motor-Circuit Overcurrent Protection, provides the following:

(A) General provides that each motor circuit must include coordinated protection to automatically interrupt overload and fault currents in the motor, the motor-circuit conductors, and the motor control apparatus. An exception states that where a motor is critical to an operation and the motor should operate to failure if necessary to prevent a greater hazard to persons, the sensing device(s) are permitted to be connected to a supervised annunciator or alarm instead of interrupting the motor circuit.

(B) Overload Protection contains four provisions:

(1) Type of Overload Device states that each motor is to be protected against dangerous heating owing to motor overloads and failure to start by a thermal protector integral with the motor or external current-sensing devices or both. Protective-device settings for each motor circuit are to be determined under engineering supervision.

(2) Wound-Rotor Alternating-Current Motors states that the secondary circuits of wound-rotor ac motors, including conductors, controllers, and resistors rated for the application, are to be considered as protected against overcurrent by the motor overload-protection means.

(3) Operation of the overload-interrupting device is to simultaneously disconnect all ungrounded conductors.

(4) Automatic Reset states that overload-sensing devices must not reset automatically after trip unless resetting of the overload-sensing device does not cause automatic restarting of the motor or there is no hazard to persons created by automatic restarting of the motor and its connected machinery.

(C) Fault-Current Protection states, in (1), that such protection is to be provided in each motor circuit as specified by either (a) or (b) below:

    a. A circuit breaker of suitable type and rating arranged so that it can be serviced without hazard. The circuit breaker must simultaneously disconnect all ungrounded conductors. The circuit breaker is permitted to sense the fault current by means of integral or external sensing elements.

    b. Fuses of a suitable type and rating placed in each ungrounded conductor. Fuses are to be used with suitable disconnecting means, or they are to be of a type that can also serve as the disconnecting means. They are to be arranged so that they cannot be serviced while they are energized.

(2) Reclosing states that fault-current interrupting devices are not to reclose the circuit automatically. An exception provides that automatic reclosing of a circuit is permitted where the circuit is exposed to transient faults and where automatic reclosing does not create a hazard to persons.

(3) Combination Protection states that overload protection and fault-current protection are permitted to be provided by the same device.

Section 430.226, Rating of Motor Control Apparatus, provides that the ultimate trip current of overcurrent (overload) relays or other motor-protective devices used must not exceed 115 percent of the controller's continuous current rating. Where the motor branch-circuit disconnecting means is separate from the controller, the disconnecting means current rating is not to be less than the ultimate trip setting of the overcurrent relays in the circuit.

Section 430.227, Disconnecting Means, states that the disconnecting means of the controller must be capable of being locked in the open position. The provision for locking or adding a lock to the disconnecting means is to be installed on or at the switch or circuit breaker used as the disconnecting means and must remain in place with or without the lock installed.

Part XII, Protection of Live Parts—All Voltages, provides the following:

Section 430.231, General, states that Part XII specifies that live parts are to be protected in a manner judged adequate for the hazard involved.

Section 430.232, Where Required, states that exposed live parts of motors and controllers operating at 50 volts or more between terminals are to be guarded against accidental contact by enclosure or by location as follows:

- By installation in a room or enclosure that is accessible only to qualified persons
- By installation on a suitable balcony, gallery, or platform elevated and arranged so as to exclude unqualified persons
- By elevation 8 feet or more above the floor

An exception provides that live parts of motors operating at more than 50 volts between terminals do not require additional guarding for stationary motors that have commutators, collectors, and brush rigging located inside of motor-end brackets and not conductively connected to supply circuits operating at more than 150 volts to ground.

Section 430.233, Guards for Attendants, states that where live parts of motors or controllers operating at over 150 volts to ground are guarded against accidental contact only by location, and where adjustment or other attendance may be necessary during operation of the apparatus, suitable insulating mats or platforms are to be provided so that the attendant cannot readily touch live parts unless standing on the mats or platforms.

Part XIII, Grounding—All Voltages, provides the following:

Section 430.241, General, states that Part XIII specifies the grounding of exposed non-current-carrying metal parts, likely to become energized, of motor and controller frames to prevent a voltage above ground in the event of accidental contact between energized parts and frames. Insulation, isolation, and guarding are suitable alternatives to grounding of motors under certain conditions.

Section 430.242, Stationary Motors, provides that the frames of stationary motors are to be grounded under any of the following conditions:

- Where supplied by metal-enclosed wiring
- Where in a wet location and not isolated or guarded
- In a hazardous (classified) location
- If the motor operates with any terminal at over 150 volts to ground

Where the frame of the motor is not grounded, it is to be permanently and effectively insulated from ground. The most common scenario is for a motor to be connected to the electrical supply by metal raceway wiring or metal-sheathed cable. These serve as grounding means in addition to any equipment-grounding conductor in the form of a wire that may be pulled through the raceway or may be a part of the cable. Grounding of the metal frame of a motor prevents it from becoming energized with respect to ground owing to a ground fault, provided that the grounding means is of sufficiently low impedance. Beware of raceway connections that may become loose or detached at the motor owing to vibration.

Section 430.243, Portable Motors, states that the frames of portable motors that operate over 150 volts to ground are to be guarded or grounded, with two exceptions:

Number 1: Listed motor-operated tools, listed motor-operated appliances, and listed motor-operated equipment are not required to be grounded where protected by a system of double insulation or its equivalent. Double-insulated equipment is to be distinctively marked.

Number 2: Listed motor-operated tools, listed motor-operated appliances, and listed motor-operated equipment connected by a cord and attachment plug other than those required to be grounded in accordance with 250.114, Equipment Connected by Cord and Plug, are not required to be grounded.

Section 430.244, Controllers, states that controller enclosures are to be connected to the equipment-grounding conductor regardless of voltage. Controller enclosures must have means for attachment of an equipment-grounding conductor termination. An exception provides that enclosures attached to ungrounded portable equipment are not required to be grounded.

Section 430.245, Method of Grounding, states that connection to the equipment-grounding conductor is to be done in the manner specified in Part VI of Article 250.

(A) Grounding Through Terminal Housings states that where the wiring to motors is metal-enclosed cable or in metal raceways, junction boxes to house motor terminals are to be provided and the armor of the cable or the metal raceways are to be connected to them in the manner specified in 250.96(A) and 250.97.

(B) Separation of Junction Box from Motor states that the junction box required by Section 430.245(A) above is permitted to be separated from the motor by not more than 6 feet, provided that the leads to the motor are stranded conductors within Type AC cable, interlocked metal-tape Type MC cable where listed and identified in accordance with 250.118(10)(a), or armored cord or are stranded leads enclosed in liquid-tight flexible metal conduit, flexible metal conduit, intermediate metal conduit, rigid metal conduit, or electrical metallic tubing not smaller than trade size 3/8, the armor or raceway being connected both to the motor and to the box.

Liquid-tight flexible nonmetallic conduit and rigid nonmetallic conduit are permitted to enclose the leads to the motor, provided that the leads are stranded and the required equipment-grounding conductor is connected to both the motor and the box.

Where stranded leads are used, protected as specified earlier, each strand within the conductor may not be larger than 10 AWG and must comply with other requirements of the Code for conductors to be used in raceways.

(C) Grounding of Controller-Mounted Devices states that instrument transformer secondaries and exposed non-current-carrying metal or other conductive parts or cases of instrument transformers, meters, instruments, and relays are to be grounded as specified in Sections 250.170 through 250.178.

The tables in Part XIV, Tables, have been placed in a separate part at the end of the article because reference is made to them throughout. They are quite familiar to anyone who has studied Article 430 in detail. Owing to their great importance, we'll take a final look at them before leaving Article 430.

Tables 430.247 through 430.250 are similar in arrangement and application. They provide full-load current in amperes as a function of horsepower for various voltage levels.

Table 430.247 is for dc motors.

Table 430.248 is for single-phase ac motors.

Table 430.249 is for two-phase ac motors—rarely used.

Table 430.250 is for three-phase ac motors.

These tables are used in most cases, though not always, as we have seen, to size motor branch-circuit conductors. Rather than using the nameplate values, you use the full-load current from these tables based on the motor nameplate horsepower. To size the overload protection, however, use the nameplate value, not the table.

Tables 430.251(A) and (B) are essentially the same, except that (A) is for single-phase motors and (B) is for polyphase Design B, C, and D motors. These are conversion tables that give maximum locked-rotor current for selection of disconnecting means and controllers. The currents are given in amperes for various voltage levels.

This concludes our review of Article 430, Motors, Motor Circuits and Controllers. The Code article is well organized and an excellent guide for compliance when designing or installing motors. For an open-book electrician's licensing exam, answers to questions on motors, circuits, and controllers can be accessed easily if you know the structure of the article. If you are not sure of the exact location of a Code reference within Article 430, try the Contents first and then the Index. Avoid Code hopping because that consumes valuable time.

This also concludes our review of Chapter 4, Equipment for General Use. Rather than reviewing each article, we have focused exclusively on Article 430, Motors, Motor Circuits, and Controllers, because it is of such central concern. This is not to diminish the importance of other articles, but it is not possible to cover everything in this book. Furthermore, Chapter 4 is highly user-friendly in the sense that each article directly addresses a single category of equipment. Most of the articles are much briefer than Article 430, and after our study of Article 430, they should be easy to understand. Article 450, Transformers and Transformer Vaults (Including Secondary Ties), may seem a little difficult, and Section 450.6(B), Overcurrent Protection for Secondary Connections, at first may seem daunting, but after working through Article 430, it will soon become familiar.

NEC® Chapter 5,
Special Occupancies

W e have completed a fairly intensive article-by-article examination of the first four chapters of the 2011 National Electrical Code (NEC). As stated in the Code Introduction, these
chapters apply generally to all electrical installations. Therefore, a
detailed knowledge of them is fundamental to the profession. Starting with Chapter 5, the Code shifts gears and becomes much more
specific. The knowledge in Chapters 5, 6, and 7 is more specialized, as
reflected in their titles, Special Occupancies, Special Equipment, and
Special Conditions, respectively. As the Code continues its progress
from the general to the more specific, the content of these three chapters will be seen to become rather complex and technical. For most of
us, this material is impossible to memorize—there is just too much of
it. Fortunately, though, for those who are faced with a licensing exam
or the real-world choices that have to be made in electrical design and
installation, the material is fairly easy to access, so on an open-book
basis, correct answers are available at all times. Once you compre-
hend the Code structure, the answer to any question that may arise is
right in front of you.

Notice that these three chapter titles all start with the word *special*.
These three chapters, as noted in NEC Article 90, Introduction, supplement or modify Chapters 1 through 4. The mandates in the first
four chapters are still applicable, except where specifically ruled out
in specific sections of Chapters 5 through 7.

In contrast, Chapter 8, Communications Systems, stands alone in
that it is not subject to the requirements in Chapters 1 through 4,
unless specifically referenced. The requirements in Chapters 1 through
4 are applicable in Chapters 5 through 7 unless specifically ruled out,
but in Chapter 8, they are applicable only if referenced; otherwise,
they are not applicable. This distinction is critical in interpreting these
chapters and mention of it will be made again.

As the Code changes its rules of the game, we shall shift gears
as well. Beginning with Chapter 5, we'll look at Code material from

a somewhat more distant perspective. Instead considering individual articles in detail, we'll consider the overall structure and in that way learn to quickly access individual sections on a case-by-case basis.

As mentioned, Chapter 5 is highly technical and detailed but not at all beyond reach. Look for an overall understanding and comprehension of the meaning of terms, but don't expect to retain the whole thing at once because this is not, in most cases, likely to happen.

Chapter 5 generally is considered to be the place where wiring in locations characterized by explosive gases and liquids is covered. We think of a large gasoline refinery erupting into a massive fireball owing to one small spark produced by arcing contacts in a switch or relay. But Chapter 5 has other content, some of it addressing locations that may be considered hazardous in varying degrees and some of it concerned with relatively benign settings such as Article 547, Agricultural Buildings.

Article 500

Article 500, Hazardous (Classified) Locations, Classes I, II, and III, Divisions 1 and 2, consists of an overview of these environments, differentiating them from each other and from nonhazardous locations. (The terms *hazardous* and *classified* are used interchangeably. Some building codes use one term; some, the other. It is current NEC practice to use the word *hazardous* followed by *classified* in parentheses.) Following Section 500.2, Definitions, which contains definitions of 17 terms specific to Article 500, an important section appears. It is Section 500.5, Classification of Locations, and it provides critical information for determining the classification of specific sites and areas within sites. The basis for properly classifying a location is twofold:

- The properties of the flammable gas, flammable liquid-produced vapor, combustible liquid-produced vapors, combustible dusts or fibers/flyings that may be present

- The likelihood that a flammable or combustible concentration or quantity is present

The Code lays out the classification parameters for various locations. It is the responsibility of owners and on-site personnel to apply these parameters and map out the boundaries of each classified area within the premises or determine whether they exist within the premises.

Classes I, II, and III correspond to the nature of the material present at the location. Divisions 1 and 2 correspond to the degree of likelihood that these materials will pose a threat in the form of fire or explosion or some other threat, such as dust that might collect on a motor housing, cause overheating, and damage to the equipment even if an actual fire or explosion did not occur.

An Informational Note observes that through the exercise of ingenuity in the layout of electrical installations for hazardous locations, it is frequently possible to locate much of the equipment in a reduced level of classification or in an unclassified location and thus reduce the amount of special equipment required.

Explosion-proof equipment, typically cast aluminum and gasketed, is much more expensive than sheet-metal equipment suitable for unclassified areas, and the labor involved in installing wiring within a classified area is much greater (Figure 5-1).

Class I, II, and III hazardous locations are not to be confused with Class 1, 2, and 3 remote-control signaling and power-limited circuits.

Class I locations generally are considered to be the most hazardous, and the degree of hazard is less in Class II and least in Class III. Notwithstanding, powerful explosions and fiery infernos may occur in any of these locations, accompanied by massive loss of life and damage to property. Each class is subdivided into Division 1 and Division 2, with Division 1 being the more hazardous.

Class I locations are those in which flammable gases, flammable liquid-produced vapors, or combustible liquid-produced vapors are or may be present in the air in quantities sufficient to produce explosive or ignitable mixtures. Locations that fall within this class include propane filling stations, automotive gas stations, and stations that dispense compressed natural gas (CNG).

FIGURE 5-1 Don't be fooled! This rugged looking box is not marked explosion-proof and is not suitable for Class I, Division 1 locations. The gasketed area is not wide enough to ensure that flame could not get through.

Class I locations are divided into Division 1 and Division 2 locations. A Class I, Division 1 location is one in which

- Ignitable concentrations of flammable gases, flammable liquid-produced vapors, or combustible liquid-produced vapors can exist under normal operating conditions.

- Ignitable concentrations of such flammable gases, flammable liquid-produced vapors, or combustible liquids above their flash points can exist frequently because of repair or maintenance operations or because of leakage.

Breakdown or faulty operation of equipment or processes can release ignitable concentrations of flammable gases, flammable liquid-produced vapors, or combustible liquid-produced vapors and also can cause simultaneous failure of electrical equipment in such a way as to directly cause the electrical equipment to become a source of ignition.

Class I, Division 1 locations are characterized by the presence of these materials in the air or confined in such a manner that they may be released frequently into the air or in case of breakdown or faulty operation may be released or may cause hazardous equipment failure.

Class I, Division 2 locations are characterized by the same materials, but the possibility of them being released into the air is less likely but still a possibility. Specifically, a Class I, Division 2 location is a location in which

- Flammable gases, flammable liquid-produced vapors, or combustible liquid-produced vapors are handled, processed, or used but in which the liquids, vapors, or gases normally will be confined within closed containers or closed systems from which they can escape only in case of accidental rupture or breakdown of such containers or systems or in case of abnormal operation of equipment.

- Ignitable concentrations of flammable gases, flammable liquid-produced vapors, or combustible liquid-produced vapors are normally prevented by positive mechanical ventilation and might become hazardous through failure or abnormal operation of the ventilating equipment.

- A Class I, Division 2 location is adjacent to ignitable concentrations of flammable gases, flammable liquid-produced vapors, or combustible liquid-produced vapors above their flash points that might be communicated occasionally unless such communication is prevented by adequate positive-pressure ventilation from a source of clean air and effective safeguards against loss of ventilation are provided.

The same pattern of division differentiation is found in all three classes. The classes apply to the type of material involved, and the divisions apply to the immediacy of the hazard.

Class II locations are those which are hazardous because of the presence of combustible dust. A Class II, Division 1 location is a location in which

- Combustible dust is in the air under normal operating conditions in quantities sufficient to produce explosive or ignitable mixtures.

- Mechanical failure or abnormal operation of machinery or equipment might cause such explosive or ignitable mixtures to be produced and also might provide a source of ignition through simultaneous failure of electrical equipment, failure of protection devices, or from other causes.

- Group E combustible dusts may be present in quantities sufficient to be hazardous.

An Informational Note states that dusts containing magnesium or aluminum are particularly hazardous, and the use of extreme precautions is necessary to avoid ignition and explosion.

A Class II, Division 2 location is a location in which

- Combustible dust owing to abnormal operations may be present in the air in quantities sufficient to produce explosive or ignitable mixtures.

- Combustible dust accumulations are present but are normally insufficient to interfere with normal operation of electrical equipment or other apparatus but could, as a result of infrequent malfunctioning of handling or processing equipment, become suspended in the air.

- Combustible dust accumulations on, in, or in the vicinity of the electrical equipment could be sufficient to interfere with the safe dissipation of heat from electrical equipment or could be ignitable by abnormal operation or failure of electrical equipment.

An Informational Note states that the quantity of combustible dust that may be present and the adequacy of dust-removal systems are factors that merit consideration in determining the classification and may result in an unclassified area.

A second Informational Note states that where products such as seed are handled in a manner that produces low quantities of dust, the amount of dust deposited may not warrant classification.

Class III locations are those which are hazardous because of the presence of easily ignitable fibers or where materials producing combustible flyings are handled, manufactured, or used but in which such fibers/flyings are not likely to be in suspension in the air in quantities sufficient to produce ignitable mixtures.

This definition indicates that the materials involved may be the same as those in Class II but less finely divided. Larger but fewer particles would have less surface area and less exposure to oxygen in the surrounding air. For these reasons, Class III locations are less hazardous than Class II locations, and as we shall see, the wiring requirements are less onerous.

A Class III, Division 1 location is a location in which easily ignitable fibers/flyings are handled, manufactured, or used.

A Class III, Division 2 location is a location in which easily ignitable fibers/flyings are stored or handled other than in the process of manufacture. As in the earlier two classes, the materials are the same in both divisions, but the hazard is greater in Division 1 because the exposure to any source of ignition is greater or more immediate.

The whole layout of classes and divisions is given in this early part of Article 500. To review, there are three classes and two divisions within each class, each beginning with the more hazardous. The first step in designing an electrical installation in a facility that may contain hazardous areas is to delineate those areas with precise boundaries between them and make a map showing necessary details. In every case, it is beneficial to locate electrical wiring and equipment in a higher-numbered class or division or outside the classified areas altogether. (For the purposes of the NEC, *hazardous* and *classified* are synonymous.)

Section 500.6, Material Groups, adds another element to the mix. This consideration is separate from the primary classification procedure, but it is necessary in order to continue the process of designing a hazardous location electrical installation.

Various air mixtures have to be grouped in accordance with (A) and (B):

Class I Group Classifications provides the following:

- Group A: Acetylene. Anyone who has done much work with an oxyacetylene torch knows that a small amount of pure acetylene released into the air ignites very easily with a loud bang.

- Group B: Flammable gas, flammable liquid-produced vapor, or combustible liquid-produced vapor mixed with air that may explode or burn with a maximum experimental safe gap (MESG) of less or equal to 0.45 mm or a minimum igniting current ratio (MIC ratio) of less than or equal to 0.40. An Informational Note states that a typical Class I, Group B material is hydrogen.

- Group C: Flammable gas, flammable liquid-produced vapor, or combustible liquid-produced vapor mixed with air that may burn or explode with an MESG of less than or equal to 0.45 mm and a minimum MIC ratio of less than or equal to 0.80. An Informational Note states that a typical Class I, Group C material is ethylene.

- Group D: Flammable gas, flammable liquid-produced vapor, or combustible liquid-produced vapor mixed with air that may burn or explode with an MESG of less than or equal to 0.75 mm and a minimum MIC ratio of less than or equal to 0.80.

Here again, the earlier groups are the more hazardous because they have lower MESG and MIC numbers. A less dangerous burning or exploding material will be extinguished passing through a larger gap, for example, between two parts of a metal enclosure and also takes more electric current to be ignited.

The National Fire Protection Association's (NFPA) 2011 *NEC® Handbook* contains an extensive list of materials in the four groups for Class I and the seven groups for Class II.

The Class I material groups were listed earlier. The Class II material groups are more numerous and can be accessed from the Code as needed. There are no Class III material groups. These material groups are used to select proper equipment for use in the areas involved.

Article 500 continues the overview of hazardous area electrical design work by listing specialized protection techniques along with a statement of where these techniques are required in terms of class and division. It is important to assimilate this information and learn where to access it. You never know when a licensing exam will ask a question on one or more of these points. Moreover, most professional electricians have occasion to do some wiring, if not a complete job, within a commercial garage, hospital, or other facility that may contain a hazardous area. Here are the protection techniques, as given in Section 500.7:

(A) Explosion-proof Equipment: Permitted in Class I, Division 1 or 2 locations.

(B) Dust Ignition-proof: Permitted in Class II, Division 1 or 2 locations.

(C) Dust-tight: Permitted in Class II, Division 1 or 2 locations.

(D) Purged and Pressurized: Permitted in any hazardous location for which it is identified.

(E) Intrinsic Safety: Permitted in any hazardous area. Articles 501 through 503 and 510 through 516 are not applicable to such installation, except as required by Article 504, and installation of intrinsically safe apparatus and wiring must be in accordance with Article 504.

(F) Nonincendive Circuit: Permitted in Class I, Division 2; Class II, Division 2; or Class III, Division 1 or 2 locations.

(G) Nonincendive Equipment: Same as (F).

(H) Nonincendive Component: Same as (F).

(I) Oil Immersion: Permitted for Class I, Division 2 locations.

(J) Hermetically Sealed: Same as (F).

(K) Combustible Gas-Detection System: Permitted as a means of protection in industrial establishments with restricted public access and where the conditions of maintenance and supervision ensure that only qualified persons service the installation. Where such a system is installed, equipment specified in (1), (2), or (3) below is permitted:

(1) Inadequate Ventilation states that in a Class I, Division 1 location that is so classified owing to inadequate ventilation, electrical equipment suitable for Class I, Division 2 locations is permitted. Combustible gas-detection equipment must be listed for Class I, Division 1, for the appropriate material group, and for the detection of the specific gas or vapor to be encountered.

(2) Interior of a Building states that in a building located in or with an opening into a Class I, Division 2 location where the interior does not contain a source of flammable gas or vapor, electrical equipment for unclassified locations is permitted. Combustible gas-detection equipment is to be listed for Class I, Division 1 or Class I, Division 2, for the appropriate material group, and for the detection of the specific gas or vapor to be encountered.

(3) Interior of a Control Panel states that in the interior of a control panel containing instrumentation using or measuring flammable liquids, gases, or vapors, electrical equipment suitable for Class I, Division 2 locations is permitted. Combustible gas-detection equipment is to be listed for Class I, Division 1, for the appropriate material group, and for the detection of the specific gas or vapor to be encountered.

(L) Other Protection Techniques specifies the techniques used in equipment identified for use in hazardous locations.

Section 500.8, Equipment, notes that Articles 500 through 504 require equipment construction and installation that ensure safe performance under conditions of proper use and maintenance. It is provided in this section, moreover, in (A), that suitability of identified equipment is to be determined by

(1) Equipment listing or labeling

(2) Evidence of equipment evaluation from a qualified testing laboratory or inspection agency concerned with product evaluation

(3) Evidence acceptable to the authority having jurisdiction (AHJ), such as manufacturer's self-evaluation or an owner's engineering judgment

(B) Approval for Class and Properties provides in (1) that equipment is to be identified not only for the class of location but also for the explosive, combustible, or ignitable properties of the specific gas, vapor, dust, or fiber/flyings that will be present. In addition, Class I equipment may not have any exposed surface that operates at a temperature in excess of the ignition temperature of the specific gas or vapor. Class II equipment may not have an external temperature higher than specified in Section 5008(D)(2) and the associated table, Class II ignition temperatures. This table is divided into two parts, one for equipment not subject to overloading and one for equipment subject to overloading, such as motors.

(2) Equipment that is identified for a Division 1 location is permitted in a Division 2 location of the same class, group, and temperature class and must comply with (a) and (b) below:

    a. An intrinsically safe apparatus having a control drawing requiring the installation of associated apparatus for a Division 1 installation is permitted to be installed in a Division 2 location if the same associated apparatus is used for the Division 2 installation.

    b. Equipment that is required to be explosion-proof must incorporate seals in accordance with Section 501.15(A) or (D) when the wiring methods of Section 501.10(B) are employed.

This paragraph refers ahead to the sections on conduit seals in Article 501, Class 1 Locations.

Article 500 concludes with (C) Marking and (D) Temperature, which contains the table mentioned earlier.

To summarize, Article 500 provides an overview of the three hazardous location classes and the two divisions into which each of the classes is divided according to the immediacy or likelihood of ignition. The article also contains an overview of the measures that may be taken to mitigate these hazards.

The next three articles cover the three classes of hazardous locations, the numbering of the articles indicating the class being considered.

Article 501

Article 501, Class I Locations, covers these locations in greater detail than we saw in the overview in Article 500.

Part I, General, is very brief. It contain a statement of scope, followed by Section 501.5, Zone Equipment, which notes that equipment listed and marked for use in Zone 0, 1, or 2 locations is permitted in Class I, Division 2 locations for the same gas and with a suitable temperature class. Similarly, equipment listed and marked for use in Zone 0 locations is permitted in Class I, Division 1 or Division 2 locations for the same gas and with a suitable temperature class.

This introduces the concept of zone classification of hazardous areas, which is covered in detail in Article 505, Zone 0, 1, and 2 Locations. The zone classification is an optional alternative system that replaces division classifications and is applicable only to Class I, where fire or explosion hazards may exist owing to flammable gases, vapors, or liquids. The zone classification system is based on International Electrotechnical Commission (IEC) standards for area classification. The electrician should become adept at working with both standards.

Part II, Wiring, discusses wiring methods for both divisions within Class I.

(A) Class I, Division 1 Wiring Methods states in (1) General that the following wiring methods are permitted in these most hazardous of locations:

a. Threaded rigid metal conduit or threaded steel intermediate metal conduit. An exception permits polyvinyl chloride (PVC) and reinforced thermosetting resin conduit (RTRC) conduit where encased in 2 inches of concrete and additionally covered with 24 inches of backfill measured from the top of the conduit to grade. In this case, the metal conduit specified above must constitute the final 24 inches of the underground run, and an equipment-grounding conductor must be included.

b. Type MI cable terminated with fittings listed for the location.

c. In industrial establishments with restricted public access where conditions of maintenance and supervision ensure that only qualified persons service the installation, Type MC-HL cable listed for use in Class I, Zone 0 or Division 1 locations, with a gas/vapor-tight continuous corrugated metallic sheath, an overall jacket of suitable polymeric material, and a separate equipment-grounding conductor(s) terminated with fittings listed for the application. This statement answers the frequently asked question of whether Type MC cable is permitted in Class I, Division 1 locations. The answer is that it is permitted, but not the ordinary Type MC that we use frequently on industrial and commercial job sites.

d. With similar restrictions, Type ITC-HL cable.

(2) Flexible Connections provides that where necessary to employ flexible connections, as at motor terminals, flexible fittings listed for the location or flexible cord terminated with cord connectors listed for the location is permitted.

(3) Boxes and Fittings states that boxes and fittings must be approved for Class I, Division 1. These are very robust, explosion-proof boxes with threaded hubs and a threaded opening for a cover that will withstand an internal explosion if gas or vapor infiltrates and ignition occurs.

(B) Class I, Division 2 lists in (1) the permitted wiring methods for Class I, Division 2:

- All wiring methods permitted in Class I, Division 1
- Enclosed gasketed busways and wireways
- Type PLTC and Type PLTC-ER cable, including installation in cable tray systems, terminated with listed fittings
- Type ITC and Type ITC-ER cable terminated with listed fittings
- Type MC-MV or Type TC cable, including installation in cable tray systems, terminated with listed fittings
- In industrial establishments with restricted public access, where conditions of maintenance and supervision ensure that only qualified persons service the installation and where metallic conduit does not provide sufficient corrosion resistance, listed RTRC, factory elbows, and associated fittings, all marked with the suffix –XW, and Schedule 80 PVC conduit, factory elbows, and associated fittings

(2) Flexible Connections provides that where provision must be made for limited flexibility, one or more of the following are permitted:

- Listed flexible metal fittings
- Flexible metal conduit with listed fittings
- Liquid-tight flexible metal conduit with listed fittings
- Liquid-tight flexible nonmetallic conduit with listed fittings
- Flexible cord listed for extra hard usage and terminated with listed fittings (A connector for use as an equipment-grounding conductor must be included in the flexible cord.)

(3) Nonincendive Field Wiring states that such wiring is permitted using any of the wiring methods permitted for unclassified locations. Nonincendive field wiring systems are to be installed in accordance with the control drawing(s). Simple apparatus, not shown on the control drawing, are permitted in a nonincendive field wiring circuit, provided that the simple apparatus does not interconnect the nonincendive field wiring circuit to any other circuits. Separate nonincendive field wiring circuits are to be installed in accordance with one of the following:

- In separate cables
- In multiconductor cables where the conductors of each circuit are within a grounded metal shield
- In multiconductor cables or in raceways where the conductors of each circuit have insulation with a minimum thickness of 0.01 inch

(4) Boxes and Fittings states that boxes and fittings are not required to be explosion-proof except as required by Sections 501.105(B), 501.115(B)(1), and 501.150(B)(1).

Boxes and fittings do not have to be explosion-proof in Class I, Division 2 locations if they contain no arcing devices, provided that the maximum operating temperature of exposed surfaces does not exceed 80 percent of the ignition temperature for the material involved.

Section 501.15, Sealing and Drainage, covers the important area of conduit seals for Class I, where the conduits enter enclosures and also where the conduits cross classification boundaries. Sealing is accomplished by introducing the sealing compound into the raceway system at appropriate locations using sealing fittings that are installed in advance as the raceway system is installed (Figure 5-2).

The key concept is that flammable gases and vapors must not be allowed to migrate through the conduit system into areas where they can become hazardous. Even though explosion-proof enclosures are able to withstand an explosion from within if an arcing device ignites a material that has infiltrated, it is better to prevent that infiltration in the first place. Furthermore, flammable gases and vapors must be prevented from migrating across classification boundaries where the appropriate protection may not be in place (Figure 5-3).

This is the section that must be accessed in order to answer exam questions regarding Class I sealing and drainage.

Section 501.30, Grounding and Bonding, discusses specialized grounding and bonding techniques that are required in Class I locations.

Figure 5-2 Conduit sealing fittings are designed to allow introduction of sealing material at class and division boundaries as required after wires have been pulled.

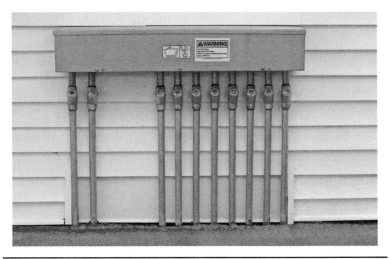

Figure 5-3 A commercial garage with conduit seals where raceways emerge from underground from gasoline dispensing pumps.

(A) Bonding emphasizes that the ordinary method of double locknuts and locknut-bushing combination is not sufficient in Class I, Division 1 and 2 locations (Figure 5-4).

Bonding jumpers with proper fittings or other approved (by the AHJ) means of bonding are to be employed. These means are to be

Figure 5-4 Bonding fittings are required at boxes in hazardous locations, as opposed to the two locknuts shown on the left. These are required straight back to the grounding lug or other point of grounding, even outside the hazardous location.

employed at all intervening raceways, fittings, boxes, enclosures, and so forth within the Class I location regardless of division. Furthermore, even after the wire run has left the classified area, the specialized bonding must be maintained until the wire run ends at a point of grounding for service equipment or separately derived system. An exception provides that the specific bonding means is required only to the nearest point where the grounded circuit and the grounding electrode are connected together on the line side of the building or structure disconnecting means, provided that the branch-circuit overcurrent protection is located on the load side of the disconnecting means.

Installation of a wire-type equipment-grounding conductor within the raceway does not eliminate the special bonding method required for the conduit.

Section 501.40, Multiwire Branch Circuits, specifies that these are not permitted within a Class I, Division 1 location except where the disconnect device(s) for the circuit opens all ungrounded conductors of the multiwire circuit simultaneously.

Part III, Equipment, covers use of various types of equipment within Class I areas. Included are transformers and capacitors, meters, instruments and relays, switches, circuit breakers, motor controllers and fuses, control transformers and resistors, motors and generators, luminaires, utilization equipment (such as heaters), flexible cords, receptacles and attachment plugs and signaling, and alarm, remote-control, and communications systems.

In all cases it is preferable to locate electrical equipment outside the classified area, if possible. If this is not possible, the more expensive equipment and wiring methods we have discussed must be used.

Article 502

Article 502, Class II Locations, is similar to Article 501 in scope and structure. Since it applies to Class II locations, as defined in Article 500, the wiring methods are somewhat less stringent, although considerably beyond ordinary Chapter 3 techniques. It must be noted, however, that even though Class II locations may be considered less hazardous than Class I locations, this is not always the case. In addition to creating an explosive atmosphere, Class II environments are also prone to an additional hazard, which is that dust may settle onto a motor or other surface that is supposed to dissipate heat. The temperature rise can damage the equipment, causing it to generate still more heat, and the end result could be fire. Then there is another issue regarding electrically conductive particles. These may settle on terminals, shorting them out to ground or to each other where they are at different voltage levels. Communication and data transmission may be disrupted, and higher power levels can produce arcing.

The bottom line is that some Class I wiring and equipment may not be suitable for Class II, as opposed to Division 1 wiring and equipment, which is always suitable for Division 2 usage. In all cases, it is necessary to look at the listing regarding both classification and the specific materials involved. (There are dual-rated enclosures that are suitable for more than one class, within limitations.)

Section 502.5, Explosion-proof Equipment, makes the point that explosion-proof equipment and wiring are not required and are not acceptable in Class II locations unless also identified for such locations.

Part II, Wiring, states, in Section 502.10, Wiring Methods, that wiring methods must comply with (A) or (B) below:

(A) Class II, Division 1 states that permitted wiring methods include

(1) Threaded rigid metal conduit or threaded steel intermediate metal conduit

(2) Type MI cable with termination fittings listed for the location

(3) In industrial establishments with limited public access, where the conditions of maintenance and supervision ensure that only qualified persons service the installation, Type MC-HL cable, listed for use in Class II, Division 1 locations, with a gas/vapor-tight continuous corrugated metallic sheath, an overall jacket of suitable polymeric material, a separate equipment-grounding conductor(s), and provided with termination fittings listed for the application

(4) Fittings and boxes are to be provided with threaded bosses for connection to conduit or cable terminations and must be dust-tight. Fittings or boxes in which taps, joints, or terminal connections are made or that are used in Group E locations are to be identified for Class II locations.

(B) Class II, Division 2 states that permitted wiring methods include

(1) All wiring methods permitted for Class II, Division 1

(2) Rigid metal conduit, intermediate metal conduit, electrical metallic tubing, dust-tight wireways

(3) Type MC or Type MI cable with listed termination fittings

(4) Type PLTC and Type PLTC-ER cable with the provisions of Article 725, including termination in cable tray systems (The cable is to be terminated with listed fittings.)

(5) Type ITC and Type ITC-ER cable terminated with listed fittings

(6) Type MC, Type MI, or Type TC cable installed in ladder, ventilated trough, or ventilated channel cable trays in a single layer, with a space not less than the larger cable diameter between the two adjacent cables

(7) In industrial establishments with restricted public access where the conditions of maintenance and supervision ensure that only qualified persons service the installation and where metallic conduit does

not provide sufficient corrosion resistance, RTRC factory elbows, and associated fittings, all marked with the suffix –XW, and Schedule 80 PVC conduit, factory elbows, and associated fittings [An exception states that Type MC cable listed for use in Class II, Division 1 locations is permitted to be installed without the spacings required by (6).]

Nonincendive field wiring is permitted using any of the wiring methods permitted for unclassified locations. All boxes and fittings must be dust-tight.

Section 502.15, Sealing, Class II, Divisions 1 and 2, provides that where a raceway provides communication between an enclosure that is required to be dust ignition-proof and one that is not, suitable means must be provided to prevent the entrance of dust into the dust ignition-proof enclosure through the raceway.

Section 502.30, Grounding and Bonding, Class II, Divisions 1 and 2, states that the locknut-bushing and double-locknut types of contacts are not to be depended on for bonding purposes, but bonding jumpers with proper fittings or other approved means of bonding are to be used.

The specialized bonding means required for Class II locations are identical to those for Class I locations.

Section 502.40, Multiwire Branch Circuits, states that such circuits are not permitted with the same exception as allowed for Class I locations.

Part III, Equipment for Class II Locations, is similar to Part III of Article 501 except that the emphasis is on dust-tightness as a requirement. The sections, as in Article 501, are divided into subsections for Divisions 1 and 2.

Article 503

Article 503, Class III Locations, covers the requirements for electrical equipment and wiring for all voltages in Class III, Division 1 and 2 locations, where fire or explosion hazards may exist owing to ignitable fibers/flyings.

Textile mills and some sawmills are included in this classification. If sawdust is finely divided, the location is Class II. Class III, Division 1 is applicable if the material is handled, manufactured, or used. The location is Division 2 if the hazard is less immediate, for example, if the material is stored and less likely to be in the air.

Section 503.5, General, states that equipment installed in Class III locations must be able to function at full rating without developing surface temperatures high enough to cause excessive dehydration or gradual carbonization of accumulated fibers/flyings. Organic material that is carbonized or excessively dry is highly susceptible to spontaneous ignition. The maximum surface temperature under operating conditions is not to exceed 165°C for equipment that is not subject to overloading and 120°C for equipment (such as motors or power transformers) that may be overloaded.

Part II, Wiring, provides, in Section 503.10, that wiring methods are to comply with (A) or (B) below:

(A) Class III, Division 1 provides the following:

(1) Rigid metal conduit, PVC conduit, RTRC conduit, intermediate metal conduit, electrical metallic tubing, dust-tight wireways, or Type MC or MI cable with listed termination fittings

(2) Type PLTC and Type PLTC-ER cable, including installation in cable tray systems, terminated with listed fittings

(3) Type ITC and Type ITC-ER cable terminated with listed fittings

(4) Type MC, Type MI, or Type TC cable installed in ladder, ventilated trough, or ventilated channel cable trays in a single layer, with a space not less than the larger cable diameter between two adjacent cables.

All boxes and fittings must be dust-tight. Nonincendive field wiring is permitted using any of the wiring methods indicated for unclassified locations.

(B) Class III, Division 2 states that the wiring methods must comply with (A) above.

Section 503.30, Grounding and Bonding—Class III, Divisions 1 and 2, provides that the requirements for grounding and bonding are the same as for Classes I and II.

Part III, Equipment, is structured like Part III of the two preceding articles. In Class III, there is less difference in requirements between the two divisions.

If you remember that Articles 501, 502, and 503 are structured similarly, exam answers will be very easy to access. Article 500 defines the classes and divisions, and the next three articles provide specific requirements for each class and division.

Article 504

Article 504, Intrinsically Safe Systems, provides details of this useful wiring method. As we have seen, intrinsically safe wiring may be used to good advantage in Class I, Class II, and Class III hazardous areas.

Section 504.1, Scope, states that the article covers the installation of intrinsically safe apparatus, wiring, and systems for Class I, II, and III locations.

Section 504.2, Definitions, provides an excellent overview. An intrinsically safe system goes beyond being nonincendive because it safeguards itself from infiltration of dangerous levels of electrical energy from outside sources owing to any fault condition that may occur.

Associated apparatus is apparatus in which the circuits are not necessarily intrinsically safe themselves but that influence the energy in the intrinsically safe circuits and so must be relied on to maintain intrinsic safety. Associated apparatus may be either of the following:

- Electrical apparatus that has an alternative-type protection for use in the appropriate hazardous location
- Electrical apparatus not so protected that is not to be used within a hazardous location

An Informational Note states that associated apparatus must have intrinsically safe connections for intrinsically safe apparatus and also may have connections for non–intrinsically safe apparatus.

Another Informational Note states that an example of associated apparatus is an intrinsic safety barrier, which is a network designed to limit the energy (voltage and current) available to the protected circuit in the hazardous location under specified fault conditions.

- *Control drawing* is a drawing or other document provided by the manufacturer of the intrinsically safe or associated apparatus or of the nonincendive field wiring apparatus or associated nonincendive field wiring apparatus that details the allowed interconnections between the intrinsically safe and associated apparatus or between the nonincendive field wiring apparatus and associated nonincendive field wiring apparatus.
- *Different intrinsically safe circuits* are intrinsically safe circuits in which the possible interconnections have not been evaluated and identified as intrinsically safe.
- *Intrinsically safe apparatus* is apparatus in which all the circuits are intrinsically safe.
- *Intrinsically safe circuit* is a circuit in which any spark or thermal effect is incapable of causing ignition of a mixture of flammable or combustible material in air under prescribed test conditions.
- *Intrinsically safe system* is an assembly of interconnected intrinsically safe apparatus, associated apparatus, and interconnecting cables in that those parts of the system that may be used in hazardous locations are intrinsically safe circuits.
- *Simple apparatus* is an electrical component or combination of components of simple construction with well-defined electrical parameters that does not generate more than 1.5 volts, 100 mA, and 25 mW or a passive component that does not dissipate more than 1.3 watts and is compatible with the intrinsic safety of the circuit in which it is used.

Section 504.10, Equipment Installation, provides in (A) that intrinsically safe apparatus, associated apparatus, and other equipment is to be installed in accordance with the control drawing(s). Compliance with the control drawings is essential if the level of safety is to be achieved. An improper hookup could allow infiltration of non–intrinsically safe levels of electrical energy into a hazardous area, resulting in an explosion and loss of life.

(B) Location states that intrinsically safe apparatus is permitted to be installed in any hazardous location for which it has been identified. General-purpose enclosures are permitted for intrinsically safe apparatus.

Associated apparatus is permitted to be installed in any hazardous location for which it has been identified or, if protected by other means, permitted by Articles 501 through 503 and Article 505.

A simple apparatus is permitted to be installed in any hazardous location in which the maximum surface temperature of the simple apparatus does not exceed the ignition temperature of the flammable gases or vapors, flammable liquids, combustible dusts, or ignitable fibers/flyings present.

Section 504.20, Wiring Methods, states that any of the wiring methods suitable for unclassified locations, including those covered by Chapters 7 and 8, are permitted for installation in intrinsically safe apparatus.

Section 504.30, Separation of Intrinsically Safe Conductors, provides that conductors of intrinsically safe circuits are not to be placed in any raceway, cable tray, or cable with conductors of any non–intrinsically safe circuit. There are some exceptions to this very definitive mandate, but the basic idea is that even though the intrinsically safe circuit may be nonincendive, it is necessary to ensure that power from other circuits does not infiltrate the intrinsically safe wiring so that it constitutes a hazard on entering the classified location.

Along these lines, it is further provided that within enclosures, conductors of intrinsically safe circuits are to be secured so that any conductor that might become loose from a terminal is unlikely to come into contact with another terminal.

Section 504.50, Grounding, provides that intrinsically safe apparatus, enclosures, and raceways, if of metal, are to be connected to the equipment-grounding conductor.

Section 504.60, Bonding, provides the following:

(A) In hazardous locations, intrinsically safe apparatus is to be bonded in the hazardous location in accordance with Section 250.100, Bonding in Hazardous Locations.

(B) In unclassified locations, where metal raceways are used for intrinsically safe system wiring in hazardous locations, associated apparatus is to be bonded in accordance with hazardous location sections as applicable.

Specialized bonding means must be in place outside the hazardous location if the enhanced bonding requirements for inside the hazardous location are to be meaningful.

Section 504.70, Sealing, states that conduits and raceways required to be sealed by previous articles are to be sealed where part of an intrinsically safe system. These seals are not required to be explosion-proof or flameproof, and they are not required for enclosures that contain only intrinsically safe apparatus.

Section 504.80, Identification, provides that terminals and wiring for intrinsically safe circuits are to be identified to prevent unintentional interference during testing and servicing. This identification must extend into the unclassified areas to ensure that non–intrinsically safe wire will not be added inadvertently to existing raceways at a later date.

Color coding is permitted to identify intrinsically safe conductors. Light blue is to be used for conductors, raceways, cable trays, and junction boxes.

To summarize, intrinsically safe wiring, with reduced energy levels and isolation from power wiring, is permitted in all hazardous locations. Where feasible, intrinsically safe wiring can greatly reduce the cost of wiring hazardous areas. As always, the strategy of choice is to locate wiring outside the hazardous areas or to install it in a Division 2 location instead of a Division1 location.

Article 505

Article 505, Zone 0, 1, and 2 Locations, provides that zone classification is an optional alternative to division classification. Zones 0, 1, and 2 correspond to Divisions 1 and 2 within Class I. Since there are three zones rather than two divisions, there is not an exact correspondence between the two systems. Zone 0 is more severe than Division 1, even though there is a partial overlap. Zone 0 is so hazardous that the only wiring method permitted therein is intrinsically safe wiring. To emphasize, explosion-proof motors, luminaires, and other explosion-proof equipment are not permitted in Class I, Zone 0 locations.

Zone classification is based on International Electrotechnical Commission (IEC) standards. Founded in 1906, the IEC is today a worldwide body headquartered in Geneva, Switzerland, that mandates electrical standards in a manner similar to the NEC. Much of the language is similar, although there are significant differences. Throughout NEC 2011, reference is made to IEC mandates, and at some time in the future, the two standards may merge. At present, zone classification for hazardous areas is the most prominent part of the NEC where IEC influence appears.

Many independent electricians and smaller electrical contracting firms stick to the division method of classification because it has worked for them in the past. Larger engineering and design firms may figure large projects on the basis of both systems and choose the one that is less expensive or easier to implement.

Licensing exams are more likely to focus on the division system. It is important to know the structure and location of zone language within the Code so that if the need arises, information can be accessed easily. Article 505 parallels Articles 500 and 501. It begins with a delineation of the zones, then lists protection techniques, and goes on to discuss use of equipment in Zones 0, 1, and 2.

Article 506

Article 506, Zone 20, 21, and 22 Locations for Combustible Dusts or Ignitable Fibers/Flyings, states that these zones are applicable to both Class II and Class III and that they are an optional alternative to the division categories. Combustible metallic dusts are not included in Article 506.

The article begins with definitions, proceeds to cover classification of locations, lists protective techniques, mandates wiring methods, and ends with a now-familiar grounding and bonding discussion.

Article 510

Article 510, Hazardous (Classified) Locations—Specific, is a brief statement of scope and contents. It looks ahead to Articles 511 through 517, which cover occupancies or parts of occupancies that are or may be hazardous because of atmospheric concentrations of flammable liquids, gases, or vapors or because of deposits or accumulations of materials that may be readily ignitable.

In contrast, Articles 518 through 590 do not address hazardous locations but do address locations that, because of the nature of their usage, require special wiring methods and materials above and beyond those required in the first four chapters of the Code.

Article 511

Article 511, Commercial Garages, Repair, and Storage, is an important article because, in a vastly automotive world, commercial garages are everywhere, and as a consequence, they comprise a large proportion of the electrician's workload. It is to be noted that the type of garage covered in this article does not include attached and detached residential garages associated with many residences. What about the backyard garage that aspires to provide work and part-time income for the owner? Trunk slammers commonly wire a facility of this sort with Type NM cable, flying through the air to power fluorescent fixtures hanging from the ceiling. This is definitely noncompliant. A garage of this sort typically has many older power tools and is filled with gasoline vapors, a recipe for disaster.

Section 511.1, Scope, states that these occupancies include locations used for service and repair operations in connection with self-propelled vehicles (including, but not limited to, passenger automobiles, buses, trucks, and tractors) in which volatile flammable liquids or flammable gases are used for fuel or power.

Section 511.2, Definitions, contains these terms:

- *Major repair garage.* A building or portions of a building where major repairs, such as engine overhauls, painting, body and fender work, and repairs that require draining of the motor vehicle fuel tank, are performed on motor vehicles, including associated floor space used for offices, parking, or showrooms.

- *Minor repair garage.* A building or portions of a building used for lubrication, inspection, and minor automotive maintenance work, such as engine tune-ups, replacement of parts, fluid changes (e.g., oil, antifreeze, transmission fluid, brake fluid, and air-conditioning refrigerants), brake system repairs, tire rotation, and similar routine maintenance work, including associated floor space used for offices, parking, or showrooms.

Section 511.3, Area Classification, General, provides that where Class I liquids or gaseous fuels are stored, handled, or transferred, electrical wiring and electrical utilization equipment is to be designed in accordance with the requirements for Class I, Division 1 or 2 hazardous locations. A Class I location does not extend beyond an unpierced wall, roof, or other solid partition that has no openings.

Class I liquids are flammable liquids, such as gasoline. Diesel fuel, with a flash point above 38°C, is a Class II combustible liquid. This terminology has no relation to and is not to be confused with Class I and Class II hazardous locations.

(A) Parking Garages states that garages used for parking or storage are unclassified.

(B) Repair Garages, with Dispensing, states that major and minor repair garages that dispense motor fuels into the fuel tanks of vehicles, including flammable liquids having a flash point below 38°C, such as gasoline, or gaseous fuels such as natural gas, hydrogen, or liquid petroleum gas (LPG), are to have the dispensing functions and components classified in accordance with Table 514.3(B)(1) in addition to any classification required by this section. Where Class I liquids other than fuels are dispensed, the area within 3 feet of any fill or dispensing point, extending in all directions, is a Class I, Division 2 location.

Table 514.3(B)(1) is a large, multipage table titled Class I Locations—Motor Fuel Dispensing Facilities. This is an example of how two articles interact. The requirements of both Article 511 and Article 514, Motor Fuel Dispensing Facilities, must be incorporated into the design of many commercial garages. There are, however, garages that do not dispense fuel and fuel dispensing facilities that do not service vehicles (Figure 5-5).

FIGURE 5-5   Emergency pump shutoff more than 20 feet but less than 100 feet from the dispensers, as required in Article 514, Motor Fuel Dispensing Facilities. This is an example of how Code articles interact.

(C) Major Repair Garages states that where flammable liquids having a flash point below 38°C, such as gasoline or gaseous fuels such as natural gas, hydrogen, or LPG, will not be dispensed, but repair activities that involve the transfer of such fluids or gases are performed, the classification rules in (1), (2), and (3) apply:

(1) Floor Areas provides the following:

    a. Ventilation Provided: The floor area is unclassified where there is mechanical ventilation providing a minimum of four air changes per hour. Ventilation must provide for air exchange across the entire floor area, and exhaust air is to be taken at a point within 12 inches of the floor.

    b. Ventilation Not Provided: The entire floor area up to a level of 18 inches above the floor is to be classified as Class I, Division 2.

(2) Ceiling Areas states that where lighter-than-air gaseous–fueled vehicles, such as vehicles fueled by natural gas or hydrogen, are repaired or stored, the area within 18 inches of the ceiling is to be considered for classification in accordance with (a) or (b) below:

    a. Ventilation Provided: The ceiling area is unclassified where ventilation is provided from a point not more than 18 inches from the highest point in the ceiling to exhaust the ceiling area at a rate of not less than 1 cubic foot per minute for each square foot of ceiling area at all times that the building is occupied or when vehicles using lighter-than-air gaseous fuels are parked below this area.

b. Ventilation Not Provided: Ceiling areas that are not ventilated are classified as Class I, Division 2.

(3) Pit Areas in Lubrication or Service Room states that any pit, below-grade work area, or subfloor work area is classified as in (a) or (b) below:

a. Ventilation Provided: Where ventilation is provided to exhaust the pit area at a rate of not less than 1 cubic foot per minute per square foot of floor area at all times that the building is occupied or when vehicles are parked in or over this area and where exhaust air is taken from a point within 12 inches of the floor of the pit, below-grade work area, or subfloor work area, the pit is unclassified.

b. Ventilation Not Provided: Where ventilation is not provided, any pit or depression below floor level is a Class I, Division 2 location that extends up to the floor level.

(E) Modifications to Classification provides the following:

(1) Specific Areas Adjacent to Classified Locations states that areas adjacent to classified locations in which flammable vapors are not likely to be released, such as stock rooms, switchboard rooms, and other similar locations, are unclassified where mechanically ventilated at a rate of four or more air changes per hour or designed with positive air pressure or where effectively cut off by walls or partitions.

(2) Alcohol-Based Windshield Washer Fluid states that the area used for storage, handling, or dispensing into motor vehicles of alcohol-based windshield washer fluid in repair garages is unclassified.

These are the rules that govern levels of classification in commercial garages. As can been seen, it is frequently possible and cost-effective to mitigate the hazard and alter or eliminate a classified designation by providing ventilation that meets Code specifications. Moreover, savings are possible by locating wiring outside a classified area. In a commercial garage, power receptacles should be located above bench level rather than close to the floor, as in a residence. Beware, however, that there are mandates regarding wiring installed above a Class I area, as we shall see.

Section 511.4, Wiring and Equipment in Class I Locations, notes that wiring and equipment classified in Article 511 must conform to applicable provisions of Article 501. This may seem obvious, but it is stated for the record.

Where fuel-dispensing units (other than LPG, which is prohibited) are located within buildings, the requirements of Article 514, Motor Fuel Dispensing Facilities, are applicable. If mechanical ventilation is provided in the dispensing area, the control is to be interlocked so that the dispenser cannot operate without ventilation.

Portable lighting equipment must be equipped with handle, lampholder, hook, and substantial guard attached to the lampholder or handle. All exterior surfaces that might come in contact with battery terminals, wiring terminals, or other objects are to be of nonconducting material or effectively protected with insulation. Lampholders are to be of an unswitched type and may not provide means for plug-in of attachment plugs. The outer shell is to be of molded composition or other suitable material. Unless the lamp and its cord are supported or arranged in such a manner that they cannot be used in the locations classified in Section 511.3, they are to be of a type identified for Class I, Division 1 locations.

Section 511.7, Wiring and Equipment Installed Above Class I Locations, provides that all fixed wiring above Class I locations must be in metal raceways; rigid nonmetallic conduit; electrical nonmetallic tubing; flexible metal conduit; liquid-tight flexible metal conduit; liquid-tight flexible nonmetallic conduit; Types MC, AC, and MI manufactured wiring systems; Type PLTC cable; Type TC cable; or Type ITC cable.

Electrical equipment in a fixed position must be located above the level of any defined Class I location or must be identified for the location.

Section 511.9, Sealing, notes that seals complying with provisions in Article 501 are to be provided and are required for both horizontal and vertical boundaries of defined Class I locations.

Section 511.12, Ground-Fault Circuit-Interrupter Protection for Personnel, provides that all 125-volt, single-phase, 15- and 20-ampere receptacles installed in areas where electrical diagnostic equipment, electrical hand tools, or portable lighting equipment are to be used must have ground-fault circuit-interrupter (GFCI) protection for personnel.

Section 511.16, Grounding and Bonding Requirements, reminds us that grounding in Class I locations is to comply with the specialized requirements that we discussed in connection with Section 501.30.

We have looked at the article on commercial garage wiring in detail because it is a frequent location for electrical work, both new and repair. Because floor slabs may be damp or have standing water, the potential for electrical shock is great. The rough environment demands robust protection for wiring and equipment. Add to the mix the presence of flammable liquids and hydrogen released from rapidly charging batteries, and the result is a sensitive setting to say the least. The Code helps us meet this challenge in a highly effective manner so that safety standards may be met.

Articles 513 through 516

Articles 513, Aircraft Hangars, through Article 516, Spray Application, Dipping, and Coating Processes, are referenced less frequently by working electricians. These articles are short and succinct, all organized similarly, and easy to understand. It you encounter an exam question on one of these topics, or if information is needed for a real-life job, it is just a matter of looking up the relevant section. To minimize Code hopping, start with the Contents and proceed to navigate the individual article as needed.

Article 517

Article 517, Health Care Facilities, is one of the more complex and challenging of Code topics. Electrical work in a hospital is difficult primarily because many of the clientele are disabled to varying degrees and unable to care for themselves. In case of an emergency, many of them would not be able to exit the facility without help. They may be more prone to injury in the event of fire or shock. Those undergoing surgery or on life support may not remain alive in the event of loss of electrical power, so that cannot be tolerated. Certain medical procedures involve contact with the bloodstream, which, being a saline liquid, constitutes a low-impedance conductor directly to the heart, where very low voltage can be fatal. Add to this the fact that there may be explosive gases present, and the entire setting becomes sensitive to mishap.

Health care facility wiring is difficult but rewarding for those who are able to assimilate the large quantity of information and acquire the expertise needed. You may not be ready to do the electrical installation for a new hospital construction project, but health care facilities often call in a local electrician to do troubleshooting and repair. These institutions are always doing renovations and expansion projects. The way to get started is to study the material carefully, perform every job, no matter how minor, in an impeccable and conscientious fashion, scrutinize your own work at every stage, and study some more. As always, the place to start is the NEC.

If you learn the contents and structure of Article 517, you certainly will score some licensing exam points. At first, the material may seem daunting, but with familiarity, it will be seen to be well organized and accessible.

Two major areas to focus on are definitions and grounding/bonding.

The definitions are particularly important in understanding the various subcategories of health care facilities, which is essential in applying mandates.

Following the overall Code template, Part I is titled General.

Section 517.1, Scope, states that the provisions of Article 517 apply to electrical construction and installation criteria in health care facilities that provide services to human beings. (A veterinary clinic or pet hospital would not be included.)

The requirements in Parts II and III not only apply to single-function buildings but are also intended to be applied individually to their respective forms of occupancy within a multifunction building (e.g., a doctor's examining room located within a limited-care facility would be required to meet the provisions of Section 517.10, Applicability).

As usual the NFPA offers other standards that are referenced in Article 517. These include NFPA 99, *Standard for Health Care Facilities*, NFPA 101, *Life Safety Code*, and NFPA 20, *Standard for the Installation of Stationary Pumps for Fire Protection*.

Section 517.2, Definitions, contains terms that are specific to this article. Here are the definitions that pertain to delineating various health care locations and branches of the overall electrical system:

- *Ambulatory health care occupancy.* A building or portion thereof used to provide services or treatment simultaneously to four or more patients that provides, on an outpatient basis, one or more of the following:

  1. Treatment for patients that renders the patients incapable of taking action for self-preservation under emergency conditions without the assistance of others

  2. Anesthesia that renders the patients incapable of taking action for self-preservation under emergency conditions without the assistance of others

  3. Emergency or urgent care for patients who, owing to the nature of their injury or illness, are incapable of taking action for self-preservation under emergency conditions without the assistance of others

- *Anesthetizing location.* Any area of a facility that has been designated to be used for the administration of any flammable or nonflammable inhalation anesthetic agent in the course of examination or treatment, including the use of such agents for relative analgesia.

- *Critical branch.* A subsystem of the emergency system consisting of feeders and branch circuits supplying energy to task illumination, special power circuits, and selected receptacles serving areas and functions related to patient care and that are connected to alternate power sources by one or more transfer switches during interruption of the normal power source.

- *Emergency system.* A system of circuits and equipment intended to supply alternate power to a limited number of prescribed functions vital to the protection of life and safety.

- *Equipment system.* A system of circuits and equipment arranged for delayed, automatic, or manual connection to the alternate power source and that serves primarily three-phase equipment.

- *Essential electrical system.* A system comprised of alternate sources of power and all connected distribution systems and ancillary equipment designed to ensure continuity of electrical power to designated areas and functions of a health care facility during disruption of normal power sources and also to minimize disruption within the internal wiring system.

- *Flammable anesthetizing location.* Any area of the facility that has been designated to be used for the administration of any flammable inhalation anesthetic agents in the normal course of examination or treatment. (These agents are not normally used anymore, but the Code retains coverage of the location.)

- *Health care facilities.* Buildings or portions of buildings in which medical, dental, psychiatric, nursing, obstetrical, or surgical care is provided. Health care facilities include, but are not limited to, hospitals, nursing homes, limited-care facilities, clinics, medical and dental offices, and ambulatory care centers, whether permanent or movable.

- *Hospital.* A building or portion thereof used on a 24-hour basis for the medical, psychiatric, obstetrical, or surgical care of four or more inpatients.

- *Isolated power system.* A system comprising an isolating transformer or its equivalent or a line-isolation monitor and its ungrounded circuit conductors.

- *Life safety branch.* A subsystem of the emergency system consisting of feeders and branch circuits meeting the requirements of Article 700 and intended to provide adequate power needs to ensure safety to patients and personnel and that is automatically connected to alternate power sources during interruption of the normal power source.

- *Limited-care facility.* A building or portion thereof used on a 24-hour basis for the housing of four or more persons who are incapable of self-preservation because of age, physical limitation owing to accident or illness, or limitations such as mental retardation/developmental disability, mental illness, or chemical dependency.

- *Nurses' stations.* Areas intended to provide a center of nursing activity for a group of nurses serving bed patients, where patient calls are received, nurses are dispatched, nurses' notes are written, inpatient charts are prepared, and medications are prepared for distribution to patients. Where such activities are carried on in more than one location within a nursing unit, all such separate areas are considered a part of the nurses' station.

- *Nursing home.* A building or portion of a building used on a 24-hour basis for the housing and nursing care of four or more persons who, because of mental or physical incapacity, might be unable to provide for their own needs and safety without the assistance of another person.

- *Patient bed location.* The location of a patient sleeping bed or the bed or procedure table of a critical-care area.

- *Patient-care area.* Any portion of a health care facility wherein patients are intended to be examined or treated. Areas of a health care facility in which patient care is administered are classified as general-care areas or critical-care areas. The governing body of the facility designates these areas in accordance with the type of patient care anticipated and with the following definitions of the area classification:

  1. *General care areas.* Patient bedrooms, examining rooms, treatment rooms, clinics, and similar areas in which it is intended that the patient will come in contact with ordinary appliances such as a nurse call system, electric beds, examining lamps, telephones, and entertainment devices.

  2. *Critical-care areas.* Those special-care units, intensive-care units, coronary-care units, angiography laboratories, cardiac catheterization laboratories, delivery rooms, operating rooms, and similar areas in which patients are intended to be subjected to invasive procedures and connected to line-operated, electromedical devices.

  3. *Wet procedure locations.* Those spaces within patient-care areas where a procedure is performed that is normally subject to wet conditions while patients are present. This includes standing fluids on the floor or drenching of the work area, either of which condition is intimate to the patient or staff. Routine housekeeping procedures and incidental spillage of liquids do not define a wet procedure location.

- *Patient-care vicinity.* In an area in which patients are normally cared for, the patient-care vicinity is the space with surfaces likely to be contacted by the patient or an attendant who can touch the patient. Typically in a patient room, this encloses a space within the room not less than 6 feet beyond the perimeter of the bed in its nominal location and extending vertically not less than 7½ feet above the floor.

- *Psychiatric hospital.* A building used exclusively for the psychiatric care on a 24-hour basis of four or more inpatients.

Part II, Wiring and Protection, covers wiring methods required to be used in patient-care areas of all health care facilities. It is here we find the specifications for redundant grounding. Part II does not apply to business offices, corridors, waiting rooms, and the like in clinics, medical and dental offices, and outpatient facilities. It also does not apply to areas of nursing homes and limited-care facilities where these areas are used exclusively as patient sleeping areas.

To emphasize, Part II, with its stringent, redundant grounding requirements, applies to patient-care areas of all health care facilities but not to areas that are not designated as patient-care areas.

Section 517.11, General Installation—Construction Criteria, states that the purpose of Article 517 is to specify the installation criteria and wiring methods that minimize electrical hazards by the maintenance of adequately low-potential differences only between exposed conductive surfaces that are likely to become energized and could be contacted by a patient.

Section 517.13, Grounding of Receptacles and Fixed Electrical Equipment in Patient Care Areas, contains two subsections that together make up the protocol for redundant grounding in the areas where it is required.

(A) Wiring Methods provides that all branch circuits serving patient-care areas are to be provided with an effective ground-fault current path by installation in a metal raceway system, or a cable having a metallic cable armor, or sheath assembly. This must itself qualify as an equipment-grounding conductor.

Accordingly, Type NM (trade name Romex) is absolutely excluded from use in patient-care areas. The permitted types enclose the conductors in metal and thereby provide a higher level of protection, but this is not the purpose of the requirement in this section. The purpose is that the metal raceway or metal sheath constitutes one part of the redundant grounding designed to provide enhanced reliability for grounding in patient-care areas. The second part of the redundant grounding system is as follows:

(B) Insulated Equipment Grounding Conductor provides the following:

(1) General states that the following must be connected directly to an insulated copper equipment-grounding conductor that is installed with the branch-circuit conductors:

- The grounding terminals of all receptacles
- Metal boxes and enclosures containing receptacles
- All non-current-carrying conductive surfaces of fixed electrical equipment likely to become energized that are subject to personal contact and operate at over 100 volts

An exception allows an insulated equipment bonding jumper that connects directly to the equipment-grounding conductor, which is permitted to connect the box and receptacle(s) to the equipment-grounding conductor. Metal faceplates, luminaires more than 7½ feet above the floor, and switches located outside the patient vicinity are exempted from redundant grounding requirements.

It is further provided that equipment-grounding conductors and bonding jumpers are to be sized in accordance with Section 250.122.

Section 517.14, Panelboard Bonding, states that the equipment-grounding terminal buses of the normal and essential branch-circuit panelboards serving the same individual patient-care vicinity are to be connected together with an insulated continuous copper conductor not smaller than 10 American Wire Gauge (AWG). Where two or more panelboards serving the same individual patient-care vicinity are served from separate transfer switches on the emergency system, the equipment-grounding terminal buses of those panelboards are to be connected together with an insulated continuous copper conductor not smaller than 10 AWG. This conductor is permitted to be broken in order to terminate on the equipment-grounding terminal bus in each panelboard.

Section 517.18, General Care Areas, provides in (A) that each patient bed location is to be supplied by at least two branch circuits, one from the emergency system and one from the normal system. All branch circuits from the normal system must originate in the same panelboard.

The branch circuit serving patient bed locations may not be part of a multiwire branch circuit. There are three exceptions:

Number 1: Branch circuits serving only special-purpose outlets or receptacles, such as portable x-ray outlets, are not required to be served from the same distribution panel or panels.

Number 2: Requirements of Section 517.18(A) do not apply to patient bed locations in clinics, medical and dental offices, and outpatient facilities; psychiatric, substance abuse, and rehabilitation hospitals; and sleeping rooms of nursing homes and limited-care facilities.

Number 3: A general-care patient bed location served from two separate transfer switches on the emergency system is not required to have circuits from the normal system.

(B) Patient Bed Location Receptacles states that each patient bed location is to be provided with a minimum of four receptacles. They are permitted to be single, duplex, or quadruplex type or any combination of the three. All receptacles, whether four or more, are to be listed "hospital grade" and so identified (Figure 5-6).

The grounding terminal of each receptacle is to be connected to an insulated copper equipment-grounding conductor sized in accordance with Table 250.122. There are two exceptions:

FIGURE 5-6   Duplex hospital-grade receptacle required for patient bed locations.

Number 1: These requirements do not apply to psychiatric, sub-stance abuse, and rehabilitation hospitals.

Number 2: Psychiatric security rooms are not required to have receptacle outlets installed in the room.

(C) Pediatric Locations states that receptacles located within the rooms, bathrooms, playrooms, activity rooms, and patient-care areas of designated pediatric locations are to be listed tamper resistant or are to employ a listed tamper-resistant cover.

Section 517.19, Critical Care Areas, provides that branch-circuit requirements are the same as for patient bed locations in general-care areas. However, each patient bed location in these areas is required to have at least six receptacles, as opposed to the four required for each patient bed location in general-care areas. Additionally, the grounding terminal of each receptacle is required to be connected to the reference grounding point by means of an insulated copper equipment-grounding conductor.

Section 517.20, Wet Procedure Locations, provides that wet procedure location patient areas are to be provided with special protection against electric shock by one of the following means:

- Power distribution system that inherently limits the possible ground-fault current owing to a first fault to a low value without interrupting the power supply
- Power distribution system in which the power supply is interrupted if the ground-fault current does in fact exceed a value of 6 mA

Section 517.21, Ground-Fault Circuit-Interrupter Protection for Personnel, provides that GFCI protection for personnel is not required for receptacles installed in critical-care areas where the toilet and basin are installed within the patient room.

GFCI nuisance tripping could be problematic for patients requiring electrical power for treatment or monitoring. A toilet and basin in the patient room would define the room as a bathroom, with GFCI required if this section did not provide otherwise.

Part III, Essential Electrical System, describes the overall structure of various health care facility electrical systems, including switching and assigning loads. Power for a health care facility typically consists of a utility-supplied normal source and an alternate source most commonly supplied by an onsite diesel generator, although any number of other configurations are possible.

The load is divided into nonessential loads and the essential electrical system. The essential electrical system is divided into the equipment system and the emergency system. The emergency system is divided into the life-safety branch and the critical branch. All these are connected to transfer switches that connect them to the normal power source and the alternate power source, except for the nonessential loads, which are connected directly (with no transfer switch) to the normal power source. This is the minimum setup for systems greater than 150 kVA.

(A) Applicability states that the requirements of Part III, Sections 517.30 through 517.35, apply to hospitals where an essential electrical system is required. Later sections apply to nursing homes and limited-care facilities.

(B) General contains information on hospital loads and transfer switches.

(1) Separate Systems states that essential electrical systems for hospitals are comprised of two separate systems capable of supplying a limited amount of lighting and power service that is considered essential for life safety and effective hospital operation during the time the normal electrical service is interrupted for any reason. These two systems are the emergency system and the equipment system.

(2) Emergency systems are limited to circuits essential to life safety and critical patient care. These are designated the life-safety branch and the critical branch.

(3) Equipment System supplies major electrical equipment necessary for patient care and basic hospital operation.

(4) Transfer Switches states that the number of transfer switches to be used is based on reliability, design, and load considerations. Each branch of the emergency system and each equipment system is to have one or more transfer switches. One transfer switch is permitted to serve one or more branches or systems in a facility with a maximum demand on the essential electrical system of 150 kVA.

(C) Wiring Requirements provides the following:

(1) Separation from Other Circuits states that the life-safety branch and critical branch of the emergency system are to be kept entirely independent of all other wiring and equipment and may not enter the same raceways, boxes, or cabinets with each other or with other wiring.

Where general-care locations are served from two separate transfer switches on the emergency system, the general-care circuits from the two separate systems must be kept independent of each other.

Where critical-care locations are served from two separate transfer switches on the emergency system, the critical-care circuits from the two separate systems are to be kept independent of each other.

Under certain limited conditions, wiring of the life-safety branch and the critical branch are permitted to occupy the same raceways, boxes, or cabinets of other circuits not part of the branch where such wiring

1. Is in transfer-equipment enclosures

2. Is in exit or emergency luminaires supplied from two sources

3. Is in a common junction box attached to exit or emergency luminaires supplied from two sources

4. Is for two or more emergency circuits supplied from the same branch and same transfer switch

The wiring of the equipment system may occupy the same raceways, boxes, or cabinets of other circuits that are not part of the emergency system.

It is further provided that wiring of the emergency systems in hospitals is to be mechanically protected. Where installed as branch circuits in patient-care areas, the installation is to comply with the requirements of Sections 517.13(A) and (B), which provide for redundant grounding. The following wiring methods are permitted:

1. Nonflexible metal raceways, Type MI cable, or Schedule 80 PVC conduit (Nonmetallic raceways may not be used for branch circuits that supply patient-care areas.)

2. Where encased in not less than 2 inches of concrete, Schedule 40 PVC conduit, flexible nonmetallic or jacketed metallic raceways, or jacketed metallic cable assemblies listed for installation in concrete

3. Listed flexible metal raceways and listed metal-sheathed cable assemblies in any of the following:

   a. Where used in listed prefabricated medical headwalls

   b. Where used in listed office furnishings

    c. Where fished into existing walls or ceilings not otherwise accessible and not subject to physical damage

    d. Where necessary for flexible connection to equipment

  4. Flexible power cords of appliances or other utilization equipment connected to the emergency system

  5. Cables for Class 2 or Class 3 systems permitted by Part VI of Article 517, with or without raceways

(D) Capacity of Systems states that the essential electrical system is to have adequate capacity to meet the demand for all functions and equipment to be served by each system and branch.

Demand calculations for sizing of the generator set(s) are to be based on any of the following:

  1. Prudent demand factors and historical data

  2. Connected load

  3. Feeder calculation procedures described in Article 220

  4. Any combination of the above

When it comes to generator sizing for health care facility essential electrical systems, the Code does not employ its usual formulaic mandate mode. Instead, it defers to the judgment of the designers, who make the decision based on experience and other factors of their choosing. Oversizing a generator for this function can create problems. A diesel engine runs poorly if it is lightly loaded a large proportion of the time.

(E) Receptacle Identification provides that the cover plates for the electrical receptacles or the receptacles themselves supplied by the emergency system must have a distinctive color or marking so as to be readily identifiable. The color is not specified, but red is used commonly (Figure 5-7).

Section 517.31, Emergency System, states that those functions of patient care depending on lighting or appliances that are connected to the emergency system are divided into two mandatory branches: the life-safety branch and the critical branch. The branches of the emergency system are to be installed and connected to the alternate power source so that all functions specified for the emergency system will be restored automatically to operation within 10 seconds of interruption of the normal source.

Section 517.32, Life Safety Branch, specifies that no functions other than those listed in (A) through (H) below are to be connected to the life-safety branch:

(A) Illumination of means of egress, such lighting for corridors, passageways, stairways, and landings at exit doors and all necessary ways of approach to exits

Figure 5-7 Hospital receptacle supplied by the emergency system, with distinctive color so as to be readily identifiable.

(B) Exit signs

(C) Alarm and alerting systems, including fire alarms

(D) Communications systems, where used for issuing instructions during emergency conditions

(E) Generator and transfer switch locations, necessary for maintenance during an outage

(F) Generator set accessories required for operation

(G) Elevators

(H) Automatic doors, used for building egress

It is important that the life-safety branch be limited so that the power supply and wiring are absolutely reliable.

Section 517.33, Critical Branch, states that it is necessary to supply the following loads, listed in (A):

1. Critical-care areas that use anesthetizing gases—task illumination, selected receptacles, and fixed equipment

2. Isolated power systems in special environments

3. Patient-care areas—task illumination and selected receptacles in the following:

   a. Infant nurseries

   b. Medication preparation areas

   c. Pharmacy dispensing areas

   d. Selected acute nursing areas

   e. Psychiatric bed areas (omit receptacles)

   f. Ward treatment rooms

g. Nurses stations (unless adequately lighted by corridor luminaires)

4. Additional specialized patient-care task illumination and receptacles, where needed

5. Nurse call systems

6. Blood, bone, and tissue banks

7. Telephone equipment rooms and closets

8. Task illumination, selected receptacles, and selected power circuits for the following:

   a. General-care beds (at least one duplex receptacle in each patient bedroom)

   b. Angiographic labs

   c. Cardiac catheterization labs

   d. Coronary-care units

   e. Hemodialysis rooms or areas

   f. Emergency room treatment areas (selected)

   g. Human physiology labs

   h. Intensive-care units

   i. Postoperative recovery rooms (selected)

9. Additional task illumination, receptacles, and selected power circuits needed for effective hospital operation (Single-phase fractional horsepower motors are permitted to be connected to the critical branch.)

(B) Subdivision of the Critical Branch states that it is permitted to subdivide the critical branch into two or more branches.

An Informational Note states that it is important to analyze the consequences of supplying an area with only critical-care branch power when failure occurs between the area and the transfer switch. Some proportion of normal and critical power from separate transfer switches may be appropriate.

Section 517.34, Equipment System Connection to Alternate Power Source, provides that the equipment system is to be installed and connected to the alternate power source such that the equipment is restored automatically to operation at appropriate time-lag intervals following the energizing of the emergency system. An exception states that for essential electrical systems under 150 kVA, deletion of the time-lag intervals feature for delayed automatic connection to the equipment system is permitted.

The section continues by listing equipment for delayed automatic connection and delayed automatic or manual connection.

Section 517.35, Sources of Power, states that the alternate source of power is to be one of the following:

1. Generator(s) driven by some form of prime mover(s) and located on the premises
2. Another generating unit(s) where the normal source consists of a generating unit(s) located on the premises
3. An external utility service when the normal source consists of a generating unit(s) located on the premises
4. A battery system located on the premises

As can be seen, a number of configurations are possible, although by far the most common is utility power for the normal source with the alternate source consisting of an on-site diesel generator. In the future, it is probable that solar power with on-site storage will play a role.

Section 517.40, Essential Electrical Systems for Nursing Homes and Limited Care Facilities, takes note of the fact that these institutions may or may not have patients or residents who may need to be sustained by electrical life-support equipment.

For nursing homes and limited-care facilities that admit patients who need to be sustained by electrical life-support equipment, the essential electrical system from the source to the portion of the facility where such patients are treated must comply with the requirements of Part III, Sections 517.30 through 517.35.

Section 517.41, Essential Electrical Systems, states that essential electrical systems for nursing homes and limited-care facilities are to be comprised of two separate branches capable of supplying a limited amount of lighting and power service that is considered essential for the protection of life safety and effective operation of the institution during the time normal electrical service is interrupted for any reason. These two separate branches are to be the life-safety branch and the critical branch.

The number of transfer switches to be used is to be based on reliability, design, and load considerations. Each branch of the essential electrical system is to be served by one or more transfer switches. One transfer switch is permitted to serve one or more branches or systems in a facility with a maximum demand on the essential electrical system of 150 kVA.

Section 517.42, Automatic Connection to Life Safety Branch, provides that the life-safety branch is to be installed and connected to the alternate source of power so that all functions specified in this section are restored automatically to operation within 10 seconds of the interruption of the normal source. No function other than those listed may be connected to the life-safety branch. The list is the same as for hospitals, except that there is no mention of automatic doors.

Section 517.43, Connection to Critical Branch, provides that the critical branch is to be installed and connected to the alternate power source so that the equipment listed for delayed automatic connection is restored automatically to operation at appropriate time-lag intervals following restoration of the life-safety branch to operation.

Section 517.44, Sources of Power, provides that essential electrical systems are to have a minimum of two independent sources of power: a normal source supplying the entire electrical system and one or more alternate sources for use when the normal source is interrupted. The alternate source of power is to be a generator driven by an on-site prime mover. An exception allows the normal source to consist of generating units on the premises and the alternate source either another generator or an external utility service. Another exception allows certain nursing homes and limited-care facilities to use a battery system or self-contained battery integral with the equipment.

At one time flammable inhalation anesthetics such as ether were used widely for surgical procedures in health care facilities. Their use has become obsolete in the United States, but the Code covers electrical installations where they are present in Part IV, Inhalation Anesthetizing Locations. Licensing exams are not likely to focus on this material, but if reference to it is required, the mandates are simple and well organized in Part IV.

Part V, X-Ray Installations, stipulates that nothing in Part V is to be construed as specifying safeguards against the useful beam or stray x-ray radiation.

Section 517.71, Connection to Supply Circuit, provides in (A) that fixed and stationary x-ray equipment is to be connected to the power supply by means of a wiring method complying with applicable requirements of NEC Chapters 1 through 4, as modified by Article 517. An exception states that equipment properly supplied by a branch circuit rated at not over 30 amperes is permitted to be supplied through a suitable attachment plug and hard-service cable or cord.

(B) Portable, Mobile, and Transportable Equipment states that individual branch circuits are not required for portable, mobile, and transportable medical x-ray equipment requiring a capacity of not over 60 amperes.

(C) Over 600-Volt Supply provides that circuits and equipment operated on a supply circuit of over 600 volts must comply with Article 490.

Section 517.72, Disconnecting Means, states in (A) that a disconnecting means of adequate capacity for at least 50 percent of the input required for the momentary rating or 100 percent of the input required for the long-time rating of the x-ray equipment, whichever is greater, is to be provided for the supply circuit.

(B) Location provides that the disconnecting means is to be operable from a location readily accessible from the x-ray control.

(C) Portable Equipment states that for equipment connected to a 120-volt branch circuit of 30 amperes or less, a grounding-type attachment plug and receptacle of proper rating is permitted to serve as the disconnecting means.

Section 517.73, Rating of Supply Conductors and Overcurrent Protection, is divided into two subsections by equipment function:

(A) Diagnostic Equipment provides the following:

(1) Branch Circuits states that the ampacity of supply branch-circuit conductors and the current rating of overcurrent protective devices is to be not less than 50 percent of the momentary rating or 100 percent of the long-time rating, whichever is greater.

(2) Feeders states that the ampacity of supply feeders and the current rating of overcurrent protective devices supplying two or more branch circuits supplying x-ray units must not be less than 50 percent of the momentary demand rating of the largest unit plus 25 percent of the momentary demand rating of the next largest unit plus 10 percent of the momentary demand rating of each additional unit. Where simultaneous biplane examinations are undertaken with x-ray units, the supply conductors and overcurrent protective devices must be 100 percent of the momentary demand rating of each x-ray unit.

An Informational Note states that the minimum conductor size for branch and feeder circuits is also governed by voltage regulation requirements. For a specific installation, the manufacturer usually specifies minimum distribution transformer and conductor sizes and rating of disconnecting means and overcurrent protection.

(B) Therapeutic Equipment states that the ampacity of conductors and rating of overcurrent protective devices are not to be less than 100 percent of the current rating of medical x-ray equipment.

Section 517.74, Control Circuit Conductors, provides the following:

(A) Number of Conductors in Raceway states that this number is to be determined in accordance with Section 300.17, Number and Size of Conductors in Raceway.

(B) Minimum Size of Conductors provides that size 18 AWG or 16 AWG fixture wires and flexible cords are permitted for the control and operating circuits of x-ray and auxiliary equipment where protected by not larger than 20-ampere overcurrent devices.

Section 517.77, Installation of High-Tension X-Ray Cables, provides that cables with grounded shields connecting x-ray tubes and image intensifiers are permitted to be installed in cable trays or cable troughs along with x-ray equipment control and power conductors without the need for barriers to separate the wiring.

Section 517.78, Guarding and Grounding, states in (A) that all high-voltage parts, including x-ray tubes, are to be mounted within grounded enclosures. Air, oil, gas, or other suitable insulating media are to be used to insulate the high-voltage from the grounded enclosure. The connection from the high-voltage equipment to x-ray tubes and other high-voltage components is to be made with high-voltage shielded cables.

(B) Low-Voltage Cables states that such cables connecting to oil-filled units that are not completely sealed, such as transformers, condensers, oil coolers, and high-voltage switches, are to have insulation of the oil-resistant type.

(C) Non-Current-Carrying Metal Parts of X-Ray and Associated Equipment (controls, tables, x-ray tube supports, transformer tanks, shielded cables, x-ray tube heads, etc.) states that such parts are to be connected to an equipment-grounding conductor in the manner specified in Part VII of Article 250, as modified by Sections 517.13(A) and (B).

Part VI, Communications, Signaling Systems, Data Systems, Fire Alarm Systems, and Systems Less than 120 Volts, Nominal, contains information on these so-called low-voltage systems.

Section 517.80, Patient Care Areas, provides that equivalent insulation and isolation to that required for the electrical distribution systems in patient-care areas must be provided for communications, signaling system, data system circuits, fire alarm systems, and systems less than 120 volts, nominal.

Class 2 and Class 3 signaling and communications systems and power-limited fire alarm systems are not required to comply with the grounding requirements of Section 517.13, to comply with the mechanical protection requirements of Section 517.30(C)(3)(5), or to be enclosed in raceways unless otherwise specified by Chapter 7 or 8.

Secondary circuits of transformer-powered communications or signaling systems are not required to be in raceways unless otherwise specified by Chapter 7 or 8.

Section 517.81, Other Than Patient Care Areas, states that installations are to be in accordance with the applicable provisions of other parts of the Code.

Section 517.82, Signal Transmission Between Appliances, states that permanently installed signal cabling from an appliance in a patient location to remote appliances must employ a signal-transmission system that prevents hazardous grounding interconnection of the appliances.

Common signal grounding wires (i.e., the chassis ground for single-ended transmission) are permitted to be used between appliances in all located within the patient-care vicinity, provided that the appliances are served from the same reference grounding point.

We have covered Class I, Class II, and Class III hazardous areas and health care facilities in detail because the wiring and equipment requirements are very exacting. These articles require careful scrutiny. It is certain that some licensing exam questions will focus on these areas.

The remaining articles in Chapter 5 do not address hazardous locations per se, but extra care in designing and installing wiring for such areas is called for. The answers to exam questions on these locations are easy to locate because the articles are brief and are all organized in the same manner. It is recommended that you know the locations in this article in order to avoid code hopping.

NEC® Chapter 6, Special Equipment

S ome types of electrical equipment fall into a category that is not particularly hazardous but nevertheless, because of their unique construction and use, require special National Electrical Code (NEC) treatment. Each of these is discussed in one of the 25 articles that comprise Chapter 6, Special Equipment. Most of these articles are fairly brief, and we shall not examine them individually. The requirements are easy to find because each article is organized in the same manner, beginning with a statement of scope and definitions and ending with a discussion of grounding or some other subtopic that is peculiar to the equipment under consideration. At all times, the Code proceeds from the more general to the more specific.

What might be difficult, given the time constraints of an open-book exam, is determining where to look for a particular type of equipment. Will it be in Chapter 4, Equipment for General Use, or in Chapter 6, Special Equipment? You could consult the Contents, but a better approach is to consider whether the equipment is, in fact, for general use, such as switches and appliances, or more specialized, such as welders and elevators.

Since the material is easy to access on an as-needed basis, we shall examine only two articles because they are of great interest at this time.

Article 690

Article 690, Solar Photovoltaic (PV) Systems, pertains to all nonutility PV installations, whether stand-alone or interactive (Figure 6-1).

Part I, General, provides an introduction to and overview of PV system structures and requirements.

Section 690.1, Scope, states that the provisions of Article 690 apply to solar PV electrical energy systems, including the array circuit(s), inverter(s), and controller(s) for such systems. Solar PV systems may be interactive with other electrical power production sources or

Figure 6-1 This solar array provides power for a dwelling and feeds excess into the grid through an inverter that provides synchronized alternating current (ac) to the utility.

stand-alone, with or without electrical energy storage such as batteries. These systems may have alternating-current (ac) or direct-current (dc) output for utilization.

Section 690.2, Definitions, defines terms that are specific to Article 690 and do not appear in Article 100 definitions:

Alternating-current PV module—A complete, environmentally protected unit consisting of solar cells, optics, inverter, and other components, exclusive of tracker, designed to generate ac power when exposed to sunlight.

Array—A mechanically integrated assembly of modules or panels with a support structure and foundation, tracker, and other components, as required, to form a dc power-producing unit.

Bipolar PV array—A PV array that has two outputs, each having opposite polarity to a common reference point or center tap.

Blocking diode—A diode used to block reverse flow of current into a PV source circuit.

Building-integrated photovoltaics—PV cells, devices, modules, or modular materials that are integrated into the outer surface or structure of a building and serve as the outer protective surface of that building.

Charge controller—Equipment that controls dc voltage or dc current or both that is used to charge a battery.

Diversion charge controller—Equipment that regulates the charging process of a battery by diverting power from energy storage to dc or ac loads or to an interconnected utility service.

Electrical production and distribution network—A power production, distribution, and utilization system, such as a utility system and connected loads, that is external to and not controlled by the PV power system.

Hybrid system—A system comprised of multiple power sources. These power sources may include PV, wind, micro hydro generators, engine-driven generators, and others but do not include electrical production and distribution network systems. Energy storage systems, such as batteries, do not constitute a power source for the purpose of this definition.

Interactive system—A solar PV system that operates in parallel with and may deliver power to an electrical production and distribution network. For the purpose of this definition, an energy storage subsystem of a solar PV system, such as a battery, is not another electrical production source.

Inverter—Equipment that is used to change voltage level or waveform or both of electrical energy. Commonly, an inverter is a device that changes dc input to ac output. Inverters also may function as battery chargers that use ac from another source and convert it into dc for charging batteries.

Inverter input circuit—Conductors between the inverter and the battery in stand-alone systems or the conductors between the inverter and the PV output circuits for an electrical production and distribution network.

Inverter output circuit—Conductors between an inverter and an ac panelboard for stand-alone systems or the conductors between the inverter and service equipment or another electrical power production source, such as a utility, for an electrical production and distribution network.

Module—A complete, environmentally protected unit consisting of solar cells, optics, and other components, exclusive of tracker, designed to generate dc power when exposed to sunlight.

Monopole subarray—A PV subarray that has two conductors in the output circuit, one positive and one negative. Two monopole PV subarrays are used to form a bipolar PV array.

Panel—A collection of modules mechanically fastened together, wired, and designed to provide a field-installable unit.

PV output circuit—Circuit conductors between the PV source circuit(s) and the inverter or dc utilization equipment.

PV power source—An array or aggregate of arrays that generates dc power at system voltage and current.

PV source circuit—Circuits between modules and from modules to the common connection point(s) of a dc system.

PV system voltage—The dc voltage of any PV source or PV output circuit. For multiwire installations, the PV system voltage is the highest voltage between any two dc conductors.

Solar cell—The basic PV device that generates electricity when exposed to light.

Solar PV system—The total components and subsystems that, in combination, convert solar energy into electrical energy suitable for connection to a utilization load.

Stand-alone system—A solar PV system that supplies power independently of an electrical production and distribution network.

Subarray—An electrical subset of a PV array.

Section 690.3, Other Articles, specifies the relationship between Article 690 and other portions of the Code. It states that wherever the requirements of other NEC articles and Article 690 differ, the requirements of Article 690 apply, and if the system is operated in parallel with a primary source of electricity, the requirements in Article 705, Interconnected Electric Power Production Sources, apply.

An exception states that solar PV systems, equipment, or wiring installed in a hazardous location also must comply with the applicable portions of Articles 500 through 516.

Section 690.4, Installation, covers basic requirements, notably connections among parts of the system and among multiple systems.

(A) PV Systems provides that such systems are permitted to supply a building or other structure in addition to any other electricity supply system(s).

(B) Identification and Grouping provides that PV source circuits and PV output circuits are not to be contained in the same raceway, cable tray, cable, outlet box, junction box, or similar fitting as conductors, feeders, or branch circuits of other non-PV systems unless the conductors of the different systems are separated by a partition. PV system conductors are to be identified and grouped as required by (1) through (4) below. The means of identification may be by separate color coding, marking tape, tagging, or other approved means.

(1) PV Source Circuits states that such circuits are to be identified at all points of termination, connection, and splices.

(2) PV Output and Inverter Circuits states that these conductors are to be identified at all points of termination, connection, and splices.

(3) Conductors of Multiple Systems states that where these occupy the same junction box, raceway, or equipment, the conductors of each system are to be identified at all termination, connection, and splice points.

(4) Grouping states that where the conductors of more than one PV system occupy the same junction box or raceway with a removable cover(s), the ac and dc conductors of each system are to be grouped separately by wire ties or similar means at least once and then are to be grouped at intervals not to exceed 6 feet. An exception states that the requirement for grouping does not apply if the circuit enters from a cable or raceway unique to the circuit that makes the grouping obvious.

(C) Module Circuit Arrangement provides that the connection to a module or panel is to be arranged so that removal of a module or panel from a PV source circuit does not interrupt a grounded conductor to other PV source circuits.

(D) Equipment provides that inverters, motor generators, PV modules, PV panels, ac PV modules, source-circuit combiners, and charge controllers intended for use in PV power systems are to be identified and listed for the application.

(E) Wiring and Connections states that the preceding equipment and all associated wiring and interconnections are to be installed only by qualified persons.

(F) Circuit Routing states that PV source and PV output conductors, in and out of conduit and inside of a building or structure, are to be routed along building structural members such as beams, rafters, trusses, and columns where the location of those structural members can be determined by observation. Where circuits are imbedded in built-up, laminate, or membrane roofing materials in roof areas not covered by PV modules and associated equipment, the location of circuits must be clearly marked.

(G) Bipolar PV Systems states that where the sum, without consideration of polarity, of the PV system voltages of the two monopole subarrays exceeds the rating of the conductors and connected equipment, monopole subarrays in a bipolar PV system are to be physically separated, and the electrical output circuits from each monopole subarray are to be installed in separate raceways until connected to the inverter. The disconnecting means and overcurrent protective devices for each monopole subarray output are to be in separate enclosures. All conductors from each separate monopole subarray are to be routed in the same raceway.

(H) Multiple Inverters states that a PV system is permitted to have multiple utility-interactive inverters installed in or on a single building or structure. Where the inverters are remotely located from each other, a directory in accordance with Section 705.10, Directory (for Interconnected Electric Power Production Sources), is to be installed at each dc PV system–disconnecting means and at the main service–disconnecting means showing the location of all ac and dc PV system–disconnecting means in the building.

Section 690.5, Ground-Fault Protection, specifies that grounded dc PV arrays are to be provided with dc ground-fault protection meeting the requirements of (A) through (C) below to reduce fire hazards.

(A) Ground-Fault Detection and Interruption states that the ground-fault protection device or system is to be capable of detecting a ground-fault current, interrupting the flow of the fault current, and providing an indication of the fault. Automatically opening the grounded conductor of the faulted circuit to interrupt the ground-fault current path is permitted. If a grounded conductor is opened to interrupt the ground-fault current path, all conductors are to be opened automatically and simultaneously. Manual operation of the main PV dc disconnect is not to activate the ground-fault protection device or result in grounded conductors becoming ungrounded.

(B) Isolating Faulted Circuits states that the faulted circuits are to be isolated by one of the two following methods:

(1) The ungrounded conductors of the faulted circuit are disconnected automatically.

(2) The inverter or charge controller fed by the faulted circuit ceases to supply power to output circuits.

(C) Labels and Markings states that such labels and markings must appear on the utility-interactive inverter or be applied by the installer near the ground-fault indicator at a visible location stating the following: "WARNING—Electric shock hazard if a ground fault is indicated; normally grounded conductors may be ungrounded and energized."

When the PV system also has batteries, the same warning also must be applied by the installer in a visible location at the batteries.

When it comes to solar PV equipment, no assumptions should be made. As soon as an array is unpacked and exposed to ambient light, its output becomes live and may be a shock hazard.

Section 690.6, Alternating-Current Modules, provides in (A) that the requirements of Article 690 pertaining to PV source circuits do not apply to ac modules. The PV source circuit, conductors, and inverters are to be considered as internal wiring of an ac module.

(B) Inverter Output Circuit states that the output of an ac module is considered to be an inverter output circuit.

(C) Disconnecting Means states that a single disconnecting means is permitted for the combined ac output of one or more ac modules. Additionally, each ac module in a multiple ac module system is to be provided with a connector, bolted, or terminal-type disconnecting means.

(D) Ground-Fault Detection states that ac module systems are permitted to use a single detection device to detect only ac ground faults and to disable the array by removing ac power to the ac module(s).

(E) Overcurrent Protection states that the output circuits of ac modules are permitted to have overcurrent protection and conductor sizing in accordance with Section 240.5(B)(2), Fixture Wire, which permits smaller conductors to be connected to branch circuits of a greater rating.

Part II, Circuit Requirements, is important in sizing and adequately protecting PV circuits.

Section 690.7, Maximum Voltage, mentions for the first time in this article that PV output is highly dependent on ambient temperature. You would expect that output might decrease as the ambient temperature drops, but actually, the reverse is true. The electrical energy produced by a solar cell, which is a diode, is greatest on a cold, sunny day in winter.

(A) In a dc PV source circuit or output circuit, the maximum PV system voltage for that circuit is to be calculated as the sum of the rated open-circuit voltage of the series-connected PV modules corrected for the lowest expected ambient temperature. For crystalline and multicrystalline silicon modules, the rated open-circuit voltage is to be multiplied by the correction factors provided in Table 690.7. This voltage is to be used to determine the voltage rating of cables, disconnects, overcurrent devices, and other equipment. Where the lowest expected ambient temperature is below –40°F, (–40°C) or where other than crystalline or multicrystalline PV modules are used, the system voltage adjustment is to be made in accordance with manufacturer's instructions.

When open-circuit voltage temperature coefficients are supplied in the instructions for listed PV modules, they are to be used to calculate the maximum PV system voltage as required by Section 110.3(B) (which states that listed or labeled equipment is to be installed and used in accordance with any instructions included in the listing and labeling) instead of using Table 690.7.

That table, titled Voltage Correction Factors for Crystalline and Multicrystalline Silicon Modules, has Celsius ambient temperatures in the left column and Fahrenheit in the right column. The center column has the factors, which range from 1.02 for 76 to 68°F (24 to 20°C) to 1.25 for –32 to –40°F (–36 to –40°C). The procedure is to multiply the rated open-circuit voltage by the appropriate correction factor and use that figure to select cables, disconnects, overcurrent devices, and other equipment.

(B) Direct-Current Utilization Circuits states that the voltage of these must conform to Section 210.6, Branch-Circuit Voltage Limitations, which provides that various occupancies are permitted to contain equipment operating at specified maximum voltages. For example, circuits exceeding 600 volts, nominal, between conductors are permitted to supply utilization equipment in installations where conditions of maintenance and supervision ensure that only qualified persons service the installation.

(C) PV Source and Output Circuits states that in one- and two-family dwellings, PV source circuits and PV output circuits that do not include lampholders, fixtures, or receptacles are permitted to have a maximum PV system voltage of up to 600 volts. Other installations with a maximum PV system voltage of over 600 volts must comply with Article 690, Part IX, Systems Over 600 Volts (for PV systems).

(D) Circuits over 150 Volts to Ground states that in one- and two-family dwellings, live parts in PV source circuits and PV output circuits over 150 volts to ground are not to be accessible to other than qualified persons while energized.

(E) Bipolar Source and Output Circuits provides that for two-wire circuits connected to bipolar systems, the maximum system voltage is to be the highest voltage between the conductors of the two-wire circuit if all the following conditions apply:

(1) One conductor of each circuit of a bipolar subarray is solidly grounded.

(2) Each circuit is connected to a separate subarray.

(3) The equipment is clearly marked with a label as follows: "WARNING—Bipolar Photovoltaic array. Disconnection of neutral or grounded conductors may result in overvoltage on array or inverter."

Section 690.8, Circuit Sizing and Current, contains requirements for sizing calculations.

(A) Calculation of Maximum Circuit Current provides that the maximum current for the specific circuit is to be calculated in accordance with (1) through (4) below:

(1) PV Source Circuit Currents states that the maximum current is to be the sum of parallel-module rated short-circuit currents multiplied by 125 percent.

(2) PV Output Circuit Currents states that the maximum current is to be the sum of parallel-source-circuit maximum currents.

(3) Inverter Output Circuit Current states that the maximum current is the inverter continuous output current rating.

(4) Stand-Alone Inverter Input Circuit Current states that the maximum current is the stand-alone continuous inverter input current rating when the inverter is producing rated power at the lowest input voltage.

(B) Ampacity and Overcurrent Device Ratings states that PV system currents are considered to be continuous.

(1) Overcurrent Devices states that where required, these devices are to be rated according to (a) through (d) below:

a. They are to carry not less than 125 percent of the maximum currents calculated in (A) above.

b. Terminal temperature limits are to be in accordance with Sections 110.3(B), Installation and Use, and 110.14(C), Temperature Limitations.

c. Where operated at temperatures greater than 104°F (40°C), the manufacturer's temperature correction factors apply.

d. The rating or setting of overcurrent devices is permitted to be in accordance with Section 240.4(B), (C), and (D), which contain rules for permitted use of the next-higher rating.

(2) Conductor Ampacity states that circuit conductors are to be sized to carry not less than the larger of (a) or (b) below:

a. One hundred and twenty-five percent of the maximum currents calculated in Section 690.8(A) above without any additional correction factors for conditions of use.

b. The maximum currents calculated in Section 690.8(A) after conditions of use have been applied.

c. The conductor selected, after application of conditions of use, is to be protected by the overcurrent protective device where required.

(C) Systems with Multiple Direct-Current Voltages provides that for a PV power source that has multiple output circuit voltages and employs a common-return conductor, the ampacity of the common-return conductor is to be not less than the sum of the ampere ratings of the overcurrent devices of the individual output circuits.

(D) Sizing of Module Interconnection Conductors states that where a single overcurrent device is used to protect a set of two or more parallel-connected module circuits, the ampacity of each of the module interconnection conductors is not to be less than the sum of the rating of the single fuse plus 125 percent of the short-circuit current from the other parallel-connected modules.

Section 690.9, Overcurrent Protection, provides the following:

(A) Circuits and Equipment states that PV source circuit, PV output circuit, inverter output circuit, and storage battery circuit conductors and equipment are to be protected in accordance with the requirements of Article 240. Circuits connected to more than one electrical source are to have overcurrent devices located so as to provide overcurrent protection from all sources.

(B) Power Transformers states that overcurrent protection for a transformer with a source(s) on each side is to be provided by considering first one side of the transformer and then the other side of the transformer as the primary.

(C) PV Source Circuits states that branch-circuit or supplementary-type overcurrent devices are permitted to provide overcurrent protection in PV source circuits. The overcurrent devices are to be accessible but are not required to be readily accessible. Standard values of supplementary overcurrent devices are to be in 1-ampere size increments starting at 1 ampere up to and including 15 amperes.

Higher standard values above 15 amperes for supplementary over-current devices are to be based on the standard sizes provided in Section 240.6(A).

(D) Direct-Current Rating states that overcurrent devices, either fuses or circuit breakers, used in any dc portion of a PV power system are to be listed for use in dc circuits and are to have the appropriate voltage, current, and interrupt ratings.

(E) Series Overcurrent Protection states that in PV source circuits, a single overcurrent protection device is permitted to protect the PV modules and the interconnecting conductors.

Section 690.10, Stand-Alone Systems, provides that the premises wiring system is to be adequate to meet NEC requirements for a similar installation connected to a service. The wiring on the supply side of the building or structure disconnecting means must comply with the Code except as modified by (A) through (E) below:

(A) Inverter Output states that the ac output from a stand-alone inverter(s) is permitted to supply ac power to the building or structure disconnecting means at current levels less than the calculated load connected to that disconnect. The inverter output rating or the rating of an alternate energy source is to be equal to or greater than the load posed by the largest single utilization equipment connected to the system. Calculated general lighting loads are not considered as a single load.

(B) Sizing and Protection states that the circuit conductors between the inverter output and the building or structure disconnecting means are to be sized based on the output rating of the inverter. These conductors are to be protected from overcurrents in accordance with Article 240. The overcurrent protection is to be located at the output of the inverter.

(C) Single 120-Volt Supply states that the inverter output of a stand-alone solar PV system is permitted to supply 120 volts to single-phase, three-wire, 120/240-volt service equipment or distribution panels where there are no 240-volt outlets and where there are no multiwire branch circuits. In all installations, the rating of the over-current device connected to the output of the inverter is to be less than the rating of the neutral bus in the service equipment. This equipment is to be marked with the following words or equivalent: "WARNING—Single 120-volt supply. Do not connect multiwire branch circuits!"

(D) Energy Storage or Backup Power Systems states that such systems are not required.

(E) Back-Fed Circuit Breakers states that plug-in-type back-fed circuit breakers connected to a stand-alone inverter output in either stand-alone or utility-interactive systems are to be secured in accordance with Section 408.36(D), Back-Fed Devices, which states that plug-in-type overcurrent protection devices or plug-in-type main lug assemblies that are back-fed and used to terminate field-installed ungrounded supply conductors are to be secured in place by an additional fastener that requires other than a pull to release the device from the mounting means on the panel. Circuit breakers that are marked "Line" and "Load" may not be back-fed.

Section 690.11, Arc-Fault Circuit Protection (Direct Current), provides that PV systems with dc source circuits, dc output circuits, or both, on or penetrating a building operating at a PV system maximum system voltage of 80 volts or greater, are to be protected by a listed (dc) arc-fault circuit interrupter, PV type, or other system components listed to provide equivalent protection. The PV arc-fault protection means must comply with the following requirements:

(1) The system is to detect and interrupt arcing faults resulting from a failure in the intended continuity of a conductor, connection, module, or other system component in the dc PV source and output circuits.

(2) The system is to disable or disconnect one of the following:

   a. Inverters or charge controllers connected to the fault circuit when the fault is detected

   b. System components within the arcing circuit

(3) The system must require that the disabled or disconnected equipment be restarted manually.

(4) The system must have an annunciator that provides a visual indication that the circuit interrupter has operated. This indication is not to reset automatically.

Part III, Disconnecting Means, provides the following:

Section 690.13, All Conductors, states that means must be provided to disconnect all current-carrying dc conductors of a PV system from all other conductors in a building or other structure. A switch, circuit breaker, or other device is not to be installed in a grounded conductor if its operation leaves the marked, grounded conductor in an ungrounded and energized state.

Section 690.14, Additional Provisions, provides the following:

(A) The disconnect is not required to be suitable as service equipment and must comply with Section 690.17, Switch or Circuit Breaker, below.

(B) Equipment such as PV source-circuit isolating switches, overcurrent devices, and blocking diodes are permitted on the PV side of the PV system–disconnecting means.

(C) Requirements for Disconnecting Means states that means are to be provided to disconnect all conductors in a building or other structure from the PV system conductors.

(1) Location states that the PV system–disconnecting means are to be installed at a readily accessible location either on the outside of a

building or structure or inside nearest the point of entrance of the system conductors. The PV system–disconnecting means must not be installed in bathrooms.

(2) Marking states that each PV system–disconnecting means is to be permanently marked to identify it as a PV system disconnect.

(3) Suitable for Use states that each PV system–disconnecting means must be suitable for the prevailing conditions. Equipment installed in hazardous locations must comply with the requirements of Articles 500 through 517.

(4) Maximum Number of Disconnects states that the PV system–disconnecting means is to consist of not more than six switches or six circuit breakers mounted in a single enclosure, in a group of separate enclosures, or in or on a switchboard.

(5) Grouping states that the PV system–disconnecting means is to be grouped with other disconnecting means for the system to comply with Section 690.14(C)(4) above. A PV system–disconnecting means is not required at the PV module or array location.

(D) Utility-Interactive Inverters Mounted in Not-Readily-Accessible Locations states that utility-interactive inverters are permitted to be mounted on roofs or other exterior areas that are not readily accessible. These installations must comply with (1) through (4) below:

(1) A dc PV system–disconnecting means must be mounted within sight of or in the inverter.

(2) An ac disconnecting means must be mounted within sight of or in the inverter.

(3) The ac output conductors from the inverter and an additional ac disconnecting means for the inverter must comply with Section 690.14(C)(1), Location.

(4) A plaque is to be installed in accordance with Section 705.10, Directory (for Interconnected Electric Power Production Services).

Section 690.15, Disconnection of PV Equipment, states that means are to be provided to disconnect equipment such as inverters, batteries, charge controllers, and the like from all ungrounded conductors of all sources.

Section 690.16, Fuses, provides the following:

(A) Disconnecting Means states that such means are to be provided to disconnect a fuse from all sources of supply if the fuse is energized from both directions. Such a fuse in a PV source circuit must be capable of being disconnected independently of fuses in other PV source circuits.

(B) Fuse Servicing provides that disconnecting means are to be installed on PV output circuits where overcurrent devices (fuses) must be serviced that cannot be isolated from energized circuits. The disconnecting means are to be within sight of and accessible to the location of the fuse or integral with the fuseholder and must comply with 690.17, Switch or Circuit Breaker. Where the disconnecting means are located more than 6 feet from the overcurrent device, a directory showing the location of each disconnect is to be installed at the overcurrent device location. Non-load-break-rated disconnecting means are to be marked, "Do not open under load."

Section 690.17, Switch or Circuit Breaker, states that the disconnecting means for ungrounded conductors must consist of a manually operable switch(es) or circuit breaker(s) complying with the following requirements:

1. Located where readily accessible
2. Externally operable without exposing the operator to contact with live parts
3. Plainly indicating whether in the open or closed position
4. Having an interrupting rating sufficient for the nominal circuit voltage and the current that is available at the line terminals of the equipment.

Where all terminals of the disconnecting means may be energized in the open position, a warning sign is to be mounted on or adjacent to the disconnecting means. The sign is to be clearly legible and have the following words or equivalent: "WARNING—Electric shock hazard. Do not touch terminals. Terminals on both the line and load sides may be energized in the open position."

Section 690.18, Installation and Service of an Array, states that open circuiting, short circuiting, or opaque covering is to be used to disable an array or portions of an array for installation and service.

Part IV, Wiring Methods for Solar PV Systems, provides the following:

Section 690.31, Methods Permitted, states in (A) that all raceway and cable wiring methods included in the Code and other wiring systems and fittings specifically intended and identified for use on PV arrays are permitted. Where wiring devices with integral enclosures are used, sufficient length of cable is to be provided to facilitate replacement. Where PV source and output circuits operating at maximum system voltages greater than 30 volts are installed in readily accessible locations, circuit conductors are to be installed in a raceway.

(B) Single-Conductor Cable states that Type USE-2 and single-conductor cable listed and labeled as PV wire are permitted in exposed outdoor locations in PV source circuits for PV module interconnections within the PV array. For most applications, single-conductor cable is not used outside of raceways, but in the case of exposed outdoor locations in PV source circuits for module interconnections within the array, it is the method of choice.

(C) Flexible Cords and Cables states that where used to connect the moving parts of tracking PV modules, such cords and cables must comply with Article 400 and must be of a type identified as a hard service cord or portable power cable. They are to be suitable for extra-hard usage, listed for outdoor use, water resistant, and sunlight resistant. Allowable ampacities are to be in accordance with Section 400.5, Allowable Ampacity for Flexible Cords and Cables, Based on Ambient Temperature of 86°F (30°C). For ambient temperatures exceeding 86°F (30°C), the ampacities are to be derated by the factors given in Table 690.31(C), which follows. This gives derating factors, ranging from 1.00 to 0.33, for various temperature-rated conductors based on ambient temperatures from 86 to 176°F (30 to 80°C).

(D) Small-Conductor Cables states that single-conductor cables listed for outdoor use that are sunlight resistant and moisture resistant in sizes 16 American Wire Gauge (AWG) and 18 AWG are permitted for module interconnections where such cables meet the ampacity requirements of Section 690.8. Section 310.15 is to be used to determine the cable ampacity adjustment and correction factors.

(E) Direct-Current PV Source and Output Circuits Inside a Building states that where a dc PV source or output circuits from a building-integrated or other PV system are run inside a building or structure, they are to be contained in metal raceways, Type MC cable, or metal enclosures from the point of penetration of the surface of the building or structure to the first readily accessible disconnecting means.

The wiring methods must comply with the additional installation requirements in (1) through (3) below:

(1) Beneath Roofs states that wiring methods are not to be installed within 10 inches of the roof decking or sheathing except where directly below the roof surface covered by PV modules and associated equipment. Circuits are to be run perpendicular to the roof penetration point to supports a minimum of 10 inches below the roof decking. An Informational Note points out that the 10-inch requirement is to prevent accidental damage from saws used by fire fighters for roof ventilation during a structure fire.

(2) Flexible Wiring Methods states that where flexible metal conduit (FMC) smaller than trade size ¾ or Type MC cable smaller than 1 inch in diameter containing PV power circuit conductors is installed across ceilings or floor joists, the raceway or cable is to be protected by substantial guard strips that are at least as high as the raceway or cable. Where they are run exposed, other than within 6 feet of their connection to equipment, these wiring methods must closely follow the building surface or be protected from physical damage by an approved means.

(3) Marking or Labeling Required states that the following wiring methods and enclosures that contain PV power-source conductors

are to be marked with the wording "Photovoltaic Power Source" by means of permanently affixed labels or other approved permanent marking:

    a. Exposed raceways, cable trays, and other wiring methods.

    b. Covers or enclosures of pull boxes and junction boxes.

    c. Conduit bodies in which any of the available conduit openings are unused.

    d. The labels or markings must be visible after installation. PV power circuit labels must appear on every section of the wiring system that is separated by enclosures, walls, partitions, ceilings, or floors. Spacing between labels or markings is not to be more than 10 feet.

(F) Flexible, Fine-Stranded Cables states that such cables are to be terminated only with terminals, lugs, devices, or connectors in accordance with Section 110.14(A), Terminals, which states that connection of conductors to terminal parts must ensure a thoroughly good connection without damaging the conductors.

Section 690.32, Component Interconnections, states that fittings and connectors that are intended to be concealed at the time of on-site assembly, where listed for such use, are permitted for on-site connection of modules or other array components. Such fittings and connectors are to be equal to the wiring method employed in insulation, temperature rise, and fault-current withstand and are to be capable of resisting the effects of the environment in which they are used.

Section 690.33, Connectors, states that connectors must comply with (A) through (E) below:

(A) Configuration states that the connectors are to be polarized and have a configuration that is noninterchangeable with receptacles in other electrical systems on the premises.

(B) Guarding states that the connectors are to be constructed and installed so as to guard against inadvertent contact with live parts by persons.

(C) Type states that the connectors are to be of the latching or locking type. Connectors that are readily accessible and that are used in circuits operating at over 30 volts, nominal, maximum system voltage for dc circuits, or 30 volts for ac circuits, must require a tool for opening.

(D) Grounding Member states that the grounding member shall be the first to make and last to break contact with the mating connector.

(E) Interruption of Circuit states that connectors must either

    (1) Be rated for interrupting current without hazard to the operator.

(2) Be a type that requires the use of a tool to open and be marked "Do Not Disconnect Under Load" or "Not for Current-Interrupting."

Section 690.34, Access to Boxes, states that junction, pull, and outlet boxes located behind modules or panels are to be so installed that the wiring contained in them can be rendered accessible directly or by displacement of a module(s) or panel(s) secured by removable fasteners and connected by a flexible wiring system.

Section 690.35, Ungrounded PV Power Systems, provides that PV power systems are permitted to operate with ungrounded PV source and output circuits where the system complies with (A) through (G) below:

(A) Disconnects states that all PV source and output circuit conductors are to have disconnects complying with Article 690, Part III, Disconnecting Means.

(B) Overcurrent Protection states that all PV source and output circuit conductors are to have overcurrent protection complying with Section 690.9.

(C) Ground-Fault Protection states that all PV source and output circuits are to be provided with a ground-fault protection device or system that complies with (1) through (3) below:

(1) Detects a ground fault

(2) Indicates that a ground fault has occurred

(3) Automatically disconnects all conductors or causes the inverter or charge controller connected to the faulted circuit to cease supplying power to output circuits automatically

(D) The PV source conductors are to consist of the following:

(1) Nonmetallic jacketed multiconductor cables

(2) Conductors installed in raceways, or

(3) Conductors listed and identified as PV wire installed as exposed single conductors

(E) The PV power system dc circuits are permitted to be used with ungrounded battery systems complying with 690.71(G).

(F) The PV power source is to be labeled with the following warning at each junction box, container box, disconnect, and device where energized, ungrounded circuits may be exposed during service: "WARNING—Electric Shock Hazard. The dc conductors of this photovoltaic system are ungrounded and may be energized."

(G) The inverters or charge controllers used in systems with ungrounded PV source and output circuits are to be listed for the purpose.

Part V, Grounding, contains provisions for grounding of PV systems.

Section 690.41, System Grounding, provides that for a PV power source, one conductor of a two-wire system with a PV system voltage over 50 volts and the reference (center tap) conductor of a bipolar system must be solidly grounded or use other methods that accomplish equivalent protection in accordance with Section 250.4(A), Grounded Systems, and that use equipment listed and identified for the use.

Section 690.42, Point of System Grounding Connection, provides that the dc circuit-grounding connection is to be made at any single point on the PV output circuit. An Informational Note states that locating the grounding connection point as close as practicable to the PV source better protects the system from voltage surges owing to lightning.

Section 690.43, Equipment Grounding, states that equipment-grounding conductors and devices must comply with (A) through (F) below:

(A) Equipment Grounding Required states that exposed non-current-carrying metal parts of PV module frames, electrical equipment, and conductor enclosures are to be grounded regardless of the voltage.

(B) Equipment Grounding Conductor Required states that an equipment-grounding conductor between a PV array and other equipment is required.

(C) Structure as Equipment Grounding Conductor states that devices listed and identified for grounding the metallic frames of PV modules or other equipment are permitted to bond the exposed metal surfaces or other equipment to mounting structures. Metallic mounting structures, other than building steel, used for grounding purposes are to be identified as equipment-grounding conductors or are to have identified bonding jumpers or devices connected between the separate metallic sections and are to be bonded to the grounding system.

(D) PV Mounting Systems and Devices states that devices and systems used for mounting PV modules that are also used to provide grounding of the module frames must be identified for the purpose of grounding PV modules.

(E) Adjacent Modules states that devices identified and listed for bonding the metallic frames of PV modules are permitted to bond the exposed metallic frames of PV modules to the metallic frames of adjacent PV modules.

(F) All Conductors Together provides that equipment-grounding conductors for the PV array and structure (where installed) must be contained within the same raceway or cable or otherwise run with the PV array circuit conductors when those circuit conductors leave the vicinity of the PV array.

Section 690.45, Size of Equipment Grounding Conductors, provides that for PV source and output circuits, the equipment-grounding conductors are to be sized in accordance with (A) or (B) below:

(A) General states that equipment-grounding conductors in PV source and output circuits are to be sized in accordance with Table 250.122. Where no overcurrent protective device is used in the circuit, an assumed overcurrent device rated at the PV-rated short-circuit current is to be used in that table. Increase in equipment-grounding conductor size to address voltage-drop considerations is not required. The equipment-grounding conductors are to be no smaller than 14 AWG.

(B) Ground-Fault Protection Not Provided states that for other than dwelling units where ground-fault protection is not provided in accordance with Section 690.5(A) through (C), each equipment-grounding conductor is to have an ampacity of at least two times the temperature- and conduit fill–corrected circuit conductor ampacity.

Section 690.46, Array Equipment Grounding Conductors, provides that equipment-grounding conductors for PV modules smaller than 6 AWG must comply with Section 250.120(C), which requires raceway protection or cable armor unless installed within hollow spaces of framing members.

Section 690.47, Grounding Electrode System, contains provisions for both ac and dc systems. Of course, all solar PV systems start off by producing dc. Small stand-alone systems may employ dc to power utilization equipment. This works fine for incandescent light bulbs of the proper voltage, and many types of equipment are available that operate on 12 volts dc, such as televisions, water pumps, and fans. A wider range of equipment is available for 120-volt ac systems, and such an installation more closely resembles utility power. Either way, proper grounding is essential for a safe, efficient installation.

Grounding for ac systems and for dc PV systems follows Code grounding protocols for ac and dc systems found in Article 250. Systems with ac and dc grounding requirements are used widely. If the PV system has ac and dc circuits with no direct connection between the dc grounded conductor and the ac grounded conductor, it must have a dc grounding system. The dc grounding system is to be bonded to the ac grounding system by one of the methods in (1) through (3) below:

(1) Separate Direct-Current Grounding Electrode System Bonded to the Alternating-Current Grounding Electrode System states that a separate dc grounding electrode or system is to be installed, and it is to be bonded directly to the grounding-electrode system. The size of any bonding jumper(s) between the ac and dc systems is to be based on the larger size of the existing ac grounding-electrode conductor or the size of the dc grounding-electrode conductor specified by Section 250.166. The dc grounding-electrode system conductor(s) or the bonding jumpers to the ac grounding-electrode system must not be used as a substitute for any required ac equipment-grounding conductors.

(2) Common Direct-Current and Alternating-Current Grounding Electrode states that a dc grounding-electrode conductor of the size specified by Section 250.166 is to be run from the marked dc grounding-electrode connection point to the ac grounding electrode. Where

an ac grounding electrode is not accessible, the dc grounding-electrode conductor is to be connected to the ac grounding-electrode conductor. This dc grounding-electrode conductor is not to be used as a substitute for any required ac equipment-grounding conductors.

(3) Combined Direct-Current Grounding Electrode Conductor and Alternating-Current Equipment Grounding Conductor states that an unspliced or irreversibly spliced combined grounding-electrode conductor is to be run from the marked dc grounding-electrode conductor connection point along with the ac circuit conductors to the grounding busbar in the associated ac equipment. This combined grounding conductor is to be the larger of the sizes specified by Sections 250.122 or 250.166 and is to be installed in accordance with Section 250.64(E), Enclosures for Grounding Electrode Conductors.

Section 690.48, Continuity of Equipment Grounding Systems, states that where the removal of equipment disconnects the bonding connection between the grounding-electrode conductor and exposed conducting surfaces in the PV source or output circuit equipment, a bonding jumper must be installed while the equipment is removed.

Section 690.49, Continuity of PV Source and Output Circuit Grounded Conductors, provides that where removal of the utility-interactive inverter or other equipment disconnects the bonding connection between the grounding-electrode conductor and the PV source and/or PV output-circuit grounded conductor, a bonding jumper is to be installed to maintain the system grounding while the inverter or other equipment is removed.

Section 690.50, Equipment Bonding Jumpers, states that such jumpers, if used, are to comply with Section 250.120(C), which contains the requirement concerning equipment-grounding conductors smaller than 6 AWG being in raceways unless protected.

Part VI, Marking, contains a number of requirements for marking PV systems. Some of these are the responsibility of the installer; others are placed on equipment by manufacturers prior to the installation. Modules, including ac PV modules, are to be marked in a variety of ways, such as operating voltage and maximum power.

Section 690.54, Interactive System Point of Interconnection, provides that all interactive system(s) points of interconnection with other sources are to be marked at an accessible location at the disconnecting means as a power source and with the rated ac output current and the nominal operating ac voltage.

Section 690.55, PV Power Systems Employing Energy Storage, states that PV power systems employing energy storage must be marked with the maximum operating voltage, including any equalization voltage, and the polarity of the grounded-circuit conductor.

Additional marking requirements concern identification of power sources. Permanent plaques are required for facilities with stand-alone systems and facilities with utility services and PV systems.

Part VII, Connection to Other Sources, contains safeguards aimed at preventing unintended energizing of output conductors of an electric utility. It provides that a load disconnect that has multiple sources of power is to disconnect all sources when in the off position.

It further provides that an inverter or an ac module in an interactive solar PV system must automatically deenergize its output to the connected electrical production and distribution network on loss of voltage in that system and must remain in that state until the electrical production and distribution network voltage has been restored.

A normally interactive solar PV system is permitted to operate as a stand-alone system to supply loads that have been disconnected from electrical production and distribution network sources.

Article 705, Interconnected Electric Power Production Sources, applies to PV systems interconnected with other power production systems.

Part VIII, Storage Batteries, provides in (A) that storage batteries in a solar PV system are to be installed in accordance with the provisions of Article 480, which covers these installations for general applications.

(B) Dwellings contains provisions specific to PV systems:

(1) Operating Voltage states that storage batteries for dwellings are to have the cells connected so as to operate at less than 50 volts, nominal. Lead-acid storage batteries for dwellings are to have no more than twenty-four 2-volt cells connected in series (48 volts, nominal). An exception allows higher voltages where live parts are not accessible during routine maintenance.

(2) Guarding of Live Parts provides that live parts of battery systems for dwellings are to be guarded to prevent accidental contact by persons or objects regardless of voltage or battery type.

(C) Current Limiting states that a listed current-limiting overcurrent device is to be installed in each circuit adjacent to the batteries where the available short-circuit current from a battery or battery bank exceeds the interrupting or withstand ratings of other equipment in that circuit.

(D) Battery Nonconductive Cases and Conductive Racks states that flooded, vented, lead-acid batteries with more than twenty-four 2-volt cells connected in series (48 volts, nominal) must not be installed in conductive cases. Conductive racks used to support the nonconductive cases are permitted where no rack material is located within 6 inches of the tops of the nonconductive cases. This requirement does not apply to any type of valve-regulated lead-acid battery or any other types of sealed batteries that may require steel cases for proper operation.

(E) Disconnection of Series Battery Circuits provides that battery circuits subject to field servicing, where more than twenty-four 2-volt cells are connected in series (48 volts, nominal), are to have provisions to disconnect the series-connected strings into segments of 24 cells or less for maintenance by qualified persons. Non-load-break bolted or plug-in disconnects are permitted.

(F) Battery Maintenance Disconnecting Means states that battery installations where there are more than twenty-four 2-volt cells connected in series (48 volts, nominal) are to have a disconnecting means, accessible only to qualified persons, that disconnects the grounded circuit conductor(s) in the battery electrical system for maintenance. This disconnecting means may not disconnect the grounded circuit conductor(s) for the remainder of the PV electrical system. A non-load-break-rated switch is permitted to be used as the disconnecting means.

(G) Battery Systems of More Than 48 Volts states that on PV systems where the battery system consists of more than twenty-four 2-volt cells connected in series (more than 48 volts, nominal), the battery system is permitted to operate with ungrounded conductors, provided that the following conditions are met:

1. The PV array source and output circuits comply with Section 690.41, System Grounding.

2. The dc and ac load circuits are solidly grounded.

3. All main ungrounded battery input/output circuit conductors are provided with switched disconnects and overcurrent protection.

4. A ground-fault detector and indicator are installed to monitor for ground faults in the battery bank.

Section 690.72, Charge Control, states in (A) that equipment is to be provided to control the charging process of the battery. Charge control is not required where the design of the PV source circuit is matched to the voltage rating and charge current requirements of the interconnected battery cells and the maximum charging current multiplied by 1 hour is less than 3 percent of the rated battery capacity expressed in ampere-hours or as recommended by the battery manufacturer. All adjusting means for control of the charging process must be accessible only to qualified persons.

(B) Diversion Charge Controller provides the following:

(1) Sole Means of Regulating Charging states that a PV power system employing a diversion charge controller as the sole means of regulating the charging of a battery is to be equipped with a second independent means to prevent overcharging of the battery.

(2) Circuits with Direct-Current Diversion Charge Controller and Diversion Load states that such circuits must comply with the following:

- The current rating of the diversion load is to be less than or equal to the current rating of the diversion-load charge controller. The voltage rating of the diversion load is to be greater than the maximum battery voltage. The power rating of the diversion load is to be at least 150 percent of the power rating of the PV array.

- The conductor ampacity and the rating of the overcurrent device for this circuit are to be at least 150 percent of the maximum current rating of the diversion charge controller.

(3) PV Systems Using Utility-Interactive Inverters states that such inverters to control battery state of charge by diverting excess power into the utility system must comply with the following:

- These systems are not required to comply with (2) above. The charge-regulation circuits used must comply with the requirements of Section 690.8, Circuit Sizing and Current.
- These systems must have a second, independent means of controlling the battery-charging process for use when the utility is not present or when the primary charge controller fails or is disabled.

(C) Buck/Boost Direct-Current Converters states that when buck/boost charge controllers and other dc power converters that increase or decrease the output current or output voltage with respect to the input current or input voltage are installed, the requirements are to comply with (1) and (2) below:

(1) The ampacity of the conductors in output circuits is to be based on the maximum rated continuous output current of the charge controller or converter for the selected output voltage range.

(2) The voltage rating of the output circuits is to be based on the maximum voltage output of the charge controller or converter for the selected output voltage range.

Section 690.74, Battery Interconnections, provides the following:

(A) Flexible Cables, states that flexible cables, as identified in Article 400, in sizes 2/0 AWG and larger are permitted within the battery enclosure from battery terminals to a nearby junction box where they are to be connected to an approved wiring method. Flexible battery cables are also permitted between batteries and cells within the battery enclosure. Such cables are to be listed for hard service and identified as moisture resistant.

Flexible, fine-stranded cables are to be terminated only with terminals, lugs, devices, or connectors in accordance with Section 110.14(A), as we saw earlier in the context of Part IV, Wiring Methods (for PV Systems).

The requirements for solar PV systems are fairly straightforward once you assimilate the somewhat unfamiliar terminology. The starting place is the definitions, near the beginning of Article 690. Solar PV wiring is specialized and may require special training and certification depending on the jurisdiction. A feel for building construction is important for roof mounting of arrays. Since the arrays are always on when the sun is shining, special techniques are appropriate for maintenance and troubleshooting operations. Large battery installations also require care because there may be considerable available fault current.

Article 694

Article 694, Small Wind Electric Systems, is new to NEC 2011 (Figure 6-2). Many of the requirements parallel those in Article 690, Solar Photovoltaic (PV) Systems. (Article 692, Fuel Cell Systems, located between solar and wind articles, shares their attributes and is similar in structure and content. All three of these systems may be utility-interactive or stand-alone, and there is often battery backup and always grounding considerations.)

Section 694.1, Scope, notes that the provisions of Article 694 apply to small wind (turbine) electric systems that consist of one or more wind electric generators with individual generators having a rated power up to and including 100 kW.

To put this in perspective, this is enough power to operate one thousand 100-watt incandescent light bulbs simultaneously. Larger systems probably would be utility-owned and not under NEC jurisdiction.

Section 694.2, Definitions, defines several terms not covered in Article 690, Solar Photovoltaic (PV) Systems:

> *Diversion load*—A load connected to a diversion charge controller or diversion load controller, also known as a *dump load*.

FIGURE 6-2 A small wind electrical system consists of a wind turbine rated 100 kW or less. The system may be stand-alone or utility-interactive.

Diversion load controller—Equipment that regulates the output of a wind generator by diverting power from the generator to dc or ac loads or to an interconnected utility service.

Guy—A cable that mechanically supports a wind turbine tower.

Maximum output power—The maximum 1-minute average power output a wind turbine produces in normal steady-state operation (instantaneous power output can be higher).

Maximum voltage—The maximum voltage a wind turbine produces in operation, including open-circuit conditions.

Nacelle—An enclosure housing the alternator and other parts of a wind turbine.

Rated power—The wind turbine's output power at a wind speed of 24.6 mi/h. If a turbine produces more power at lower wind speeds, the rated power is the wind turbine's output power at a wind speed less than 24.6 mi/h that produces the greatest output power.

Tower—A pole or other structure that supports a wind turbine.

Wind turbine—A mechanical device that converts wind energy to electrical energy.

Wind turbine output circuit—The circuit conductors between the internal components of a small wind turbine (which might include an alternator, integrated rectifier, controller, and/or inverter) and other equipment.

Subsequent sections state that where the requirements of other Code articles and Article 694 differ, the requirements of Article 694 apply. An exception states that small wind systems, equipment, or wiring installed in a hazardous location also must comply with the applicable portions of Articles 500 through 516.

It is hard to imagine that a wind electrical system would be installed in a hazardous area, but the Code covers that eventuality.

Also paralleling solar PV system requirements, it is provided that systems covered by Article 694 are to be installed only by qualified persons.

Section 694.12, Circuit Sizing and Current, states in (A) that the maximum current for a circuit is to be calculated in (1) through (3) below:

(1) Turbine Output Circuit Currents states that the maximum current is to be based on the circuit current of the wind turbine operating at maximum output power.

(2) Inverter Output Circuit Current states that the maximum output current is the inverter continuous output current rating.

(3) Stand-Alone Inverter Input Circuit Current states that the maximum input current is the stand-alone continuous inverter input current rating of the inverter producing rated power at the lowest input voltage.

(B) Ampacity and Overcurrent Device Ratings provides the following:

(1) Continuous Current states that small wind turbine electrical system currents are considered to be continuous.

(2) Sizing of Conductors and Overcurrent Devices states that circuit conductors and overcurrent devices are to be sized to carry not less than 125 percent of the maximum current as calculated in (A) above. The rating or setting of overcurrent devices is to be as permitted in Sections 240.4(B) and (C), which allows going to the next-higher standard overcurrent device rating where the overcurrent device is rated 800 amperes or less.

Section 694.15, Overcurrent Protection, provides the following:

(A) Circuits and Equipment states that turbine output circuits, inverter output circuits, and storage battery circuit conductors and equipment are to be protected in accordance with the requirements of Article 240. Circuits connected to more than one electrical source are to have overcurrent devices located so as to provide overcurrent protection from all sources.

(B) Power Transformers states that overcurrent protection for a transformer with sources on each side is to be provided in accordance with Section 450.3, Overcurrent Protection (for Transformers), by considering first one side of the transformer and then the other side of the transformer as the primary.

(C) Direct-Current Rating provides that overcurrent devices, either fuses or circuit breakers, used in any dc portion of a small wind electrical system must be listed for use in dc circuits and must have appropriate voltage, current, and interrupting ratings.

Section 694.18, Stand-Alone Systems, contains requirements that parallel those for PV systems. Remember that if a single 120-volt supply powers the building or structure, multiwire branch circuits may not be used because both legs would be connected to the same pole, setting the stage for neutral overheating.

Part III, Disconnecting Means (for small wind electrical systems), resembles requirements for Solar PV systems with some variations.

Section 694.20, All Conductors, states that means are to be provided to disconnect all current-carrying conductors of a small wind electrical power source from all other conductors in a building or other structure. A switch, circuit breaker, or other device, either ac or dc, is not to be installed in a grounded conductor if its operation leaves the marked, grounded conductor in an ungrounded and energized state.

Section 694.22, Additional Provisions, further specifies disconnecting means requirements:

(A) Disconnecting means are not required to be suitable for service equipment. The disconnecting means for ungrounded conductors must consist of manually operable switches or circuit breakers complying with all the following requirements:

(1) They are to be located where readily accessible.

(2) They are to be externally operable without exposing the operator to contact with live parts.

(3) They are to plainly indicate whether in the open or closed position.

(4) They are to have an interrupting rating sufficient for the nominal circuit voltage and the current that is available at the line terminals of the equipment.

Where all terminals of the disconnecting means are capable of being energized in the open position, a warning sign to that effect is mandated.

(B) Equipment, such as rectifiers, controllers, output circuit isolating and shorting switches, and overcurrent devices, is permitted on the wind turbine side of the disconnecting means.

(C) Requirements for Disconnecting Means provides the following:

(1) Location states that the small wind electrical system disconnecting means is to be installed at a readily accessible location either on or adjacent to the turbine tower, on the outside of a building or structure, or inside, at the point of entrance of the wind system conductors.

A wind turbine disconnecting means is not required to be located at the nacelle or tower. As with solar PV systems, the disconnecting means may not be installed in a bathroom. (NEC prohibits overcurrent devices from being located in a bathroom in any occupancy.)

(2) Marking states that each turbine system disconnecting means is to be permanently marked to identify it as a small wind electrical system disconnect. A plaque is to be installed in accordance with Section 705.10, Directory (for Interconnected Electric Power Production Sources).

(3) Suitable for Use states that turbine system disconnecting means are to be suitable for the prevailing conditions. Equipment installed in hazardous locations must comply with the appropriate requirements of Articles 500 through 517.

(4) Maximum Number of Disconnects states that the turbine disconnecting means must consist of not more than six switches or six circuit breakers mounted in a single enclosure, in a group of separate enclosures, or in or on a switchboard.

(5) Equipment That Is Not Readily Accessible states that rectifiers, controllers, and inverters are permitted to be mounted in nacelles or other exterior areas that are not readily accessible.

Section 694.24, Disconnection of Small Wind Electrical System Equipment, states that means must be provided to disconnect equipment, such as inverters, batteries, and charge controllers, from all ungrounded conductors of all sources. If the equipment is energized from more than one source, the disconnecting means must be grouped and identified.

A single disconnecting means in accordance with Section 694.22 is permitted for the combined ac output of one or more inverters in an interactive system.

A shorting switch or plug is permitted as an alternative to a disconnect in systems that regulate the turbine speed using the turbine output circuit.

An exception provides that equipment housed in a turbine nacelle is not required to have a disconnecting means.

Section 694.26, Fuses, states that means are to be provided to disconnect a fuse from all sources of supply where the fuse is energized from both directions and is accessible to other than qualified persons. Switches, pullouts, or similar devices that are rated for the application are permitted to serve as a means to disconnect fuses from all sources of supply.

Section 694.28, Installation and Service of a Wind Turbine, states that open circuiting, short circuiting, or mechanical brakes may be used to disable a turbine for installation and service.

An Informational Note says that some wind turbines rely on the connection from the alternator to a remote controller for speed regulation. Opening turbine output circuit conductors may cause mechanical damage to a turbine and create excessive voltages that could damage equipment or expose persons to electric shock.

Part IV, Wiring Methods, provides the following:

Section 694.30, Permitted Methods, is somewhat different from the parallel section in Part IV of Article 690 because of inherent differences in the equipment and environment involved.

(A) Wiring Systems states that all raceway and cable wiring methods included in the Code and other wiring systems and fittings specifically intended for use on wind turbines are permitted. In readily accessible locations, turbine output circuits that operate at voltages greater than 30 volts are to be installed in raceways.

(B) Flexible Cords and Cables states that where used to connect the moving parts of turbines or where used for ready removal for maintenance and repair, such cords and cables must comply with Article 400 and must be of a type identified as hard-service cord or portable power cable, must be suitable for extra-hard usage, must be listed for outdoor use, and must be water resistant. Cables exposed to sunlight are to be sunlight resistant.

(C) Direct-Current Turbine Output Circuits Inside a Building states that such circuits are to be enclosed in metal raceways or installed in metal enclosures from the point of penetration of the surface of the building or structure to the first readily accessible disconnecting means.

Part V, Grounding, also differs somewhat from the corresponding part in Article 690.

Section 694.40, Equipment Grounding, provides the following:

(A) General states that exposed non-current-carrying metal parts of towers, turbine nacelles, other equipment, and conductor enclosures are to be connected to an equipment-grounding conductor regardless of voltage. Attached metal parts, such as turbine blades and tails that have no source of electrical energization, are not required to be connected to equipment-grounding conductors.

(B) Guy Wires states that such wires used to support turbine towers are not required to be connected to an equipment-grounding conductor.

(C) Tower Grounding provides the following:

(1) Auxiliary Electrodes states that a wind turbine tower is to be connected to one or more auxiliary electrodes to limit voltages imposed by lightning. Auxiliary electrodes are permitted to be installed in accordance with Section 250.54, Auxiliary Grounding Electrodes. Electrodes that are part of the tower foundation and meet the requirements for concrete-encased electrodes are acceptable. A grounded metal tower support is considered acceptable where meeting the requirements of Section 250.136(A), Equipment Secured to Grounded Metal Supports. Where installed in close proximity to galvanized foundation or tower anchor components, galvanized grounding electrodes are to be used.

In conductive soils, copper grounding electrodes close to galvanized components can cause electrolytic corrosion to occur.

(2) Equipment Grounding Conductor states that such a conductor is required between a turbine and the premises grounding system.

(3) Tower Grounding Connections provides that equipment-grounding conductors and grounding-electrode conductors, where used, are to be connected to the metallic tower by exothermic welding, listed lugs, listed pressure connectors, listed clamps, or other listed means. Devices such as connectors and lugs are to be suitable for the material of the conductor and the structure to which the devices are connected. Where practicable, contact of dissimilar metals is to be avoided anywhere in the system to eliminate the possibility of galvanic action and corrosion. All mechanical elements used to terminate these conductors must be accessible.

(4) Lightning Protection Systems states that auxiliary electrodes and grounding-electrode conductors are permitted to act as lightning-protection system components where meeting applicable requirements. If separate, the tower lightning-protection system grounding electrodes are to be bonded to the tower auxiliary grounding-electrode system. Guy wires used as a lightning-protection system grounding electrodes are not required to be bonded to the tower auxiliary grounding-electrode system.

Part VI, Marking, provides for various plaques and directories, some of which are the responsibility of the installer.

Section 694.50, Interactive System Point of Interconnection, provides that all interactive system points of interconnection with other sources are to be marked at an accessible location at the disconnecting means and with the rated ac output current and the nominal operating ac voltage.

Section 694.52, Power Systems Employing Energy Storage, states that small wind electrical systems employing energy storage are to be marked with the maximum operating voltage, any equalization voltage, and the polarity of the grounded circuit conductor.

Section 694.54, Identification of Power Sources, contains these two provisions:

(A) Facilities with Stand-Alone Systems states that any structure or building with a stand-alone system and not connected to a utility service source is to have a permanent plaque or directory installed on the exterior of the building or structure at a readily visible location. The plaque or directory is to indicate the location of system-disconnecting means and is to indicate that the structure contains a stand-alone electrical power system.

(B) Facilities with Utility Services and Small Wind Electrical Systems states that buildings or structures with both utility service and small wind electrical systems must have a permanent plaque or directory providing the location of the service-disconnecting means and the small wind electrical system–disconnecting means.

Section 694.56, Instructions for Disabling Turbine, states that a plaque is to be installed at or adjacent to the turbine location providing basic instructions for disabling the turbine.

Part VII, Connection to Other Sources, provides the following:

Section 694.60, Identified Interactive Equipment, states that only inverters listed and identified as interactive are permitted in interactive systems.

Section 694.62, Installation, states that small wind electrical systems, where connected to utility electrical sources, are to comply with the requirements of Article 705, Interconnected Electrical Power Sources.

Section 694.66, Operating Voltage Range, states that small wind electrical systems connected to dedicated branch or feeder circuits are permitted to exceed normal voltage operating ranges on these circuits, provided that the voltage at any distribution equipment supplying other loads remains within normal ranges.

Section 694.68, Point of Connection, states that the point of connection to interconnected electrical power sources must comply with Section 705.12, Point of Connection (for Interconnected Electrical Power Production Sources).

Part VIII, Storage Batteries, and Part IX, Systems Over 600 Volts, are substantially the same as the parallel parts in Article 690, Solar Photovoltaic (PV) Systems.

To summarize, NEC Chapter 6, Special Equipment, covers various types of equipment that, while not particularly hazardous, is somewhat unique owing to construction and function and therefore requires special treatment.

Looking in detail at solar PV systems and small wind electrical systems, we have endeavored to consider representative equipment that is of particular interest in today's world. For the electrician, proficiency in these areas may be a gateway into a large amount of high-quality work, as well as providing perspective on other aspects of the profession.

CHAPTER 7

NEC® Chapter 7, Special Conditions

Many electricians experience difficulty understanding National Electrical Code (NEC) Chapter 7, especially the infamous Article 725, Class 1, Class 2, and Class 3 Remote-Control, Signaling, and Power-Limited Circuits. Perhaps we can shed some light on the situation. Most of the difficulty arises from one basic misconception, which is that Article 725 is about *low-voltage wiring*. This is hardly the case, as we shall see.

Chapter 7 is the last of the three chapters whose titles begin with the word *Special*. This implies that there is something unique about the subject matter, and the fact is that these chapters stand somewhat apart from the rest of the Code.

Section 90.3, Code Arrangement, states that the Code is divided into the introduction and nine chapters. Chapters 1 through 4 apply generally; Chapters 5 through 7 apply to special occupancies, special equipment, and other special conditions. These latter chapters supplement or modify the general rules. Chapters 1 through 4 apply except as amended by Chapters 5 through 7 for the particular conditions.

Chapter 8 covers communications systems and is not subject to the requirements of Chapters 1 through 7 except where the requirements are specifically referenced in Chapter 8.

Chapter 9 consists of tables that are applicable as referenced.

Informative annexes are not part of the requirements of the Code but are included for informational purposes only.

You could say that Chapters 5 through 7 stand apart from the earlier portion of the Code and that Chapter 8 stands even farther apart. Chapters 5 through 7 may contain language that excludes Chapters 1 through 4, bypassing basic Code requirements, but in the case of Chapter 8, the basic Code mandates *including those in Chapters 5 through 7* are completely excluded unless specifically referenced as applicable.

Keeping this mechanism in mind, it is easy to access and understand any of the 10 articles in Chapter 7, with the exception of Article 725, which is somewhat more challenging.

Article 725

Article 725, Class 1, Class 2, and Class 3 Remote-Control, Signaling, and Power-Limited Circuits, becomes easier to understand if you put aside the idea that it is concerned with low-voltage wiring. In the first place, this is a nebulous term. Its meaning varies with context. For power distribution systems, it could be taken to mean less than medium voltage, which is 2,001 to 35,000 volts. Variously, it has been equated with under 1,000 volts, under 600 volts, under 50 volts, and under 12 volts. Second, Article 725 is not exclusively about any particular voltage level. It has to do with the function of the circuits involved. The best thing to do is to toss aside the term *low voltage* and delve into the actual provisions of Article 725.

Another impediment to understanding Article 725 is that the three classes are not members of the same general group. They are not organized by voltage levels—the lowest voltage level is found in Class 2. Classes 1 and 3 may contain higher voltage levels. In actual practice and for the purpose of licensing exam questions, Class 2 usually will be the main focus.

The best way to think about Article 725 is remember that it is primarily about remote-control, signaling, and power-limited, not voltage-limited, circuits.

Part I, General, contains basic definitions and requirements that apply to all three classes.

Section 725.1, Scope, states that Article 725 covers remote-control, signaling, and power-limited circuits that are not an integral part of a device or appliance.

An Informational Note states that the circuits described herein are characterized by usage and electrical power limitations that differentiate them from electric light and power circuits; therefore, alternative requirements to those of Chapters 1 through 4 are given with regard to minimum wire sizes, ampacity adjustment and correction factors, overcurrent protection, insulation requirements, and wiring methods and materials.

Section 725.2, Definitions, considers terms that are specific to Article 725:

Abandoned Class 2, Class 3, and PLTC cable—Installed Class 2, Class 3, and PLTC (Power-Limited Tray Cable) that is not terminated at equipment and not identified for future use with a tag. (This definition becomes important later when Section 725.25 requires abandoned cable to be removed.)

Circuit integrity (CI) cable—Cable(s) used for remote-control, signaling, or power-limited systems that supply critical circuits to ensure survivability for continued circuit operation for a specified time under fire conditions.

Class 1 circuit—The portion of the wiring system between the load side of the overcurrent device or power-limited supply and the connected equipment. An Informational Note refers to Section 725.41 for voltage and power limitation of Class 1 circuits.

Class 2 circuit—The portion of the wiring system between the load side of a Class 2 power source and the connected equipment. Owing to its power limitations, a Class 2 circuit considers safety from a fire-initiation standpoint and provides acceptable protection from electric shock.

Class 3 circuit—The portion of the wiring system between the load side of a Class 3 power source and the connected equipment. Owing to its power limitations, a Class 3 circuit considers safety from a fire-initiation standpoint. Since higher levels of voltage and current than for Class 2 are permitted, additional safeguards are specified to provide protection from an electric shock hazard that could be encountered.

The three classes are defined in more detail in Parts II and III and in Chapter 9, Tables 11(A) and 11(B), which give alternating-current (ac) and direct-current (dc) power source limitations for Class 2 and Class 3 circuits.

Section 725.21, Access to Electrical Equipment Behind Panels Designed to Allow Access, provides that access to electrical equipment is not to be denied by an accumulation of wires and cables that prevents removal of panels, including suspended panels. This requirement means that for an installation to be compliant, cables cannot be allowed to lie directly on the ceiling panels but may be hung from framing or attached directly to the ceiling surface above.

Section 725.24, Mechanical Execution of Work, provides that Class 1, Class 2, and Class 3 circuits are to be installed in a neat and workmanlike manner. Cables and conductors installed exposed on the surfaces of ceilings and sidewalls are to be supported by the building structure in such a manner that the cable will not be damaged by normal building use. Such cables are to be supported by straps, staples, hangers, cable ties, or similar fittings designed and installed so as not to damage the cable (Figure 7-1).

Section 725.25, Abandoned Cables, states that the accessible portion of abandoned Class 2, Class 3, and PLTC cables are to be removed. Where cables are identified for future use with a tag, the tag is to be of sufficient durability to withstand the environment involved.

In the event of fire, burning insulation from abandoned cable could produce a large quantity of toxic smoke. The requirement to

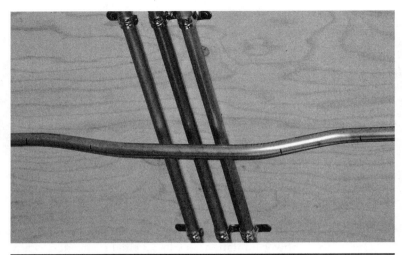

Figure 7-1   EMT (Electrical Metallic Tubing) used to protect remote-control, signaling, and power-limited circuits.

remove abandoned cable recurs in Articles 640, 645, 760, 770, 800, 820, and 830. At present, abandoned power and light conductors are not required to be removed.

Section 725.30, Class 1, Class 2, and Class 3 Circuit Identification, provides that these circuits are to be identified at terminal and junction locations in a manner that prevents unintentional interference with other circuits during testing and servicing.

Section 725.31, Safety-Control Equipment, contains information concerning Class 1 circuits.

(A) Remote-Control Circuits states that such circuits for safety-control equipment are to be classified as Class 1 if the failure of the equipment to operate introduces a direct fire or life hazard (Figure 7-2)

Room thermostats, water temperature-regulating devices, and similar controls used in conjunction with electrically controlled household heating and air conditioning are not considered safety-control equipment.

A nurse call system does not fall into the Class 1 category because in the event of failure, a hazard would not be initiated. Nurse call systems report hazards but do not initiate them. It would be possible for a nurse call system to fall into the Class 1 category, however, because of voltage and power-limitation issues.

Class 1 circuits require a higher degree of protection either because damage to them could introduce a hazard or because they have higher voltage and power limits.

(B) Physical Protection states that where damage to remote-control circuits of safety-control equipment would introduce a

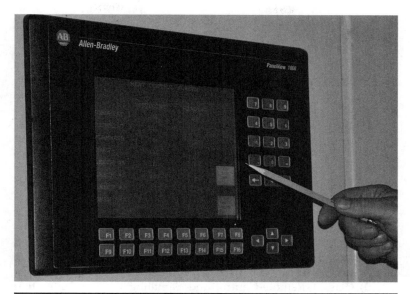

FIGURE 7-2 Output from a programmable logic controller (PLC) user-interface panel is a Class 1 circuit if failure of the equipment to operate could cause a direct fire or life hazard.

hazard, all conductors of such remote-control circuits are to be installed in rigid metal conduit, intermediate metal conduit, rigid nonmetallic conduit, electrical metallic tubing, Type MI cable, or Type MC cable, or otherwise be suitably protected from physical damage.

Section 725.35, Class 1, Class 2, and Class 3 Circuit Requirements, states that a remote-control, signaling, or power-limited circuit is to comply with the following parts of Article 725:

1. Class 1 Circuits: Parts I and II
2. Class 2 and Class 3 Circuits: Parts I and III

Part II, Class 1 Circuits, provides the following:

Section 725.41, Class 1 Circuit Classifications and Power Source Requirements, provides that Class 1 circuits are to be classified as either Class 1 power-limited circuits where they comply with the power limitations of (A) below or as Class 1 remote-control and signaling circuits where they are used for remote-control or signaling purposes and comply with the power limitations of (B) below. Notice that these power limitations are much higher than those for Class 2 and Class 3. The reason that they are permitted to be higher is that Class 1 circuits have a greater amount of protection.

(A) Class 1 Power-Limited Circuits states that these circuits are to be supplied from a source that has a rated output of not more than 30 volts and 1,000 volt-amperes (VA). In classifying Article 725 circuits, we are concerned with the rated (maximum) output of the source, not how much current the load draws.

(1) Class 1 Transformers states that transformers used to supply power-limited Class 1 circuits must comply with the applicable sections within Parts I and II of Article 450, Transformers and Transformer Vaults (Including Secondary Ties).

(2) Other Class 1 Power Sources provides that power sources other than transformers are to be protected by overcurrent devices rated at not more than 167 percent of the volt-ampere rating of the source divided by the rated voltage. The overcurrent devices may not be interchangeable with overcurrent devices of higher ratings. The overcurrent device is permitted to be an integral part of the power supply.

To comply with the 1,000-VA limitation of Section 725.41(A), the maximum output of power sources other than transformers is to be limited to 2,500 VA, and the product of the maximum current and maximum voltage is not to exceed 10,000 VA. These ratings are to be determined with any overcurrent-protective device bypassed.

(B) Class 1 Remote-Control and Signaling Circuits states that these circuits are not to exceed 600 volts. The power output of the source is not required to be limited.

Section 725.43, Class 1 Circuit Overcurrent Protection, states that overcurrent protection for conductors 14 American Wire Gauge (AWG) and larger is to be provided in accordance with the conductor ampacity without applying the ampacity and adjustment correction factors of Section 310.15 to the ampacity calculation. Overcurrent protection must not exceed 7 amperes for 18 AWG conductors and 10 amperes for 16 AWG.

Section 725.45, Class 1 Circuit Overcurrent Device Location, provides that overcurrent devices are to be located as specified in (A) through (E) below:

(A) Point of Supply states that overcurrent devices are to be located at the point where the conductor to be protected receives its supply.

(B) Feeder Taps states that Class 1 circuit conductors are permitted to be tapped, without overcurrent protection at the tap, where the overcurrent device protecting the circuit conductor is sized to protect the tap conductor.

(C) Branch-Circuit Taps states that Class 1 circuit conductors 14 AWG and larger that are tapped from the load side of the overcurrent device(s) of a controlled light and power circuit require only short-circuit and ground-fault protection and are permitted to be protected by the branch-circuit overcurrent protective device(s) where the rating of the protective device(s) is not more than 300 percent of the ampacity of the Class 1 circuit conductor.

(D) Primary Side of Transformer states that Class 1 circuit conductors supplied by the secondary of a single-phase transformer having only a two-wire (single-voltage) secondary are permitted to be protected by overcurrent protection provided on the primary side of the transformer, provided that this protection does not exceed the value determined by multiplying the secondary conductor ampacity by the secondary-to-primary transformer voltage ratio. Transformer secondary conductors other than two-wire are not considered to be protected by the primary overcurrent protection.

(E) Input Side of Electronic Power Source states that Class 1 circuit conductors supplied by the output of a single-phase, listed electronic power source, other than a transformer, having only a two-wire (single-voltage) output for connection to Class 1 circuits are permitted to be protected by overcurrent protection provided on the input side of the electronic power source, provided that this protection does not exceed the value determined by multiplying the Class 1 circuit conductor ampacity by the output-to-input voltage ratio. Electronic power source outputs, other than two-wire (single voltage), are not considered to be protected by the primary overcurrent protection.

Section 725.46, Class 1 Circuit Wiring Methods, states that Class 1 circuits are to be installed in accordance with Part I of Article 300 and with wiring methods from the appropriate articles in Chapter 3.

Section 725.48, Conductors of Different Circuits in the Same Cable, Cable Tray, Enclosure, or Raceway, states that Class 1 circuits are permitted to be installed with other circuits as specified in (A) and (B) below:

(A) Two or More Class 1 Circuits states that Class 1 circuits are permitted to occupy the same cable, cable tray, enclosure, or raceway without regard to whether the individual circuits are ac or dc, provided that all conductors are insulated for the maximum voltage of any conductor.

(B) Class 1 Circuits with Power-Supply Circuits states that Class 1 circuits are permitted to be installed with power-supply conductors as specified in (1) through (4) below:

(1) In a Cable, Enclosure, or Raceway states that Class 1 circuits and power-supply circuits are permitted to occupy the same cable, enclosure, or raceway only where the equipment powered is functionally associated.

(2) In Factory- or Field-Assembled Control Centers states that Class 1 circuits and power-supply circuits are permitted to be installed in factory- or field-assembled control centers.

(3) In a Manhole states that Class 1 circuits and power-supply circuits are permitted to be installed as underground conductors in a manhole in accordance with one of the following:

- The power-supply or Class 1 circuit conductors are in a metal-enclosed cable or Type UF cable.

- The conductors are permanently separated from the power-supply conductors by a continuous, firmly fixed nonconductor such as flexible tubing in addition to the insulation on the wire.

- The conductors are permanently and effectively separated from the power-supply conductors and securely fastened to racks, insulators, or other approved supports.

(4) In Cable Trays states that installations in cable trays are to comply with either of the following:

- Class 1 circuit conductors and power-supply conductors not functionally associated with the Class 1 circuit conductors are to be separated by a solid fixed barrier of a material compatible with the cable tray.

- Class 1 circuit conductors and power-supply conductors not functionally associated with the Class 1 circuit conductors are permitted to be installed in a cable tray without barriers where all the conductors are installed with separate multiconductor Type AC, Type MC, Type MI, or Type TC cables, and all the conductors in the cables are insulated at 600 volts.

Section 725.49, Class 1 Circuit Conductors, provides the following:

(A) Sizes and Use states that conductors of sizes 18 AWG and 16 AWG are permitted to be used, provided that they supply loads that do not exceed the ampacities given in Section 402.5, Allowable Ampacities for Fixture Wires, and are installed in a raceway, an approved enclosure, or a listed cable. Conductors larger than 16 AWG may not supply loads greater than the ampacities given in Section 310.15. Flexible cords must comply with Article 400.

(B) Insulation on conductors is to be rated for 600 volts. Conductors larger than 16 AWG are to comply with Article 310. Conductors in sizes 18 AWG and 16 AWG are to be of specified types. Conductors with other types and thicknesses of insulation are permitted if listed for Class 1 circuit use.

Section 725.51, Number of Conductors in Cable Trays and Raceway, and Ampacity Adjustment, applies exclusively to Class 1 circuits.

(A) Class 1 Circuit Conductors states that where only Class 1 circuit conductors are in a raceway, the number of conductors is to be determined in accordance with Section 300.17. The ampacity adjustment factors given in Section 310.15(B)(3)(a) apply only if such conductors carry continuous loads in excess of 10 percent of the ampacity of each conductor.

(B) Power-Supply Conductors and Class 1 Circuit Conductors states that where power-supply conductors and Class 1 circuit conductors are permitted in a raceway in accordance with Section 725.48, the number of conductors is to be determined in accordance with Section 300.17. The ampacity adjustment factors given in Section 310.15(B)(3)(a) apply as follows:

(1) To all conductors where the Class 1 circuit conductors carry continuous loads in excess of 10 percent of the ampacity of each conductor and where the total number of conductors is more than three

(2) To the power-supply conductors only where the Class 1 circuit conductors do not carry continuous loads in excess of 10 percent of the ampacity of each conductor and where the number of power-supply conductors is more than three

(C) Class 1 Circuit Conductors in Cable Trays states that where Class 1 circuit conductors are installed in cable trays, they are to comply with the provisions of Sections 392.22 and 392.80(A). This requirement refers back to Article 392, Cable Trays.

Section 725.52, Circuits Extending Beyond One Building, provides that Class 1 circuits that extend aerially beyond one building also must meet the requirements of Article 225, Outside Branch Circuits and Feeders.

Part III, Class 2 and Class 3 Circuits, is similar to Part II, but it is longer because it pertains to both Class 2 and Class 3 circuits. It also contains information on the cable-substitution hierarchy, which is important in design and installation of any size job.

Section 725, Power Sources for Class 2 and Class 3 Circuits, provides the following:

(A) Power Source for a Class 2 or Class 3 circuit is to be as specified in (1) through (5) below:

(1) A listed Class 2 or Class 3 transformer (Figure 7-3)

(2) A listed Class 2 or Class 3 power supply

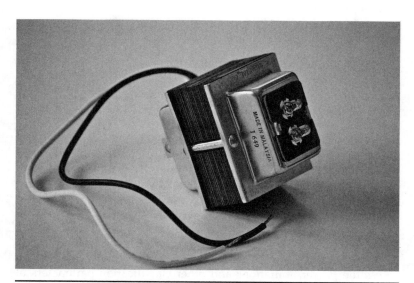

FIGURE 7-3 A Class 2 power source suitable for powering a furnace control circuit.

(3) Other listed equipment marked to identify the Class 2 or Class 3 power source

(4) Listed information technology (computer) equipment power-limited circuits

(5) A dry-cell battery is considered an inherently limited Class 2 power source, provided that the voltage is 30 volts or less and the capacity is equal to or less than that available from series-connected Number 6 carbon-zinc cells.

(B) Interconnection of Power Sources states that Class 2 or Class 3 power sources may not have the output connections paralleled or otherwise interconnected unless listed for such interconnection. Listing is not required for dry cells and thermocouples as power sources for Class 2 and Class 3 circuits.

Section 725.124, Circuit Marking, states that the equipment supplying the circuits is to be durably marked where plainly visible to indicate each circuit that is a Class 2 or Class 3 circuit.

Section 725.127, Wiring Methods on Supply Side of the Class 2 or Class 3 Power Source, provides that conductors and equipment on the supply side of the power source are to be installed in accordance with the appropriate requirements of Chapters 1 through 4. Transformers or other devices supplied from electric light or power circuits are to be protected by an overcurrent device rated not over 20 amperes.

An exception states that the input leads of a transformer or other power source supplying Class 2 and Class 3 circuits are permitted to be smaller than 14 AWG but not smaller than 18 AWG if they are not over 12 inches long and if they have insulation that complies with Section 725.49(B).

Understanding and properly applying Article 725 consists of two operations:

- Determining the correct classification for any given circuit you intend to install
- Ascertaining the correct wiring methods and materials to be used

Most Article 725 work involves Class 2 wiring. A typical Class 2 circuit is the familiar control circuit for a residential oil furnace or commercial/industrial steam heat fed through a wall-mounted room thermostat. The circuit is powered by a 24-volt (i.e., under 30 volts) transformer that is marked as a Class 2 power source and is, of course, listed by Underwriters Laboratories (UL). All the thinking is done for you. (Of course, you still have to verify that it is not necessary to install the circuit as Class 1 on the grounds that there is a safety issue.) Nevertheless, in an exam setting, it is necessary to know or be able to access the power limitations and other classification parameters of Class 1, Class 2, and Class 3 remote-control, signaling, and power-limited circuits.

Section 725.130, Wiring Methods and Materials on the Load Side of the Class 2 or Class 3 Power Source, provides that Class 2 and Class 3 circuits on the load side of the power source are permitted to be installed using wiring methods and materials in accordance with (A) or (B) below:

(A) Installation is to be in accordance with Section 725.46, Class 1 Circuit Wiring Methods. This provision is saying that Class 2 may be wired in accordance with the requirements of the more-stringent Class 1 protocol, but with the following exception, which relaxes the mandate somewhat: The ampacity adjustment factors given in Section 310.15(B)(3)(a) do not apply.

A second exception may seem baffling until you come to regard it as a consequence of (A) and nothing more profound: Class 2 and Class 3 circuits are permitted to be reclassified and installed as Class 1 circuits if the Class 2 and Class 3 markings are eliminated and the entire circuit is installed using the wiring methods and materials in accordance with Part II, Class 1 circuits.

An Informational Note states that Class 2 and Class 3 circuits reclassified and installed as Class 1 circuits are no longer Class 2 or Class 3 circuits regardless of the continued connection to a Class 2 or Class 3 power source. This exception is useful if it is necessary or desirable to place a Class 2 or Class 3 circuit in the same cable or raceway as a Class 1 circuit.

The preceding is one alternative wiring method/material mode for Class 2 and Class 3 circuits. A second is given below.

(B) Class 2 and Class 3 Wiring Methods provides that conductors on the load side of the power source are to be insulated at not less than the requirements of Sections 725.133 and 725.154. The first of these sections enables Sections 725.136 through 725.143, so there are a number of sections that come into play.

Section 725.136, Separation from Electric Light, Power, Class 1, Non-Power-Limited Fire Alarm Circuit Conductors, and Medium-Power Network-Powered Broadband Communications Cables, contains a number of provisions that must be observed.

(A) General states that cables and conductors of Class 2 and Class 3 circuits are not to be placed in any cable, cable tray, compartment, enclosure, manhole, outlet box, device box, raceway, or similar fitting with conductors of electric light, power, Class 1, non-power-limited fire alarm circuits, and medium-power network-powered broadband communications circuits unless otherwise permitted by (B) through (I) below.

(B) Separated by Barriers states that Class 2 and Class 3 circuits are permitted to be installed together with the conductors of electric light, power, Class 1, non-power-limited fire alarm, and medium-powered broadband communications circuits where they are separated by a barrier.

(C) Raceways Within Enclosures states that in enclosures, Class 2 and Class 3 circuits are permitted to be installed in a raceway to separate them from Class 1, non-power-limited fire alarm, and medium-power network-powered broadband communications circuits.

(D) Associated Systems Within Enclosures states that Class 2 and Class 3 circuit conductors in compartments, enclosures, device boxes, or similar fittings are permitted to be installed with electric light, power, Class 1, non-power-limited fire alarm, and medium-power network-powered broadband communications circuits where they are introduced solely to connect the equipment connected to Class 2 and Class 3 circuits and where (1) or (2) below applies:

(1) The electric light, power, Class 1, non-power-limited fire alarm, and medium-powered broadband communications circuit conductors are routed to maintain a minimum of 0.25 inch separation from the conductors and cables of Class 2 and Class 3 circuits.

(2) The circuit conductors operate at 150 volts or less to ground and also comply with one of the following:

    a. The Class 2 and Class 3 circuits are installed using Type CL3, CL3R, or CL3P or permitted substitute cables, provided that the Class 3 cable conductors extending beyond the jacket are separated by a minimum of 0.25 inch or by a nonconductive sleeve or nonconductive barrier from all other conductors.

    b. The Class 2 and Class 3 circuit conductors are installed as a Class 1 circuit in accordance with Section 725.41.

(E) Enclosures with Single Opening states that Class 2 and Class 3 circuit conductors entering compartments, enclosures, device boxes, outlet boxes, or similar fittings are permitted to be installed with Class 1, non-power-limited fire alarm and medium-power network-powered broadband communications systems where they are introduced solely to connect the equipment connected to Class 2 and Class 3 circuits. Where Class 2 and Class 3 circuit conductors must enter an enclosure that is provided with a single opening, they are permitted to enter through a single fitting (such as a tee), provided that the conductors are separated from the conductors of the other circuits by a continuous and firmly fixed nonconductor, such as flexible tubing.

(F) Manholes states that underground Class 2 and Class 3 circuit conductors in a manhole are permitted to be installed with Class 1, non-power-limited fire alarm and medium-power network-powered broadband communications circuits where one of the following conditions is met:

(1) The electric light, power, Class 1, non-power-limited fire alarm, and medium-power network-powered broadband communications circuit conductors are in a metal-enclosed cable or Type UF cable.

(2) The Class 2 and Class 3 circuit conductors are permanently and effectively separated from the conductors of other circuits by a continuous and firmly fixed nonconductor, such as flexible tubing, in addition to the insulation or covering on the wire.

(3) The Class 2 and Class 3 circuit conductors are permanently and effectively separated from conductors of the other circuits and securely fastened to racks, insulators, or other approved supports.

(G) Cable Trays states that Class 2 and Class 3 circuit conductors are permitted to be installed in cable trays, where the conductors of the electric light, Class 1, and non-power-limited fire alarm circuits are separated by a solid fixed barrier of a material compatible with the cable tray or where the Class 2 or Class 3 circuits are installed in Type MC cable.

(H) In Hoistways states that Class 2 or Class 3 circuit conductors are to be installed in rigid metal conduit, rigid nonmetallic conduit, intermediate metal conduit, liquid-tight flexible nonmetallic conduit, or electrical metallic tubing.

(I) Other Applications states that for other applications, conductors of Class 2 and Class 3 circuits are to be separated by at least 2 inches from conductors of any electric light, power, Class 1, non-power-limited fire alarm, or medium-power network-powered broadband communications circuits unless one of the following conditions is met:

(1) Either (a) all the electric light, power, Class 1, non-power-limited fire alarm, and medium-power network-powered broadband communications circuit conductors or (b) all the Class 2 and Class 3 circuit conductors are in a raceway or in metal-sheathed, metal-clad, nonmetallic-sheathed, or Type UF cables.

(2) All the electric light, power, Class 1, non-power-limited fire alarm, and medium-power network-powered broadband communications circuit conductors are permanently separated from all the Class 2 and Class 3 circuit conductors by a continuous and firmly fixed nonconductor, such as porcelain tubes or flexible tubing, in addition to the insulation on the conductors.

Section 725.139, Installation of Conductors of Different Circuits in the Same Cable, Enclosure, Cable Tray, or Raceway, contains the following provisions:

(A) Two or More Class 2 Circuits states that conductors of two or more Class 2 circuits are permitted within the same cable, enclosure, or raceway.

(B) Two or More Class 3 Circuits states that conductors of two or more Class 3 circuits are permitted within the same cable, enclosure, or raceway.

(C) Class 2 Circuits with Class 3 Circuits states that conductors of one or more Class 2 circuits are permitted within the same cable, enclosure, or raceway with conductors of Class 3 circuits, provided that the insulation of the Class 2 circuit conductors is at least that required for Class 3 circuits.

(D) Class 2 and Class 3 Circuits with Communications Circuits provides the following:

(1) Classified as Communications Circuits states that Class 2 and Class 3 circuit conductors are permitted in the same cable with communications circuits, in which case the Class 2 and Class 3 circuits are to be classified as communications circuits and are to be installed in accordance with the requirements of Article 800. The cables must be listed as communications cables.

(2) Composite Cables states that cables constructed of individually listed Class 2, Class 3, and communications cables under a common jacket are permitted to be classified as communications cables. The fire-resistance rating of the composite cable is to be determined by the performance of the composite cable.

(E) Class 2 or Class 3 Cables with Other Circuit Cables states that jacketed cables of Class 2 or Class 3 circuits are permitted in the same enclosure, cable tray, or raceway with jacketed cables of any of the following:

(1) Power-limited fire alarm systems in compliance with Article 760

(2) Nonconductive and conductive optical fiber cables in compliance with Article 770

(3) Communications circuits in compliance with Article 800

(4) Community antenna television and radio distribution systems in compliance with Article 820

(5) Low-power, network-powered broadband communications in compliance with Article 830

(F) Class 2 or Class 3 Conductors or Cables and Audio System Circuits states that audio system circuits described in Section 640.9(C) and installed using Class 2 or Class 3 wiring methods in compliance with Sections 725.133 and 725.154 are not permitted to be installed in the same cable or raceway with Class 2 or Class 3 conductors or cables.

A loud, sustained audio signal, especially sine wave, translates to heavy current flow and heat dissipation. (*Ac-dc* and *heavy metal* have acquired new meanings in contemporary parlance.)

Section 725.141, Installation of Circuit Conductors Extending Beyond One Building, provides that where Class 2 or Class 3 circuit conductors extend beyond one building and are run so as to be subject to accidental contact with electric light or power conductors operating at over 300 volts to ground or are exposed to lightning on interbuilding circuits on the same premises, requirements in Chapter 8 for coaxial and noncoaxial conductors apply.

Section 725.143, Support of Conductors, states that Class 2 or Class 3 circuit conductors are not to be strapped, taped, or attached by any means to the exterior of any conduit or other raceway as a means of support.

Violation of this prohibition is very common—it's fast, easy, and neat and facilitates circuit tracing. The problem is that it may compromise raceway functionality in regard to weight and heat dissipation.

Section 725.154, Application of Listed Class 2, Class 3, and PLTC Cables, lays out fundamental mandates for selecting the correct cable for remote-control, signaling, and power-limited circuits. To get started, we need to become familiar with some definitions and nomenclature. Then it is simply a matter of comprehending the very simple cable-substitution hierarchy.

The Code permits cabling technicians and electricians to substitute higher-quality (usually more expensive) cables for others that are suitable for less sensitive locations. This allows installers to stock less inventory and use short cutoffs provided that the cable-substitution hierarchy is observed.

What, in this context, is meant by *cable quality*? It has nothing to do with the electrical performance or durability of the cable. The concern is flame propagation and smoke generation. Plenum and riser environments present unique challenges regarding smoke generation and fire propagation.

Requirements for Class 2, Class 3 and PLTC cables are as follows:

(A) Plenums states that cables installed in ducts, plenums, and other spaces used for environmental air are to be Type CL2P or Type CL3P. Listed wires and cables installed in compliance with Section 300.22 are permitted. Listed plenum signaling raceways are permitted to be installed in other spaces used for environmental air, as described in Section 300.22(C). Only Type CL2P or Type CL3P cable is to be installed in these raceways. It is easy to decode the cable types. In Type CL2P, CL2 means Class 2, and P stands for plenum.

(B) Riser states that cables installed in risers are to be as described in (1) through (3) below:

(1) Cables installed in vertical runs and penetrating more than one floor or cables installed in vertical runs in a shaft are to be Type CL2R or Type CL3R. Floor penetrations requiring Type CL2R or Type CL3R are to contain only cables suitable for riser or plenum use. Listed riser signaling raceways and listed plenum signaling raceways are permitted to be installed in vertical riser runs in a shaft from floor to floor. Only Type CL2R, Type CL3R, Type CL2P, or Type CL3P cables are permitted to be installed in these raceways.

(2) Other cables as covered in Table 725.154(G) and other listed wiring methods as covered in Chapter 3 are to be installed in metal raceways or located in a fireproof shaft having fire stops at each floor.

(3) Type CL2, Type CL3, Type CL2X, and Type CL3X cables are permitted in one- and two-family dwellings. Listed general-purpose signaling raceways are permitted for use with Type CL2, Type CL3, Type CL2X, and Type CL3X cables.

(C) Cable Trays states that cables installed in cable trays outdoors are to be Type PLTC. Cables installed in cable trays indoors are to be Types PLTC, CL3P, CL3R, CL3, CL2P, CL2R, or CL2. Listed general-purpose signaling raceways are permitted for use with cable trays.

(D) Industrial Establishments states that in industrial establishments where the conditions of maintenance and supervision ensure that only qualified persons service the installation, Type PLTC cable is permitted in accordance with either (1) or (2) below:

(1) Where the cable is not subject to physical damage, Type PLTC cable that complies with the crush and impact requirements of Type MC cable and is identified as PLTC-ER for such use is permitted to be exposed between the cable tray and the utilization equipment or device. The cable is to be continuously supported and protected against physical damage using mechanical protection such as dedicated struts, angles, or channels. The cable is to be supported and secured at intervals not exceeding 6 feet.

(2) Type PLTC cable, with a metallic sheath or armor, is permitted to be installed exposed. The cable is to be continuously supported and protected against physical damage using mechanical protection such as dedicated struts, angles, or channels. The cable is to be secured at intervals not exceeding 6 feet.

(E) Other Wiring Within Buildings states that cables installed in building locations other than those covered in (A) through (D) above are to be as described in any of (1) through (6) below:

(1) General states that Type CL2 and Type CL3 are permitted.

(2) In Raceways or Other Wiring Methods states that Type CL2X and Type CL3X are permitted to be installed in a raceway or in accordance with other wiring methods covered in Chapter 3.

(3) Nonconcealed Spaces states that Type CL2X and Type CL3X are permitted to be installed where the exposed length of cable does not exceed 10 feet.

(4) One- and Two-Family Dwellings states that Type CL2X cable less than 0.25 inch in diameter and Type CL3X cable less than 0.25 inch in diameter are permitted.

(5) Multifamily Dwellings states the same as (4) above, but the cable must be in nonconcealed spaces.

(6) Under Carpets states that Type CMUC undercarpet communications wires and cables are permitted.

(F) Cross-Connect Arrays states that Type CL2 or Type CL3 conductors or cables are to be used for cross-connect arrays.

(G) Permitted Substitutions for Class 2 and Class 3 Cables provides the following:

- For CL3P, permitted substitution is CMP.
- For CL2P, permitted substitutions are CMP and CL3P.
- For CL3R, permitted substitutions are CMP, CL3P, and CMR.

- For CL2R, permitted substitutions are CMP, CL3P, CL2P, CMR, and CL3R.
- For PLTC, there are no permitted substitutions.
- For CL3, permitted substitutions are CMP, CL3P, CMR, CL3R, CMG, CM, and PLTC.
- For CL2, permitted substitutions are CMP, CL3P, CL2P, CMR, CL3R, CL2R, CMG, CM, PLTC, and CL3.
- For CL3X, permitted substitutions are CMP, CL3P, CMR, CL3R, CMG, CM, PLTC, CL3, and CMX.
- For CL2X, permitted substitutions are CMP, CL3P, CL2P, CMR, CL3R, CL2R, CMG, CM, PLTC, CL3, CL2, CMX, and CL3X.

A few observations are in order. If they are kept in mind, it will not be necessary to memorize the substitutions because they adhere to a particular logic. Type CM refers to communications wires and cables. Types CL2 and CL3 are Class 2 and Class 3 remote-control, signaling, and power-limited cables. Type PLTC is power-limited tray cable. P means plenum. R means riser. G or no letter means general purpose. X means restricted, used only in dwellings.

Communications cables are not discussed until Chapter 8, but they are part of the permitted substitutions covered in Chapter 7. Plenum is the most demanding location, followed in descending order of sensitivity by riser, general purpose, and dwelling. Class 3 may be substituted for Class 2 because the voltage and power limits are higher.

Part IV, Listing Requirements, is primarily of interest to manufacturers and listing organizations.

This completes our review of Article 725, the most challenging of the Chapter 7 topics. It may be necessary to go through Article 725 several times before it makes sense, but everything is completely logical if you think about it. First learn the meanings of the classifications; then consider the wiring methods. Despite the fact that it is considered the "low-voltage article," the voltage limit for Class 1 is 600 volts, which, along with the entirely separate safety issue, is the reason that special protection is required. The matter of reclassifying Class 2 and Class 3 circuits as Class 1 circuits may seem baffling until it is realized that the purpose of this option is to allow them to coexist within the same raceway or cable as Class 1 circuits. This is legitimate because Class 1 has a greater degree of protection.

Other articles, besides Article 725, have not been covered because they are relatively simple, and the requirements can be accessed easily.

NEC® Chapter 8, Communications Systems

C hapter 8 stands apart from the rest of the National Electrical Code (NEC) both structurally and in regard to content. With a few key differences in mind, however, it is easy to understand and implement. Licensing exam questions are not at all difficult. The material is right in front of you and easy to access if you understand the logical progression of the articles. There are not a lot of involved calculations, and circuitry is simple.

Remember that Chapter 8 is unique in that Chapters 1 through 7 are not applicable unless specifically referenced, in contrast to Chapters 5, 6, and 7, which *may* exclude themselves from the provisions in Chapters 1 through 4. Where these provisions are not specifically excluded, they are applicable.

Article 800

Article 800, Communication Circuits, presents an overview. Overhead clearances, underground conductors entering buildings, protective devices, grounding and bonding, cable hierarchy, and separation from other conductors are some of the topics. Much of this material is similar to what we encountered in Article 725, and it recurs in Articles 640, 645, 760, 770, 820, and 830.

Article 810

Article 810, Radio and Television Equipment, covers antenna systems for radio and television receiving equipment, amateur and citizen band radio transmitting and receiving equipment, and certain features of transmitter safety (Figure 8-1).

FIGURE 8-1   This TV satellite dish is a form of antenna covered in Article 810. The single coaxial transmission line is covered in Article 820.

Transmitters typically involve higher power and voltage levels than receivers. In addition to electric shock hazard, it is possible to receive radiation burns from a transmitter antenna (Figure 8-2).

Do not go near a transmitter antenna unless the power supply to the transmitter has been locked out, and there is no chance of lightning.

FIGURE 8-2   This Internet satellite dish involves both transmission and reception, hence the two coaxial cables. The cooling fins indicate the use of larger amounts of power, necessary for transmission. Power to the modem should be disconnected prior to any maintenance work to be performed on the dish.

Article 810 covers antennas such as wire-strung type, multiele-ment, vertical rod, and dish and also covers the wiring and cabling that connect them to equipment. The article does not cover equip-ment and antennas used for coupling carrier current to power-line conductors.

The Code does not look at the inner workings of factory-made transmitters and receivers, but field-fabricated equipment, at least in theory, should be approved by the authority having jurisdiction (AHJ). NEC Chapters 1 through 4 contain requirements for the power supply, studio wiring, and the like. Article 810 covers antennas, including parabolic dishes and digital satellite receiving equipment.

Article 820

Article 820, Community Antenna Television and Radio Distribution Systems, is all about coaxial cable, which is defined therein as a cylin-drical assembly composed of a conductor centered inside a metallic tube or shield, separated by a dielectric material, and usually covered by an insulating jacket. Have you wondered about characteristic impedance? Coaxial cable is labeled 75 ohms, 300 ohms, or some other value. This does not mean that it measures that resistance for any given length. In fact, characteristic impedance cannot be mea-sured by a conventional ohmmeter. We don't normally need to mea-sure it with any instrument. The manufacturer provides that and other specifications. The characteristic impedance of a cable is deter-mined by conductor sizes, spacing between them, and the type of insulation.

Article 830

Article 830, Network-Powered Broadband Communications Systems, covers systems that provide any combination of voice, audio, video, data, and interactive services through a network interface unit. An Informational Note provides a good description:

> A typical basic system configuration includes a cable supplying power and broadband signal to a network interface unit that converts the broadband signal to the component signals. Typical cables are coaxial cable with both broadband signal and power on the center conductor, composite metallic cable with a coaxial member for the broadband sig-nal and a twisted pair for power, and composite optical fiber cable with a pair of conductors for power. Larger systems may also include net-work components such as amplifiers that require network power.

Broadband and *bandwidth* are somewhat nebulous terms. In analog radio transmission, bandwidth comes into play when a low-frequency audio signal modulates a higher-frequency carrier. The modulating signal is alternately added to and subtracted from the carrier frequency, constituting bandwidth. Greater bandwidth is needed to transmit a higher-pitch audio signal.

For digital transmission, bandwidth means something quite different. It is equivalent to the amount of information that a medium can carry, which is directly related to the system's basic frequency level. Using copper as a medium, higher frequency means greater loss because of series inductance and parallel capacitance. To address this situation, broadband networks often use optical fiber.

Broadband usually equates to multiple data signals transmitted simultaneously through a medium. In this way, it is possible to achieve a greater capacity. NEC Article 830 contains provisions relating to network-powered broadband communications systems, typically taking the form of subscriber services offering voice, Internet access, television, and other products packaged and billed on a monthly basis.

Network-powered denotes that in addition to the signal, the network provides power to run amplifiers. Both power and signal are on the same conductors, a common scenario for coaxial cable. Network power may be direct current (dc) or alternating current (ac) rectified locally. The power can be applied at either end or at any point along the transmission line. Loss of either signal or network power will cause an outage, and troubleshooting involves checking for signal and power at various points along the transmission line.

Systems covered in Article 820 must have network power no higher than 60 volts. Systems covered in Article 830 may be powered at higher levels. Dc power sources exceeding 150 volts to ground but no more than 200 volts to ground, with the current to ground limited to 10 mA dc, are classified as medium-power sources.

Articles 820 and 830 contain the obligatory language regarding access to electrical equipment behind panels designed to allow access, mechanical execution of work, abandoned cables, and spread of fire or products of combustion.

NEC mandates removal of abandoned cable, and this mandate certainly has the effect of reducing fire loading, which translates into saved lives and reduced property loss. But who is responsible for this undertaking? If you do any work on a commercial building's data infrastructure, it could be construed that you are responsible for removing the accessible portion of any abandoned cable that may have accumulated in the course of previous alterations. The amount of this material often is enormous, and in the removal process, there is always the possibility of affecting an operating system, not to mention taking it down altogether. Prior to work on a large commercial building, agreements should be in place with the owners or tenants regarding the scope of the job.

Articles 820 and 830 both have a Part II, Cables Outside and Entering Buildings. Clearances above roofs and from power and light conductors, attachment to cross-arms, climbing space, interference with other communications systems, and distance from lightning conductors are covered.

Both articles also have a Part III, Protection. For Article 820, this consists of grounding the outer conductive shield of coaxial cables. Coaxial cables entering buildings or attached to buildings must comply with the provisions of Part IV, Grounding Methods.

Protection for network-powered broadband communications systems is somewhat more elaborate because it is required where these systems are neither grounded nor interrupted and are run partly or entirely in aerial cable not confined within a block. Also, primary electrical protection is to be provided on all aerial or underground network-powered broadband communications conductors that are neither grounded nor interrupted and are located within the block containing the building served so as to be exposed to lightning or accidental contact with electric light or power conductors operating at over 300 volts to ground.

Both articles have a Part IV, Grounding Methods. The major difference is that for Article 820, it is the shield of the coaxial cable that is to be bonded or grounded according to specified methods, whereas Article 830 states that for network-powered broadband communications systems, the connection is to be made at network interface units containing protectors, network interface units with metallic enclosures, primary protectors, and the metallic members of the network-powered broadband communications cable.

Otherwise, grounding provisions for both articles are substantially the same:

(A) Bonding Conductor or Grounding-Electrode Conductor contains these provisions:

(1) Insulation states that the bonding conductor or grounding-electrode conductor is to be listed and is permitted to be insulated, covered, or bare.

(2) Material states that the bonding conductor or grounding-electrode conductor is to be copper or other corrosion-resistant conductive material, stranded or solid.

(3) Size states that the bonding conductor or grounding-electrode conductor may not be smaller than 14 American Wire Gauge (AWG). It must have a current-carrying capacity not less than the outer sheath of the coaxial cable (or, for network-powered broadband communications systems, not less than the grounded metallic member and protected conductor of the cable). The bonding conductor or grounding-electrode conductor is not required to exceed 6 AWG.

(4) Length states that the grounding-electrode conductor is to be as short as practicable. In one- and two-family dwellings, the bonding conductor or grounding-electrode conductor is to be as short as practicable, not to exceed 20 feet in length.

(5) Run in Straight Line states that the bonding conductor or grounding-electrode conductor is to be run in as straight a line as practicable.

(6) Physical Protection states that bonding conductors and grounding-electrode conductors are to be protected where exposed to physical damage. Where the bonding conductor or grounding electrode is installed in metal raceway, both ends of the raceway are to be bonded to the contained conductor or to the same terminal or electrode to which the bonding conductor or grounding-electrode conductor is connected.

(B) Electrode states that the bonding conductor or grounding-electrode conductor is to be connected in accordance with (1), (2), or (3) below:

(1) In Buildings or Structures with an Intersystem Bonding Termination states that if the building or structure served has an intersystem bonding termination, the bonding conductor is to be connected to it.

(2) In Buildings or Structures with Grounding Means states that if the building or structure served has no intersystem bonding termination, the bonding conductor or grounding-electrode conductor is to be connected to the nearest accessible location on one of the following:

- The building or structure grounding-electrode system

- The grounded interior metal water piping system, within 5 feet from its point of entrance to the building

- The power service accessible means external to enclosures

- The nonflexible metallic power service raceway

- The service equipment enclosure

- The grounding-electrode conductor or the grounding-electrode metal enclosure of the power service

- The grounding-electrode conductor or the grounding electrode of a building or structure disconnecting means that is connected to an electrode

(3) In Buildings or Structures Without an Intersystem Bonding Termination or Grounding Means states that if the building or structure served has no intersystem bonding termination or grounding means, the grounding-electrode conductor is to be connected to either of the following:

- To any one of the individual electrodes described in Section 250.52(A)(1) through (4)

- To any one of the individual grounding electrodes described in Section 250.52(A)(5), (7), and (8)

(C) Electrode Connection states that such connections are to comply with Section 250.70.

(D) Bonding of Electrodes states that a bonding jumper not smaller than 6 AWG copper or equivalent is to be connected between the community antenna television system's grounding electrode (or network-powered broadband communications system's grounding electrode) and the power grounding-electrode system at the building or structure served where separate electrodes are used.

Article 820 contains one additional provision:

(E) Shield Protection Devices states that grounding of a coaxial drop cable shield by means of a protective device that does not interrupt the grounding system within the premises is permitted.

Both articles contain Part V, Installation Methods Within Buildings, which discusses cable hierarchies and separation from other conductors. These topics are similar to those in Article 725. Both articles also contain Part VI, Listing Requirements, primarily of interest to manufacturers.

Article 840

Article 840, Premises-Powered Broadband Communications Systems, is new to NEC 2011. An Informational Note included in Section 840.1, Scope, provides an overview. It states that a typical basic system configuration consists of an optical fiber cable to the premises supplying a broadband signal to an optical network terminal that converts the broadband signal into component electrical signals, such as traditional telephone, video, high-speed Internet, and interactive services. Powering of the optical network terminal typically is accomplished through a power supply and battery-backup unit that derive their power input from the available ac at the premises. The optical fiber cable is unpowered and may be nonconductive or conductive.

Provisions of Article 840 include access to electrical equipment behind panels designed to allow access, mechanical execution of work, abandoned cables, and overhead clearances.

Licensing exams may contain questions on Chapter 8, Communications Systems. The material is easy to access if you remember that Article 800, Communications Circuits, provides an overview, whereas Articles 810, 820, and 830 contain details on three major types of these systems. If you need information on coaxial cable, Article 820 is the place to look.

CHAPTER 9

NEC® Chapter 9, Tables

The main body of the National Electrical Code (NEC) concludes in a somewhat subdued manner with a chapter containing 12 tables that are "applicable as referenced." This means that some of them become mandatory and enforceable as they are referenced in Chapters 1 through 8.

Table 1, Percent of Cross Section of Conduit and Tubing for Conductors, is very brief. It gives the maximum permitted fill. If there is one conductor, the maximum fill is 53 percent. If there are two conductors, the maximum fill is 31 percent. For over two conductors, it is 40 percent. The table is very simple, but it is universally applicable when it comes to figuring conduit fill.

Table 2, Radius of Conduit and Tubing Bends, contains bend specifications for one-shot, full-shoe, and other bends. A hand or power bender produces these bends to the correct radius. Too small a radius would reduce the raceway internal diameter, causing difficulty and damage when pulling wires.

As stated earlier, conduit fill calculations are very simple when all conductors are the same size. Just consult Annex C and look up the answer. When conductors are not all the same size, it necessary to consult Table 4, Dimensions and Percent Area of Conduit and Tubing. There are separate tables for 12 types of raceway, with areas figured for 60, 53, 31, and 40 percent, as needed for different numbers of wires. Then find the areas of the various conductors as needed from Table 5. If the total cross-sectional areas are less than required, you have the correct match. All this assumes that calculations for ambient temperature and number of current-carrying conductors have been performed so that the correct size for the wires has been chosen.

Table 8, Conductor Properties, is quite useful. It gives the area in square millimeters and circular mills for uninsulated conductors, which is necessary when figuring conduit fill when a bare wire is to be pulled. It also contains information on individual strands of stranded wire and gives direct-current (dc) resistance, which is useful for calculating voltage drop for long runs.

Table 9, Alternating Current Resistance and Reactance, contains information on alternating-current (ac) resistance and reactance for 600-volt cables, three-phase, 60 Hz, at 75°C—three single conductors in conduit.

Table 10, Conductor Stranding, contains information on copper and aluminum, Class B and Class C stranding, for various conductor sizes.

Table 11(A), Class 2 and Class 3 Alternating-Current Power Source Limitations, contains ac power-source limitations for Class 2 and Class 3 circuits. Table 11(B), Class 2 and Class 3 Direct-Current Power Source Limitations, gives the same for dc circuits.

Tables 12(A), PLFA Alternating-Current Power Source Limitations, and 12(B), PLFA Direct-Current Power Source Limitations, provide the same information for power-limited fire alarm systems (Figure 9-1)

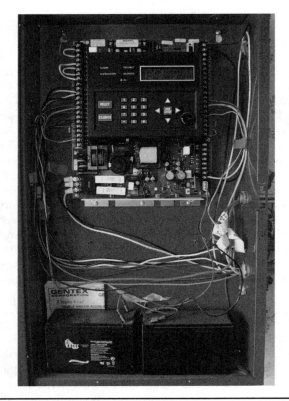

FIGURE 9-1   This fire alarm control panel with user interface qualifies as power limited. Notice the two backup 12-volt batteries hooked in series, indicating that the entire system runs at 24 volts dc.

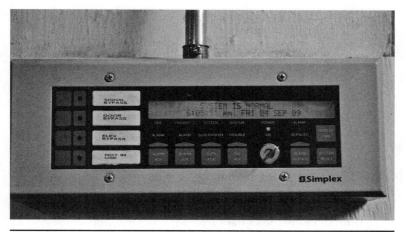

Figure 9-2 A remote subpanel showing that the system is normal. In the event of an alarm, the location will be shown, and the alarm can be acknowledged and silenced. Later, the system can be reset from this panel.

Fire alarm systems as covered in NEC Article 760 and other regulatory documents and local building codes are fairly complex integrated structures of great functionality. These are not the individual residential-type smoke detectors, even if ac powered and wired together to work in concert. They are powered and monitored by a centrally located control panel that performs a supervisory function (Figure 9-2). By *supervisory*, I am not referring to an individual who sits at a console and monitors various screens and output devises.

The supervisory role is performed by electronic components within the control panel on a continuous basis. We shall see how this works presently.

Fire alarm systems must comply with a number of separate regulatory mandates that operate together to govern the end product:

- *Life Safety Code* (NFPA 101) specifies the occupancies that are required to have fire alarm systems.

- *National Fire Alarm Code* (NFPA 72) enumerates system design requirements. Included are minimum performance requirements; operational, testing, and maintenance procedures; and system design parameters such as location and spacing of heads and pull stations.

- National Electrical Code (NFPA 70) focuses on fire alarm systems in Article 760. This article contains information on fire alarm system wiring, overcurrent protection, ampacity of power-supply conductors feeding the control panel, and zone wiring for individual initiating devices (such as pull stations and heads) and indicating appliances (such as strobes and horns). NEC also specifies and describes a variety of fire alarm functions such as fire door release, smoke doors and fan shutdown, liquefied petroleum gas shutdown, elevator capture and release, sensing of sprinkler water flow, and sprinkler supervisory capability. NEC Article 725, Class 1, Class 2, and Class 3 Remote Control, Signaling, and Power-limited Circuits, covers wiring emerging from the control panel. Power-limited circuits are classified by voltage and the amount of power that is available. They have optional less stringent requirements for overcurrent protection, insulation, minimum wire sizes, derating factors, conduit fill, and wiring methods and materials.

- Underwriters Laboratories (UL) and other inspecting agencies list fire alarm equipment, including smoke-detecting heads, control panels, batteries, horns, and pull stations.

These codes and design standards are the place to start in learning about fire alarm design and installation, but as the NEC indicates, they are not instruction manuals for untrained persons.

Manufacturers' installation and user manuals provide a good way to obtain insight and a feel for the subject. These are available as Internet downloads or as documentation accompanying existing systems.

Licensing varies from state to state. Florida, for instance, has two levels of alarm contractors' licenses. New Hampshire has voluntary fire alarm certification. Rules have been developed regarding training and testing curriculum to be used to determine minimum qualifications for certification. As of 2011, the infrastructure for implementing the system is under development.

As mentioned earlier, the essence of a fire alarm is the supervisory function performed by the control panel. To understand how this works, we need to understand that the fire alarm system's initiating devices are divided among separate zones, which may correspond to floors in a building. Any number of configurations is possible. Each zone has a number of initiating devices wired in parallel, just like receptacles in an individual branch circuit. The zone wiring consists of two wires, neither of which is grounded. They are isolated from the EMT or other raceways, which are grounded through the connectors at the control panel. Polarity is critical because the voltage, usually 24 volts dc, is required to power the solid-state circuitry within the heads (Figure 9-3).

Normally, when there is no fire, the heads and pull stations do not conduct. If smoke enters a head (some heads for use in dusty areas

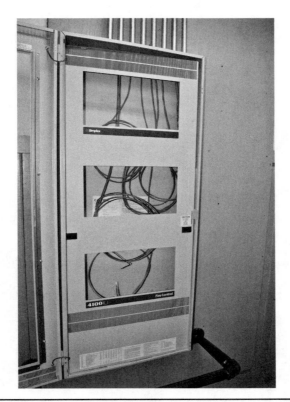

FIGURE 9-3 Fire alarm control panel under construction. The box has been mounted, EMT installed to initiating device and indicating appliance locations, and wire pulled.

are sensitive to heat only), its solid-state circuitry causes it to conduct. The control panel senses the event and goes into alarm. A pull station also causes the control panel to go into alarm (Figure 9-4).

At all times, the control panel is applying a dc voltage to each of the zones. This voltage is necessary for the solid-state circuitry in the heads to work. But it also serves another purpose as well. The control panel monitors the voltage and ascertains that the circuit is intact. This is what is meant by the *supervisory function*. But how, you might ask, is the control panel able to differentiate between a normal state (with all heads in the nonconductive mode) and an open circuit? The answer is that there is an end-of-line resistor placed across the two wires after the last head. This resistor is in the kilohm range, with the various manufacturers employing somewhat different values. Similarly, the indicating appliance zone has an end-of-line resistor so that it also can be supervised.

FIGURE 9-4 In case of a false alarm, pull stations may be checked for vandalism. This pull station has a Wiremold raceway in lieu of concealed wiring.

When the control panel sees the end-of-line resistance, the system is normal. The control panel incorporates an alphanumeric user interface that displays the state of the system.

As part of the supervisory capability, the control panel monitors itself internally. Each zone has a printed circuit card in the control panel, and if any of these develop an anomaly, or if the system loses ac power for a specified length of time, determined by programming, or if the dc voltage drops below a certain level, the system will enter the "trouble" state. The circuit boards are easy to replace.

There are three primary states: normal, alarm, and trouble. In the event of fire, the control panel, sensing that an initiating device has been activated, places the system in the "alarm" state. Horns and strobes activate so that the building may be evacuated and appropriate action may be taken (Figure 9-5). The alphanumeric display will indicate the location of the event (e.g., third floor, west wing, room 324), and onsite personnel will rush to that area. They will ascertain

FIGURE 9-5 Indicating appliance consisting of a horn and a strobe for the benefit of hearing-impaired individuals.

whether there is indeed a fire and, if so, attempt to extinguish it. In the meantime, an automated telephone link will have contacted the nearest fire department, which will have begun to deploy. If it is a false alarm, possibly caused by dust buildup in a head or vandalism at a pull station, the on-site personnel will contact an individual at the control panel who will silence the alarm. The control panel maintains the alarm status until the faulty head is removed, cleaned, or replaced, whereupon the system reset button can be pressed, and the system will return to normal.

If the control panel senses an open zone circuit, faulty telephone link, or any of a number of other aberrations that would interfere with normal operation, the system enters the "trouble" state. Audible and visual indicators activate at the panel and at any remote subpanels, and the alphanumeric display reports the problem (such as "Trouble: Zone Three Open Circuit").

Maintenance workers, often the facility electricians, disable and troubleshoot the zone, repair the fault, and reset the system.

The preceding description is typical of most systems but subject to variation. Edwards, Honeywell, and Simplex are among the many fire alarm manufacturers. Installation manuals and documentation available from them detail the installation and programming procedures. The programming is accomplished by pressing touch-screen controls at the control panel.

NEC Article 760, Fire Alarm Systems, begins, in Part I, with familiar mandates that are common to other low-voltage systems: access to electrical equipment behind panels designed to allow access, mechanical execution of work, and abandoned cable removal.

Part II deals with non-power-limited fire alarm circuits. These are relatively rare. They may have an output voltage of up to 600 volts.

Part III deals with power-limited fire alarm circuits. Power sources may be a listed power-limited Class 3 transformer, Class 3 power supply, or listed equipment marked to identify the power-limited fire alarm power source.

An important requirement, sometimes overlooked, is that the branch circuit powering the fire alarm system is to have no other loads. The location of the branch-circuit overcurrent protective device is to be permanently identified at the fire alarm control unit. The circuit-disconnecting means is to have red identification, it is to be accessible only to qualified personnel, and it is to be identified as "FIRE ALARM CIRCUIT."

Output circuits must be kept separate from electric light, power, Class 1, non-power-limited fire alarm circuits, and medium-power network-powered broadband communications circuit conductors. Separation may be maintained by barriers or raceways within enclosures.

Cable and conductors of two or more power-limited fire alarm circuits, communications circuits, or Class 3 circuits are permitted in the same cable, enclosure, cable tray, or raceway.

Conductors of one or more Class 2 circuits are permitted within the same cable, enclosure, cable tray, or raceway with conductors of power-limited fire alarm circuits, provided that the insulation of the Class 2 conductors is at least that required by the power-limited fire alarm circuits.

Low-power network-powered broadband communications circuits are permitted in the same enclosure, cable tray, or raceway with power-limited fire alarm cables.

Audio system circuits installed using Class 2 or Class 3 wiring methods are not permitted to be installed in the same cable, cable tray, or raceway with power-limited conductors or cables.

Conductors of 26 American Wire Gauge (AWG) are permitted only where spliced with a connector listed as suitable for 26 to 24 AWG or larger conductors that are terminated on equipment or where the 26 AWG conductors are terminated on equipment listed as suitable for 26 AWG conductors. Single conductors may not be smaller than 18 AWG.

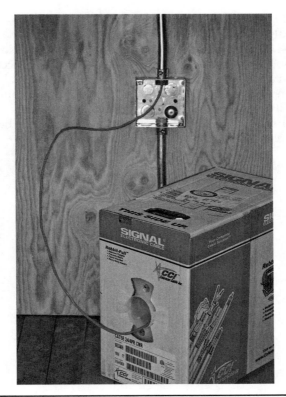

FIGURE 9-6   Fire alarm wire is being installed in widely used EMT raceway.

As always, power-limited fire alarm circuit conductors are not to be strapped, taped, or attached by any means to the exterior of any conduit or other raceway as a means of support.

As in all low-voltage wiring, there is a cable-substitution hierarchy. Plenum, riser, and other applications must be observed (Figure 9-6).

Fire alarm work is highly specialized and may require separate licensing and certification. It is a highly rewarding and recession-proof line of work and makes an excellent addition to the electrician's repertoire. Even if you don't plan to design and install these structures, it is advisable to know how to respond to alarms, disable and troubleshoot zones, and perform routine maintenance and programming.

As in all wiring, there are tremendous moral and legal obligations. It is unthinkable that a fire alarm system would fail to go into the "alarm" state in the event of fire without going into the "trouble" state first. Even if the ac power fails, there is the battery backup. And if the batteries fail, the system will report trouble.

The biggest problem with these systems is false alarms. These are problematic because, in a restaurant or hotel, they upset the guests. In an industrial facility, they cause worker downtime and loss of productivity. Moreover, repeated false alarms may cause occupants to disregard a real emergency.

When the system fails to perform properly, it is sometimes necessary to call in fire alarm technicians provided by the manufacturer or a fire alarm installation and maintenance firm. But it is better if in-house electricians can resolve the problem.

Afterword: Profitability and Ethics

We've taken a tour of the 2011 National Electrical Code (NEC). The Code does not provide a complete body of knowledge for the working electrician—more is needed. But it is always the place to start because it contains basic mandates necessary for safety. Electrical work consists of creating a part of the building project needed for the premises to be complete and usable. The end users, often with only a rudimentary knowledge of electrical and other fundamentals, depend on the construction professionals to create a safe and functional building. The overriding concern is always to make the premises safe not only short-term, but for years to come. The Code provides necessary guidance to achieve this goal. You must implement Code mandates to make the environment hazard-free for those who eventually will live there but also for your own workers and those in other trades who are on the job.

This is the first priority, but it's also good to make a fair profit in the process. The owner or tenant who is paying to have the job done has a responsibility to provide fair compensation for the work in progress. The electrician has to communicate in a timely fashion, during the course of the project, so that the owner feels confident that safety and functionality are being achieved. Ideally, there can be a degree of knowledge transfer so that when the owner finally occupies the premises, good practices will remain in place. This is part of the picture in creating a long-range hazard-free facility.

Electrical workers should ensure that everything is complete right down to the last detail. Marking and labeling need to be complete and durable for years to come. All covers should be in place and everything thoroughly tested. Even after the final payment has been issued, on a good-sized project, it is appropriate to look in one or more times to make sure that everything is functioning properly and that the owners and their workers are engaging in sound practices.

The Code has a number of mandates, such as working space, that must be continuously observed, even after the electrician has finished the job and is no longer a presence. It is important that the areas in front of entrance panels and load centers remain free of any objects that could impede ready access. If the original electrical installers have an ongoing maintenance contract, they can see to it that these and other requirements for hazard-free premises remain in place. If they do not have an ongoing maintenance contract, an example of a good fallback strategy would be to install a placard beside each panel saying, "Keep this area clear."

In addition to the NEC, local building ordinances and other juris-dictional documents must be observed. Occupational Safety and Health Administration (OSHA) regulations are applicable while work is in progress. Hard hats, scaffolding rails, adequate arc-fault mitigation procedures, and many other requirements are mandated. It is the responsibility of job supervisors and management profes-sionals to ensure that all requirements are observed. Temporary wir-ing as needed on the job must not just be thrown together to get something working but must be fully compliant with NEC, OSHA,

Code violations are plentiful. Section 230.28, Service Masts as Supports, states that only power-service drop conductors are permitted to be attached to a service mast.

and other requirements. Job-site electrocutions have been all too common, even with the introduction of ground-fault circuit interrupter (GFCI) technology.

Don't be a trunk slammer! It is in the interest of those working in the trade to comply with all applicable laws at all times. Workers' Compensation is expensive, but the cost of premiums is small compared with the liability incurred when a worker is injured. The owner of a construction firm is morally as well as legally obligated to ensure the safety of employees and, in the unfortunate event of injury, to provide insurance coverage so that the worker receives good medical care and compensatory income.

Profitability and ethical operating procedures are totally compatible. Those who work off the books or under the table are setting themselves up for not getting paid and possibly for massive liability.

I have stressed that the Code is always the place to start in achieving electrical knowledge and expertise. First, it should be assimilated to the extent that it is possible to access quickly on an open-book basis all information that is necessary to create hazard-free installations. Then knowledge and expertise may be acquired from other sources. The Internet is a great database. If a malfunctioning appliance gives a cryptic error code, type it into a search engine, and the answer will appear. Manufacturers' service information is available on the Web as well.

Electronics and troubleshooting textbooks are available from technical publishers. If you use one idea gained from a text, the book may have paid for itself. It is a good idea to keep a log of every job that you do with applicable Code references. If a complex piece of equipment is to be disassembled, a digital photograph taken in advance is of great value.

Above all, thoroughly torque all terminations!

Index

Note: An *"f"* follows page numbers referencing figures.

WITHDRAWN